Hong Kong the Super Paradox
Life after Return to China

Edited by
James C. Hsiung

St. Martin's Press
New York

#4892515

ISBN 0-312-22293-9

Library of Congress Cataloging-in-Publication Data
Hong Kong the super paradox : life after return to China/edited by
    James C. Hsiung.
        p.   cm.
    Includes bibliographical references and index.
    ISBN 0-312-22293-9
    1. Hong Kong (China)—Politics and government—1997-   2. Hong Kong
(China)—Economic conditions.   3. Hong Kong (China)—Social
conditions.   4. Hong Kong (China)—Foreign relations. I. Hsiung,
James Chieh, 1935-
DS796.H757H6595   1999 2000
951.2506—dc21
                                                                    99–40500
                                                                    CIP

Design By Letra Libre, Inc.

10 9 8 7 6 5 4 3 2 1

Dedicated to the memory of my mother
Yun-chih Hsieh Hsiung (1917–1997)

Contents

Part III
Conclusions

Foreword and Acknowledgements

The idea of this book was a long time in the coming. On a personal note, ever since 1984 and through the mid-1990s, I have been fascinated by Hong Kong's forthcoming return to China, closing a chapter on Western inroads in Chinese history dating back over a century and a half, to say the least. Furthermore, given my professional interest in international politics theory, the way Hong Kong was to exit from British colonial rule posed many tantalizing questions: What more, for example, does it tell us about decolonization, a phenomenon destined to mark off the twentieth century as distinct from the previous century of (colonial) empire-building? How is Hong Kong's exit from colonial rule different from the experiences of other former colonies of the West? If "one country, two systems," the governance structure to be installed in post-handover Hong Kong, is unprecedented in human experience, what preconditions must be present before it could succeed? Is the model really foredoomed to failure, as critics and much of the media suggested? If, on the other hand, the model should prove successful in post-colonial Hong Kong, what are the chances of its being used as an example for the remaining cases of contentious irredentism and contested secessions the world over, including but not limited to the Taiwan case, the last jigsaw puzzle in the Chinese unification bid? Then, the Sino-British disputes arising from the post-treaty British reform (changes) in Hong Kong also poked my curiosity as to their political, as well as legal, implications.

In addition to these tantalizing theoretical questions, the widespread doomsday forecasts, circulating in the months immediately before the 1 July 1997 handover day for Hong Kong, further whetted my appetite, paradoxically, for wanting to find out more about a land alleged to have no future.[1] After all, I have a perpetual interest in paradoxes and counter-intuitive findings.

I was fortunate in two ways. First, I was invited to participate in the historic handover ceremony, slated to take place at midnight of 30 June 1997, when the Union Jack was to be lowered for the last time, and Hong Kong would be returned as the proverbial "pearl" back to the "cuddle" of Mother

China's custodian hand. Second, I was offered a visiting appointment as Chair Professor and Head of the Politics-Sociology Department at an up-and-coming tertiary institution in Hong Kong, the Lingnan University (then still known by its earlier name, Lingnan College), commencing the summer of 1997. Both offers provided me with an additional incentive and the physical facility for me to come to Hong Kong and to see things for myself during a historic transition.

Hence, the idea of this book was no longer a figment of imagination, or something to groan over in a far-off observation deck in New York. Instead, it became an ongoing project from the day I arrived on the spot. I do not, however, want to take full credit for myself, as I owe an intellectual debt to many quarters for the final product. In the first place, I am indebted to my colleagues at Lingnan, especially to Dean Y. Y. Kueh of its Social Science Faculty, for encouragement and support. Professor Loksang Ho, head of the Economics Department at Lingnan, whose office was next to mine, was a walking encyclopedia on Hong Kong. In addition to their counsels, I bene-fited from two symposia, held in October 1997 and June 1998, under the auspices of the Faculty's Center for Asia and Pacific Studies (CAPS). Both had the specific charge of looking into the state of post-handover Hong Kong. The running dialogue thus started provided me a new avenue of real-istic insights into the intricacies of the newly created Special Administrative Region (SAR). These insights gave new meaning to my own observations and some of the factual data that came my way. They helped me onto a new vista in my endeavoring to decipher many complex problems involving Hong Kong's inherited legacy, new challenge, and missed opportunities, and furthermore what it entails to make the "one country, two systems" work in the new Hong Kong SAR.

Next, I am indebted to the contributors to this book. With few excep-tions, almost all of them participated in one or both of the symposia, which were used to bounce off ideas before the final drafts were written up. The first batch of the completed reports, from the first conference, was published in a special issue of *Asian Affairs: An American Review,* a learned journal pub-lished by Heldref Publications in Washington, D. C., in the winter issue of 1998. Most of the chapters in this book, on the other hand, were first dis-cussed at the second conference, some in preliminary form; but the final texts were completed, hence updated, for the purpose of this book. In addi-tion, two new chapters were solicited only after the conference, including the one by Y. Y. Kueh, which offers an up-to-date assessment of the regional financial crisis and how Hong Kong has fared under its crunch. The other one, by Danny Paau, is an essay he graciously agreed to take on, on short no-tice, even at a time of great personal hardship. I must say that almost to a person, my contributors were all extra tolerant and generous in the way they put up with my pestering them after their finished manuscripts, even when I chased them, in some cases across continents, by long-distance phone calls, faxes, and bombardments of e-mail messages.

One thing unique about the team I have assembled for this undertaking is that it includes, in addition to selected well-known academics in Hong Kong, one internationally known practicing journalist, and two former officials whose tenure (in charge of important portfolios) straddled the outgoing British administration and the incoming Chinese SAR government of Hong Kong (see more in the biographical notes provided at the end of the book). Despite their diverse backgrounds, it is amazing that what they have to say in their respective chapters, each in its own way, all support the paradox paradigm of the book, as I shall explain in the Introduction below. A note, incidentally, is in order about the spelling of the names of some of the contributors who prefer the Chinese way of putting the family name first, followed by the given name. Hence, Lau Siu-kai, Ho Loksang, and Ting Wai should really be Siu-kai Lau, Loksang Ho, and Wai Ting if spelled in the Western way.

In addition, I am indebted to Frank Ching and his wife Anna Wu for filling me in on some of the important background about Hong Kong old and new, including the nuances and hidden meanings of certain events, especially those that must be understood in perspective. Together they kept me, I think, from going astray on many issues and events that came both during the post-1984 Sino-British disputes and in the post-1997 transition. My secretary, Ivy Suk-han Tsang, at Lingnan, doubled occasionally as my research assistant and, more frequently, as my unofficial computer consultant, bringing to bear her expertise and soft touch in dealing with the latest intricate software, which is so advanced and sophisticated that it had to break down every so often. David Ji, of CAPS at Lingnan, also helped in sharing part of the tedious task of inputting onto the diskettes all necessary stylistic changes to the manuscript after it came back from the copy editor.

Last, but not least, I remain thankful to the Politics Department at New York University (NYU), my home institution, for allowing me to take off two years in a row, so that I could have the time and opportunity to do the needed field research that went into the making of this book.

Echoing a usual routine, I, alone, as architect and editor, bear the responsibility for any possible faults of the book. Breaking tradition with the routine modesty, however, I would like to make a special claim for the book *Hong Kong the Super Paradox*. While governments in Washington, London, and other capitals, the world business community, and many in academia are hankering to know what has really happened to post-1997 Hong Kong, it is nevertheless true that most publishers, especially those that have yet to recover their losses from over-stocked books about Hong Kong in the pre-handover publication glut, have lost interest in publishing anything about the place after l July 1997. This book, therefore, could very well be the first, or at least one of the very first, to offer a comprehensive assessment of the post-handover Hong Kong. Not only does it sort things out for timely analyses on the social, political, and economic fronts, but, more important, it offers a coherent framework for analyzing events in the SAR—offering not only "what" but, more important, "why"—and anticipating the future. Besides, this is most likely the first book that offers readers an opportunity to have a glimpse of how the people in Macau view Hong Kong's retrocession and what lessons they can learn for Macau's own return to China, following in Hong Kong's footprints two and a half years later, in December 1999. Furthermore, it is, I am sure, the first book that addresses the international relevance of the SAR's "one country, two systems" model, assessing its possible duplicability for the world's remaining cases of irredentism and contested secessions, ranging from Canada's Quebec, Spain's Catalonia, the United Kingdom's Northern Ireland, to Indonesia's East Timor, which are only the more celebrated cases among those that one can identify in 51 countries across four continents, outside the immediate China-Taiwan orbit.

The unanticipated Asian financial crisis, hitting the region after August 1997, has only spoiled an otherwise largely uneventful and collegial transition of Hong Kong from a British colony into a Chinese SAR. Likewise, it has created uncertainty for international investors and businesses interested in this hubbub of trade and financial activities and the largest container shipping center in Asia Pacific. The book offers not one, but three, chapters dealing in depth with Hong Kong's economy and how it has fared and pulled off relatively unscathed from under the crushing blow of the regional crisis. These chapters, written by true experts, along with the other chapters, combine to add to the book's uniqueness.

In view of the above, the publication of this book is implicitly a tribute to St. Martin's Press for its great vision, in contrast to the "once bitten, twice shy" publishers. Its senior editor, Michael Flamini, who recognized the value of the book project right away when he first saw the proposal, ought to be commended and singled out for special mention.

It had always been a fond wish of both my late parents to see the return of Hong Kong to China. While my father passed away in 1990, my mother, who turned 80 in 1997, missed the handover by just a few days. Hence, this book is dedicated to her memory.

Notes

1. On the weekend before my departure for my perch in Hong Kong in late May 1997, a multitude of friends gave me a rousing send-off party. Over 300 people came and paid for the party at a plush New York City hotel. Some, suppressing

tears, thought they were saying goodbye to me for the last time, as if they were sending some maniac off on a suicidal mission, to a place of no future (?!). I was greatly moved, but not convinced. I have survived and am at ease to tell the story about Hong Kong after the handover. I have no qualms about what I chose to do, in coming to Hong Kong. As far as I can see, this place, though temporarily distracted by the unanticipated regional financial crisis, has a post-colonial future. History will tell who has the last laugh.

INTRODUCTION

The Paradox Syndrome and Update

James C. Hsiung

A Tale of Paradoxes

"Is there life in Hong Kong after its handover to China? If so, what is it going to be like?" These were the typical questions tormenting many before the former British crown colony was returned to Chinese sovereignty on 1 July 1997. While these questions are still ringing in our ears, we can safely say, over two years later, that life in Hong Kong after reversion is paradoxically quite normal, contrary to the doomsday prophecies of many "people of little faith,"[1] as this volume demonstrates in various chapters. In contrast, as is also shown, however, it is in the other and least expected areas—to wit, Hong Kong's external aspects and its economic performance—that developments have likewise paradoxically contradicted earlier, more positive predictions.

Most dire prophecies concerned the future of domestic politics. There was near-unanimous consensus among Western media and many commentators that Hong Kong, after reversion, would not be able to enjoy, under the watchful eyes of its Chinese sovereign, its promised "high degree of autonomy." Because of its strategic commercial importance to the global economy, Hong Kong's projected handover elicited such voluminous commentaries and monographs on its future (or lack of it) that they could easily fill a small library. With few exceptions, most predictions about Hong Kong's domestic conditions following the handover were dismal and downright pessimistic. The worst scenario saw Beijing meddling in Hong Kong's politics and economic life, and trampling upon its freedoms, including freedom of the press, judicial freedom, academic freedom, and free elections. There would be corruption, nepotism, cronyism, and related plagues, brought in by the Mainland Chinese.[2]

By December 1996, when the future Chief Executive, Tung Chee-hwa, was selected by a selection committee going through rather elaborate procedures, it provided the occasion for almost universal condemnation by the U.S. media. The latter pretty much represented the judgmental sentiments of most Western commentators. The *New York Times,* in a 28 December editorial "Farewell to Hong Kong's Freedom," contended that the Chief Executive was chosen "without the slightest accommodation to democracy." As Frank Ching (1997, p.54) noted, the fact that "all previous governors of [colonial] Hong Kong had been chosen with even less 'accommodation to democracy' was not mentioned." All previous governors were appointed by London, without even a semblance of consultation with the Hong Kong people. By contrast, the Chinese selection procedures at least involved a wide spectrum of people, many of them entries in the Hong Kong *Who's Who.* The *Washington Post* condemned Tung, the Chief Executive-designate, as "Beijing's man," who was "hand-picked by a Beijing-controlled committee." Like the *New York Times* and the *Washington Post,* most of the U.S. media, not to mention members of Congress, were of the opinion that there was no future for Hong Kong after its transition from a British crown colony to a Chinese special administrative region in 1997.

The matter was of enormous interest to the American public, as the United States had over 1,100 firms, 50,000 or more American citizens, and at least $16 billion in direct investment in the British colony in transition.[3] What would happen in post-1997 Hong Kong was of crucial concern to these interests, among others. American anxiety over the transition both engineered, and was reflected in, the U.S. Hong Kong Policy Act of 1992, enacted by Congress. The main intent of the act was to mandate the Department of State to keep a close watch on Hong Kong through its reversion and beyond, and keep Congress abreast of the goings-on, more especially on the question of whether U.S. interests were protected.

Events subsequent to the handover, however, have paradoxically proven these predictions unfounded, at least during the initial period as studied by this book. Many once-pessimists now have to recant or else modify their language. In Taiwan, Hong Kong's most critical watcher, a highly placed official—Dr. King-yuh Chang, director of the Cabinet's Mainland Affairs Office—answering questions in the Legislature, gave an unequivocal, though somewhat grudging, answer that "in the nine months since the handover, Beijing did not excessively interfere, but let Hong Kong govern itself as a highly autonomous entity."[4] Even the usually skeptical *Economist* had to admit, with a straight face, that two "common predictions" about how Hong Kong would fare after leaving British rule "have refused to come true: the Communists in Beijing have refused to call the shots; and the chief executive, Tung Chee-hwa, despite being a deeply conservative fan of 'Confucian authority,' has not become an oriental despot."[5] A similar accolade came from a totally unexpected source. A spokesman for the British Prime Minister, Tony Blair, declared: "We think the pessimists have been proved wrong about the [aftermath of the] handover in Hong Kong."[6]

Despite widespread speculation that the Hong Kong Special Administrative Region (HKSAR) would be dragged down the drain of corruption under Chinese influence, a survey in April 1998 showed that the SAR's corruption during the nine months following the handover had been at its lowest level in five years.[7] The five-year period, of course, covers the last years of Governor Chris Patten's tenure.

The crime rate was also down following the handover. According to police statistics, crime rates in 1997 were 14.8 percent lower than in 1996, which was only outdone by a 22.2 percent decrease in the number of commercial crimes.[8]

In a 1 April 1998 report filed by the U.S. State Department to Congress, as required by law, one of the beginning paragraphs explicitly stated:

> There is no evidence of interference from the Chinese government in local affairs. Hong Kong's civil service remains independent, and senior officers, including those who have been critical of the PRC [China], have been retained. Hong Kong continues to play an important role as a regional finance center, actively participating in efforts to address the Asian financial crisis. The Hong Kong press remains free and continues to comment critically on the PRC and its leaders, though some self-censorship has been reported. Demonstrations— often critical of the PRC—continue to be held. Mainland Chinese companies are subject to the same laws and prudential supervision as everyone else, and the rule of law and the independent judiciary remain in place as guarantees of Hong Kong's free and open civil society.[9]

Beijing's hands-off stance has even earned the gratuitous praise of the British Foreign Office in its first report on Hong Kong after the handover (Tang 1998, p.16).

The State Department's second annual report to Congress, dated 1 April 1999, had pretty much the same to say, at times even using identical sentences. Only two noteworthy exceptions were entered: Hong Kong's economic problems and a dispute over SAR's judicial review of acts of China's parliament affecting the territory. In noting the economic problems, during the year of 1 April 1998 through 31 March 1999, the report nonetheless concluded that "overall political development [of the Hong Kong SAR] under Chinese sovereignty has gone relatively smoothly." On the next question, the report recounted the SAR Court of Final Appeal's ruling on the Chan Kam-Nga case. While "this case has raised uncertainty about how judicial independence will fare over the longer term," it declared, "for the present Hong

Kong's judicial system appears to have weathered intact the first major test of its independence." We shall have more to say in this regard in the concluding chapter of this book. For now, suffice it to quote the following statement from the State Department's 1 April 1999 report, which confirms what this book has to say on developments within the Hong Kong SAR:

> Hong Kong's political system continues to evolve. The legislature and free press have used their public fora to increase government transparency and accountability. There is vigorous public debate on the pace of democratizing elections for the legislature and chief executive. Likewise, there was controversy regarding the Government's proposal to abolish democratically elected municipal councils and to add appointed members on local district councils. Public debate on issues of democracy and law has been active and vigorous and has served to reconfirm the distinction of Hong Kong as a free society with free markets (Preface).

As to Hong Kong's international status and competence to defend its interests abroad, on the other hand, most pre-handover predictions by publicists and international law experts had been positive, some even upbeat. Almost none anticipated any significant change from how Hong Kong had fared internationally before. However, as will be shown later in this volume, these positive predictions notwithstanding, the HKSAR has encountered unforetold problems, as was most poignantly typified in the Matimak case. In it, a Hong Kong corporation was held by two U.S. federal courts to be "stateless," hence without the legal status to sue before them (see Chapter 7). The case cannot but cast doubt on the HKSAR's legal eligibility, on its own, to act internationally in defense of the interests of its natural and juridical persons.

In sum, as the chapters below demonstrate, domestic events in the SAR have fared better than predicted (see Chapters 1, 4, 5, and 6), but on the external tangent the first year's record has shown developments either unanticipated or worse off than the predictions (see Chapters 7 and 8). This contrast constitutes a giant paradox. While in social science parlance, a "paradox" is a phenomenon marked by conflict between, in game-theoretic language, dominant strategy, and optimal outcome (Quine 1965, p.145), we use the term in a more mundane sense. For our purpose, a paradox mainly denotes a discrepancy, or complex of discrepancies, between expectations and outcome. (Occasionally, the discrepancy may turn out to be between intended result and actual outcome.) In addition to the two sets of discrepancies between expectations and outcomes, just noted, there is the totally unanticipated economic downturn triggered by the post-1997 Asia Pacific regional financial crisis (see Chapters 2, 3, and 9), which stands in sharp contrast to the generally rosy forecasts about the SAR's continuing economic boom. It simply adds to the paradox syndrome. Hence, the term "super paradox" just about sums up the experience of Hong Kong during its first two years following the handover.

Deciphering the Sources of the Paradox Syndrome

Among the possible sources (i.e., origins) of the sort of paradox syndrome briefly identified above and dealt with in greater detail in the following chapters, I could think of three that need special mention. For all their importance, they seem to have been neglected in most discussions that have come to my attention. Parenthetically, I shall skip the most obvious, and overworked, explanation linked to the Tiananmen tragedy of 1989 and the world's reactions to it, including a habitual intuition to view things related to China in the most bleak light and to question the veracity of everything Beijing said or did.

Loophole in the Sino-U.K. Agreement

The first source lies in a neglected loophole in the Sino-U.K. Joint Declaration of 1984 regarding Hong Kong's return to Chinese sovereignty after one and a half centuries of British colonial rule. While the Joint Declaration provides that the "current" systems[10] of Hong Kong will remain unchanged for 50 years after reversion to China, the term "current" is nowhere defined. Does it mean, for example, the political system as it existed at the time the Declaration was signed in 1984, or when the ratified instruments were exchanged in 1985, or in 1997, the year when Hong Kong was actually retroceded to Chinese sovereignty? The definition of what was meant by the "current" system is more crucial a matter than meets the eye, because between 1984 and 1997, there was a long interregnum of 13 years. The British colonial rulers had plenty of time to change the existing (or "current") system before their exit, as indeed they did. I do not want to be too technical, but under relevant international law (i.e., "law of treaties"),[11] if no definition is given, then "current" should be coterminous with the time when the agreement is signed. And, if ratification is required, as in the present case, an agreement takes effect upon the exchange of ratifications. Hence, an alternate answer for the undefined "current" systems is what obtained in Hong Kong in 1985, when the instruments of ratification were exchanged, as was required by the Declaration (Art. 8).

All Sino-British disputes of the 1990s and the doomsday imagery they helped evoke, in a nutshell, came from the conflicting assumptions between Beijing and the British rulers, about what was the "current" system, which was not to be tampered with. To appreciate this point, which has rarely been discussed[12]—not even raised by Beijing—we have to dwell briefly on the political system of Hong Kong as it existed in 1984 and before, in contrast to what it later became between 1984 and 1997. I will try to be very brief on this enormous subject.

Until 1985 Hong Kong had a governmental structure (or political system) dominated by the executive, as symbolized by the Governor, who exercised by delegation the powers of the royal prerogative, which were characterized by one authoritative writer (Miners 1995, ch. 6) as "awesome" under the Letters Patent. These powers were comparable to those once possessed by the King of England before the coming of democracy. Until 1984, he had the power to appoint civil servants to form a majority of the seats on the Legislative Council (Legco), and they were bound to follow his instructions as to how they should vote.[13] But, between 1985 and 1997 the British, acting through their last two governors, skillfully but surely transformed Hong Kong government into a legislature-dominant structure, or a system in which the legislature was, in Norman Miners' (1995, ch. 8) words, "controlling the Administration." In order to do so, the British introduced elections for the Legco, for the first time ever in one and a half centuries of colonial rule, thereby augmenting the Legco's legitimacy and making it more powerful. It is moot to question the possible British motivations for not introducing democracy until they were about to depart from Hong Kong after 150 some years. But, the Tiananmen episode was often raised as an immediate reason why the British wanted to create democratic safeguards in Hong Kong before they were to leave (Chang and Chuang 1998, p.78). Be that as it may, the move, however laudable, nevertheless, raises an unavoidable question as to its legality and the British good faith in treaty-making. For the British-initiated change was unilateral, and subsequent to their 1984 agreement signed with the Chinese side. As will be noted below, the change made it difficult for the SAR government to maintain political predominance, such as the executive had always enjoyed under colonial rule until reversed under the last two British Governors.

It would be amusing to compare the change over time in the British attitudes toward elections for the Legco. I do not question Governor Christopher Patten's intention to use the elections as a step toward democratizing Hong Kong (Dimbleby 1997, p.140). But previous British positions on the elections were inconsistent. For instance, the first time the idea of an elected Legco came up was under Governor Sir Mark Young, after he was released from Japanese captivity and reinstated at Government House in 1946 at the end of World War II. Although the "Young Plan" had the support of the next Governor, Sir Alexander Graham, who arrived in Hong Kong in 1947, the Colonial Office in London, however, insisted that the electoral system

for the Legco should be, in Dimbleby's words (1998, p.102) "so gerryman-
dered to ensure that voters would always stack up in favour of the government." Governor Patten's elections in 1995, which produced the first fully
elected Legco in Hong Kong's history, did not follow the earlier Colonial
Office's mode, but instead used a combination of both geographical and
functional constituencies, complemented by the role of a selection committee, which picked a separate group of members. This mode, first concocted
in 1985, was used by Patten in 1995 and passed on to the post-handover era.

Chronicling the deliberate British move to transform Hong Kong's existing governmental structure, stretching over the 13-year interregnum
until 1997, is not our intent here. Nor is it a central concern of this book.
However, the end result was that the political system, as it existed in 1997,
was not the "current" system found in the colony at the time the 1984
Sino-U.K. Joint Declaration on the return of Hong Kong was signed. To be
more specific, the 1995 elected Legco became the center of the Sino-British
dispute, which came to a head as Beijing demurred on Governor Patten's
coup. It easily drew the world's attention and cast doubt on Beijing's trustworthiness, more especially after the Chinese announced in defiance that
they would replace the elected Legco with a Provisional Legislature after the
handover, pending the arrival of a legislature elected according to the Basic
Law in early 1998.

What is puzzling, however, is that Beijing never bothered to point out
the true issue in contention other than saying in generic and, hence, vague
terms, that Patten's elected legislature violated the Basic Law. The true
bone of contention in the dispute was whether, under international law,
the British had the right to unilaterally transform the "current" system as
spoken of in the 1984 Joint Declaration. In addition, the question remained whether the departing British rulers had the right to foster, in
1995, an elected legislature whose term of office would straddle the July
1997 handover date.

The Chinese argument was less than convincing because the Basic Law
was a domestic law, which, unlike the 1984 bilateral agreement, was not
binding on the British. When besieged by criticisms, the Chinese only made
the matter worse with their confounding, imprecise language, that there
would be "no through train" (*bu zhitong che*).

Consequently, the whole world was left with the unavoidable and distinct impression that, despite their pledge not to change anything for 50
years, the Chinese were, even before the takeover, contemplating changing
a major part of Hong Kong's governmental system, that is, its duly elected
legislature! The fact that it was the British who had initiated unilateral
changes first, thus provoking a Chinese backlash, was never called into
question by Beijing, nor was the possible British travesty on the sacro-

sanctity of Hong Kong's existing ("current" in 1984) system by their uni-lateral post-agreement changes.

While we are on the question of post-treaty British unilateral changes, to which Chinese backlash paradoxically boomeranged on Beijing, I might add the issue of the controversy over the Bill of Rights legislation introduced by Governor David Wilson in 1990, following the Tiananmen fiasco. According to Frank Ching (1997, p.60), Britain, as a signatory to the two mainstay international human rights conventions—International Covenant on Civil and Political Rights (ICCPR) and International Covenant on Economic, Social, and Cultural Rights (ICESCR), both signed in 1966 and entering into effect in 1968—had always maintained to the United Nations that Hong Kong needed no separate bill of rights legislation because human rights were fully protected in the colony. While the alleged protection was presumably guaranteed under the two covenants, the British had introduced no enabling legislation to make the rights enforceable in Hong Kong. Nor did the British even inform their Hong Kong subjects that they were entitled to the rights that these human rights covenants purported to protect. After the 1990 Bill of Rights Ordinance took effect in 1991, the British realized in hindsight that, despite their earlier assurances to the United Nations, dozens of Hong Kong's existing laws contravened the ICCPR covenant. Hence, beginning in 1992, several dozen laws had to be amended in what actually were post-agreement wholesale changes (Ching, p.60).

For the sake of illustration, let us look at two representative laws, the revisions of which were made necessary by the 1990 Bill of Rights Ordinance: (a) the Societies Ordinance and (b) the Public Order Ordinance—amended respectively in 1992 and 1995. The old Societies Ordinance, enacted in 1920, gave the Governor absolute discretion to declare as unlawful any society (or organization) that was used or even might be used for unlawful purposes or "purposes incompatible with the peace and good order of the colony" (Chan 1994, p.40). The law also prohibited local groups in Hong Kong from having links with foreign political organizations.[14] After the 1992 amendment, however, the Societies Ordinance permitted Hong Kong organizations to have ties with foreign political organizations. On the other hand, the old Public Order Ordinance required organizers of demonstrations to apply for police permits beforehand. The amended law made such permit no longer necessary: the organizers needed only to inform the police of their plan to hold a protest (Ching 1997, p.62).

The details of the amendments are not of direct concern to our discussion here. Nor is the true British motivation. What is relevant is the Chinese reaction and the bad publicity it brought, adding to the gloom of prophecies about Hong Kong's future after reversion. Quite typically, the Chinese objected to these belated relaxations of the old stringent laws on the ground that they defied the Basic Law; and avowed that after 1 July 1997 they would reintroduce the more stringent provisions that had been deleted or amended (Ching 1997, pp.60–62), as they later did. Again, it is puzzling that the Chinese did not point out that the British amendments of the existing laws were unilateral changes that undercut the inviolability of the "current" system guaranteed under the 1984 U.K.-PRC agreement. Equally, their failure to point out that their intended action after 1 July 1997 was simply to undo these unilateral, hence illegal, changes by the departing British and to restore *status quo ante* was amazing and beyond comprehension. No wonder the outside world's initial pessimism was magnified by the Chinese avowal to tamper with what appeared to be purely British efforts at democratizing Hong Kong.

All this boils down to the culprit alluded to earlier, that is, the loophole in the Sino-U.K. agreement of 1984, which failed to define what was the "current" system that was not to be changed for 50 years. Equally, just when the 50 years of no change should commence, and whether the British were permitted under any circumstances to make unilateral changes before the handover, were questions not nailed down. Nor were they raised by the Chinese side in the post-agreement Sino-British disputes. Under international law, good faith requires that parties to an agreement not frustrate the object and purpose of the agreement even before it enters into effect.[15] The British, because of the loophole, were able to initiate methodical changes that the Chinese found repugnant. The paradoxical result was that the entire world appeared to side with the British and to distrust the Chinese, because the latter had inscrutably failed to pinpoint the real source of the problems.

It may be too late to cry over spilled milk. But, it is an appropriate time to take due cognizance of this episode of the post-agreement Sino-British disputes, as we assess a possible rationale for the kind of paradoxical outcome on the SAR's domestic scene. Since the Chinese gave the impression—erroneously, as it turned out—that they were contemplating on meddling in Hong Kong's political system even before they took over, as just described, Western commentators as well as many local pundits and political party activists were led (or misled) to the worst scenario case in their expectations about what was to happen after 1 July 1997. Hence, the prophecies of doom took off.

Eventually, when things did not turn out as badly as expected on the SAR's political scene, the prophets of doom were at a loss to explain why. I know of one publication, written to justify its co-authors' earlier downright dismal forecasts, none of which has come true, tried instead—hitting below the belt, as it were—to make hay of the SAR's unanticipated economic downturn. The region-wide financial crisis was twisted to appear as though it were proof that life in Hong Kong after reversion was, as the co-authors had predicted, "worse off than during the last years of British colonial rule."[16] The specter thus created would, so it was hoped, justify its co-authors' previous predictions for the end of Hong Kong after the Chinese takeover, which, they had in fact claimed, would bring corruption, nepotism, cronyism, and related plagues as well as burial of the legal and judicial system and loss of freedoms.

In order to use the economic downturn as evidence that Hong Kong is going under because of the end of British colonial rule, one has to confront a methodological problem, as I shall explain below. Since there are people who have done so—and no doubt more people may be tempted to judge the Hong Kong SAR by its unanticipated economic woes—I think it necessary to address the economic downturn separately as another source of the paradox syndrome being addressed in this book.

Unanticipated Regional Financial Attack

The second possible source of the paradox surrounding Hong Kong after reversion is its unheralded economic problems, the incidence and gravity of which, again paradoxically, flew in the face of some upbeat projections (cf. Enright, Scott, and Dodwell 1997). But, before we jump to conclusions, methodological prudence dictates that we disaggregate Hong Kong from the rest of Asia Pacific, which is similarly battered by economic woes in the midst of a regional financial crisis, ranging from Japan to Indonesia, and from Singapore to Thailand. For only Hong Kong was undergoing a transition to post-colonial rule, and its exit from colonial rule coincided with the onslaught of the regional financial crisis. In fact, the crisis was brought on by the devaluation of the Thai baht on 2 July 1997, the day after British rule ended in Hong Kong.

To claim that Hong Kong's economic deterioration was due to the end of British rule, as some seem to insinuate, one would have to mount a similar prior claim, however ludicrous, that the British departure (from Hong Kong) was responsible for the Thai baht devaluation! While both domestic and external factors have been offered by analysts for the economic plague sweeping the region, one commonality was that, despite differences and the varying degrees of severity of some of the similar domestic problems,[17] all economics in the Asian Pacific region were hit by a financial crisis externally fueled by speculations by mammoth international hedge funds and competitive currency depreciations (Sung 1997).[18] To claim that Hong Kong's economic difficulties were brought on by the infelicitous departure of the British, logical consistency demands that the same claim (about British departure) be made for all the region's economies.

In addition, methodological prudence requires that we place Hong Kong's unexpected economic downturn in comparative light. Before we do so, however, let us not lose sight of the context of this discussion, which is that the sometimes rosy forecasts by respectable economists (e.g., Enright, Scott, and Dodwell 1997), echoed by the media, including the *Asian Wall Street Journal,* about Hong Kong's post-1997 continuing economic prosperity were, in retrospect, a possible "source" for the paradox surrounding the post-handover turn of events to the contrary. By our definition, this discrepancy between earlier (even though quite rosy) expectations and subsequent outcomes (the unforeseen economic upset) constituted a paradox, an auspicious one at that, as contra-distinguished from the opposite, inauspicious paradox (in which doomsday predictions were followed by surprisingly normal outcomes) mentioned above. Indisputably, the region-wide financial turbulence was totally unanticipated; so were the economic woes it triggered in Hong Kong as elsewhere. Y. Y. Kueh's chapter has more to say on this. In comparative light, Hong Kong's record in coping with the financial crisis was not bad at all.

Even if one discounts the external factors and zeroes in on the domestic problems that may have accounted for the economic downturns throughout the region in 1997–98, it has to be pointed out that Hong Kong, next only to Singapore (and possibly Taiwan), was faring much better than the rest of the region, similarly hit by the financial crisis. As the U.S. Consul General, Richard A. Boucher (1998, p.1), put it, Hong Kong's ability thus far to "have weathered the regional economic crisis is in many ways a testament to the success of tested principles: adherence to the rule of law and the rules of the market, the free flow of information, capital, and goods, and clean, efficient, and non-interventionist government." All these principles, concededly, were inherited from the British era but, to the SAR's credit, have remained intact (Huang 1998, p.45). In addition, one could add another factor contributing to Hong Kong's holding out as well as it did under the crush of the regional financial turmoil. That factor was the unequivocal assurance by Premier Zhu Rongji that the Central Government would back up the SAR, if need be, with China's massive U.S.$149.1 billion foreign reserves, on top of Hong Kong's U.S.$96.4 billion.

In this comparative light, it would be less than convincing, in fact misleading, to try to link Hong Kong's economic downturn to the departure of its former British rulers.[19] If this link could be maintained, does it mean that the solution would be to bring back British colonial rule? Even if it could be proven, as it has not been, that there is a link as such, it would not be tantamount to proving that Beijing was at fault through its (unproven) undue meddling in Hong Kong's affairs. As noted above, previous speculations about Beijing's possible high-handed interventions were the basis for most of the downcast predictions about life in Hong Kong beyond 1997. But, all sources, including the ones quoted above, point to the opposite direction, namely, that Beijing did not interfere in the SAR's socio-political life after the handover. While Ting Wai's chapter has more to say on the subject, let me cite the relevant findings from a survey conducted by the *Yazhou Zhoukan* (1998, pp.42–50). On the question of the SAR's relations with Beijing in the year following the handover, an overwhelming majority of the respondents gave a rating of 8.1, on a scale of zero to ten. Close to one-third even gave a satisfaction score of a full mark of 10 (p.42). In other words, the survey confirmed the people's gratification with Beijing's hands-off policy, which calmed earlier apprehensions.

In this respect, I would suggest that a fine distinction be made between the HKSAR government, headed by Chief Executive Tung Chee-hwa, and the Central Government in Beijing. Plenty of analysts have ascribed post-reversion Hong Kong's economic discomfiture—assuming we can momen-tarily leave out its possible external causes—to the misguided policies of the SAR government. Loksang Ho, for instance, in his chapter takes issue with Chief Executive Tung's new housing policy as the culprit for the unraveling of Hong Kong's real estate business, and with it the overall economy at large. Yet, Ho is quick to point out that the fall of the SAR's economy is unrelated to Beijing, because the Central Government has stuck to a record of leaving Hong Kong "entirely on its own." As an example, he points to the holding of demonstrations in commemoration of the 4 June incident in Victoria Park, without interference, in the summer of 1998.[20]

Earlier we quoted from the U.S. Department of State's report (as of 1 April 1998) to Congress on the state of Hong Kong after it became a Spe-cial Administrative Region of China in 1997. In the quote, we noticed the passage that "[d]emonstrations—often critical of the PRC [China]—con-tinued to be held." The report went on to say that the Hong Kong govern-ment reported that there had been over 1,000 public demonstrations in the seven months since the handover in July 1997 (p.9). As additional evidence of Beijing's non-interference in the SAR, the same U.S. State Department report noted that, despite the 8,000 People's Liberation Army (PLA) soldiers stationed in Hong Kong, as per prior agreement between U.K. and PRC, their presence is next to being unnoticeable. For, after they had quietly moved in, "symbolizing the 'retaking' of Hong Kong after a century and a half of colonial rule," the report added, "the PLA soldiers have remained out of sight and confined to barracks since July [1997]" (p.13). This part of the U.S. State Department report collided head-on with the vision once held by Martin Lee, the most noted anti-Beijing opposition leader in the Legco, who saw the PLA soldiers marching into Hong Kong on handover day, arresting innocent people, and putting them in jail without a trial (Dimbleby 1997, p.287). The Department's second report, dated 1 April 1999, as we also noted, confirmed its own report filed the previous year.

Even assuming we can establish the fault of the Tung Chee-hwa SAR administration for its misguided policy, such as its Tenant Purchase Scheme (TPS) housing policy, which according to Loksang Ho's chapter is largely responsible for triggering Hong Kong's economic setback, the question remains as to Beijing's culpability in all this. Considering all the evidence of Beijing's non-interference, the only compelling, although somewhat specious, conclusion would seem to be that the Central Government's fault was that it had remained too aloof and followed too literally a policy of allowing "Hong Kongers to run Hong Kong."[21] However, any deviation by Beijing from its hands-off policy thus far would run afoul of the Basic Law, Hong Kong's mini-constitution, and would, moreover, throw out of sync the "one country, two systems" model so thoughtfully crafted by the late Deng Xiaoping for Hong Kong in the post-colonial era.

Confusing Novelty of the "One Country, Two Systems" Model

The third possible source for the kind of paradox syndrome discussed in this book can be traced to the innovativeness, as well as ingenuity, of the " one country, two systems" formula used in accommodating Hong Kong's return to a China that has a system totally different from the former's capitalistic system.

The "one country, two systems" model, in the way it is implemented after Hong Kong's reversion, is probably without precedent in human experience. To put it succinctly and in simple constitutional-legal terms, the Hong Kong SAR is both an "inalienable part of China" and (with emphasis on "and") a distinct entity with a "high degree of autonomy."[22] But the simplicity of this statement, while absolutely accurate, is deceptive. As we shall see below, on more than one occasion, foreign judicial organs were confused as to the exact nature of the symbiotic system to be installed in the HKSAR, in their anticipatory evaluations of the latter's legal status and competence both in its own right and vis-à-vis its Chinese sovereign.

For the uninitiated, it seems, an amplification is necessary, in order to help explain why I have suggested that the confusion regarding Hong Kong's forthcoming unprecedented governance structure might have been a third source of the outside world's rampant prophecies of doom about politics and human rights in the future SAR—prophecies that have not been borne out by the unfolding reality after the handover.

Despite the model's awkward phraseology ("one country, two systems") and the deceptive simplicity in the way it can be described, as shown above, it is a most complex and elaborate conjoining of a nominally all-powerful sovereign with a small autonomous enclave, the Hong Kong SAR, which is granted plenty of wiggle room under the Basic Law and by the grace of the sovereign. The symbiosis is a seamless integral whole. Its design strikes a most carefully calibrated balance between the sovereign and the Special Administrative Region, in their respective powers and functions. The epithet "autonomy," or to be more exact, "a high degree of autonomy," connotes that the Hong Kong SAR has powers to govern itself in all aspects of *domestic* life, that is, in all aspects except in defense and foreign relations. In the context of the time-honored unitary (as opposed to federal) system that has characterized the Chinese polity since 221 B.C., I must hasten to add that the powers of the SAR's autonomy are exercised within the parameters set by the constitution of its Chinese sovereign (the People's Republic of China, or PRC), and by the wills of the sovereign's national parliament, the National People's Congress (NPC).

As such, the SAR's preexisting social and political systems, conceived in the capitalist and common law tradition from its pre-1997 colonial era, were destined to continue, at least for 50 years, after Hong Kong's return to China. Equally, all existing laws shall continue in force. The SAR's own legislations (enacted by the Legco), which stand on their own after becoming law upon signing by the Chief Executive (Art. 48(3) of the Basic Law), are to be filed with the NPC in Beijing, not for approval but for record keeping (Art. 17). Other than a fixed number of national laws passed by the NPC—as stipulated in Annex III, as amended, to the Basic Law—such as pertaining to the designation of Beijing as the national capital, the design of the national flag, national holidays, etc., no other national laws are applicable to the SAR (Art. 18).

As a part of China, on the other hand, Hong Kong's autonomy does not preclude the Chinese sovereign's power in assuming the responsibility of the SAR's foreign affairs and defense, as per Articles. 13 and 14 of the Basic Law, respectively. Later on, I am going to come back to a peculiarity in respect of the power to conduct the SAR's "foreign affairs." Let me pause now to point out that the uniqueness of the "one country, two systems" model does not lie in the "two systems" half of the schema. The conjoining of two systems into one common sovereign entity of sorts has been attempted elsewhere before. A more recent example was the United Arab Republic (UAR), comprising Egypt and Syria, which existed from 1958 through 1961, when Syria withdrew, although the name UAR continued until 1971. Other examples in our century could arguably include the dual monarchy of Austro-Hungary (1867–1918)[23] and the Czech-Slovak experience (1918–93). Still another example, though briefer in time but closer to home, was the trouble-laden union of Malaya and Singapore into Malaysia, lasting only two years (1963–65).[24]

The novelty of the "one country, two systems" model, in the final analysis, is that Hong Kong, as an inalienable part of China, is a local unit comparable to a province in the Chinese unitary system. Despite the potential conflicts which Ting Wai lays his hand on in his chapter, the Hong Kong SAR as such, within its own territorial confines, enjoys and exercises executive, legislative, and judicial powers in a way that can be described, in the de facto sense, as "independently" of the Central Government. The "one country, two systems" model demonstrates a rare "unity in diversity," to borrow words from Albrecht-Carrie (1973, p.155) in another context, which was not found in any of the other examples. As an SAR under Chinese sovereignty, Hong Kong's residents enjoy freedoms and rights protected not only by the Basic Law, but under its inherited common law tradition, and also by the relevant international treaties to which Hong Kong is a party (cf. Roda Mushkat 1997). It is possible that Hong Kong residents may be entitled to more rights and wider access to international protection, such as afforded by certain human rights treaties, than are citizens in China proper. Besides, although not endowed with the legal attribute of sovereignty, the Hong Kong SAR is empowered to act on the international plane through its membership in 51 international organizations[25] and as a party to 169 multilateral treaties, not counting the numerous bilateral treaties.

The uniqueness of the "one country, two systems" model, in short, lies rather in the "one country" half of the equation. This is particularly noteworthy, when one considers that the Chinese word for "country" (*guo*) in the "one country, two systems" scheme actually encompasses three ingredients represented in English by three separate words: state, country, and nation. Yet, while Hong Kong is an inalienable and integral part of sovereign China, the sovereign's laws, with the exception of those specified in Annex III, as amended, to the Basic Law, do not apply in the SAR. The SAR's representatives may sit on China's delegations to those international organizations of which Hong Kong is not a member. The reverse, however, is not necessarily true. For Hong Kong, in its own right, may be a member of certain other international organizations, such as the World Trade Organization (WTO), of which China is not or not yet a member, and Hong Kong is under no obligation to take on Chinese representatives as part of its delegations.

Hong Kong's relationship with its sovereign cannot be described in the normal terms of either a confederate or federal system. China has a unitary system in that, contrary to what the tenth amendment to the U.S. constitution provides for the American union, whatever power is not delegated to Hong Kong inures to the Central Government of China. I would love to see someone come up with either a historical or a contemporary example in human history that had or has a governance structure that can be compared with the "one country, two systems" arrangement accommodating Hong

Kong's relationship with China proper, following its return to Chinese sovereignty (more on this in my concluding chapter). This outstanding uniqueness, in the final analysis, quite understandably caused confusion among most commentators, including foreign judicial organs, in trying to pinpoint Hong Kong, a non-sovereign international actor, on the grid of our Westphalian system still conceived in the paradigm of sovereign states. The confusion, as such, contributed to the paradox syndrome that the book attempts to explicate.

Let us now return to the "peculiarity" of the SAR's foreign affairs. The complexity of the problem is further abetted by the peculiarity in the arrangements laid down by the Basic Law, in the area of the SAR's foreign affairs. In the first place, the word "affairs," not "relations," is used in the Basic Law; and it covers wider grounds. For instance, the issuance of travel documents, such as the SAR has the capacity to do for its residents, is a matter of "foreign affairs," not "foreign relations." Secondly, a careful perusal of the relevant provisions of the Basis Law reveals, paradoxically, that the Law has more to say on what the SAR can do, than what its sovereign (central government) can do, in regard to Hong Kong's foreign affairs. Other than a brief provision (Art. 13) that the Central Government "shall be responsible for the foreign affairs relating to the Hong Kong Special Administrative Region and that it shall have an office in Hong Kong to deal with foreign affairs," the remaining relevant provisions (Arts. 150–57) all pertain to Hong Kong's roles and initiatives in foreign affairs. They cover the SAR's participation, as members of Chinese delegations, in diplomatic negotiations affecting Hong Kong; its capacity in concluding and implementing international agreements; its participation in international organizations (though under the designation "Hong Kong, China"); its power in the issuance of passports and other travel documents to the SAR's permanent residents who are Chinese citizens; its capacity to maintain trade and consular missions in foreign countries; its discretionary power in receiving non-governmental missions from countries having no formal diplomatic relations with Beijing, etc. Although in regard to these functions the Basic Law provisions do pay lip service to either authorization by or consultation with the sovereign power in Beijing, the general thrust of Chapter 7 on Hong Kong's "Foreign Affairs" (i.e., Arts. 150–57) is the high degree of autonomy that the Law confers upon the SAR.

Here is the problem, or source of confusion, for an outside observer unversed in the tradition of Chinese politics, including the background provided above regarding the deliberate balance built into the symbiotic "one country, two systems" model. The crux of the matter is that the provisions in the Basic Law concerning Hong Kong as an "inalienable part of China" are far outnumbered and eclipsed by those pertaining to the "high degree of autonomy" enjoyed by Hong Kong. The lopsided imagery, thus conveyed, seems to have confounded external commentators before and after the inauguration of the SAR in July 1997 and, equally, the foreign judicial organs that had occasion to pronounce themselves on the legal capacity of the SAR.

In my Chapter 7, below, I examine four different cases that came before the judicial organs of three countries (United States, Britain, and Australia), in which both the novelty and the complexity (including the eclipse of Beijing's sovereign power by the detailed provisions in the Basic Law substantiating the SAR's autonomous power and status) seemed to have overwhelmed the jurists concerned. Without duplicating what I say in Chapter 7, I would like to zero in on one consequence from this novelty and complexity of the "one country, two systems" model that was installed in Hong Kong after the handover. In all four cases to varying degrees, the foreign judicial organs seemed to be highly uncertain about two possibly conflicting interpretations, i.e., to see the Hong Kong SAR as a distinct entity with a "high degree of autonomy," on the one hand, and to take it as a subordinate local unit ("an inalienable part") within the Chinese colossus upon its return from British rule, on the other hand. In one specious decision (the Matimak Trading case), two U.S. federal courts opted for the first interpretation. They defined Hong Kong, living under the "one country, two systems" structure, as a "stateless" entity, since it is endowed with no sovereignty, although it enjoys a high degree of autonomy. The net result was that the Matimak Trading Co., incorporated under the laws of Hong Kong, was likewise tainted with a "stateless" stigma, hence having no *jus standi* to sue before a U.S. federal court. As I suggest in that chapter, the two successive U.S. federal courts' decisions were egregiously erroneous, given our knowledge of the American *stare decisis* (cf. the precedent set by the Cedec Trading Co. case), and considering the import of the U.S. Hong Kong Policy Act of 1992, the intent of the Basic Law of the HKSAR, and whatever guide we can find from general international law. Without rehashing the arguments made in Chapter 7, let me reiterate that the reason for the erroneous decision in the Matimak case and, equally, for the initial consternation of the foreign judicial organs in the other three cases, stemmed from the seeming incomprehensibility of the innovative "one country, two systems" governance structure of the HKSAR.

Hence, if any commentator had been thrown off by the "inscrutability" of the model as such and had come away with forecasts of great uncertainty ahead, it should have come as no surprise. It is for this reason that I have suggested that the seeming inscrutability of the SAR's "one country, two systems" governance structure could very well have been a third source of the paradox syndrome under study in this book, arising as it did from the glaring discrepancy between the (obviously confounded) prior fore-casts and the subsequent (quite normal but contrary) outcome following Hong Kong's reversion.

Bad Luck, Mismanagement, and Unintended Landmines

In addition to the fine distinction between the SAR government and the Central Government in Beijing, which I urged should be maintained, I would suggest a further distinction: that between mismanagement by the SAR government and some unexpected turns of events that were purely nat-ural accidents (bad luck?). Both, of course, had a bearing on the ratings of Chief Executive Tung Chee-hwa's job performance and the public's confi-dence in the SAR government.[26] The two, admittedly, are related. LAU Siu-kai's chapter alludes to "a slew of mishaps whose convergence at more or less the same time was mind-boggling." More specifically, he is referring to the bird flu hitting Hong Kong in the winter of 1997–98, the series of bus crashes during the year, a string of medical accidents caused by human neg-ligence but mysteriously recurring for no good reason, and the toxic algae "red tide" in March 1998 that killed large quantities of fish, an important source of staple food for Hong Kongers. To this list can be added the advent of killer diseases such as the dengue fever epidemic that hit the territory and put the government on alert in the early part of 1998.[27] For a people who strongly believe in omens inherent in natural disasters, this slew of mishaps cannot but make them wonder, in Lau's words, "what has gone wrong after the end of colonial rule." In addition to the economic slowdown, Lau finds, these events of bad luck contributed to the people's declining confidence in Tung Chee-hwa's ability to run the SAR government. In fact, Tung and his SAR government were seen as "incompetent in handling [these] crises."

Below, I shall explain that certain changes the departing British rulers had made, while well intentioned to ensure the independence of the bureaucracy and to enhance the power of the legislature, have inadvertently turned out to be gratuitous road-blocks to Tung's executive power and his ability to function normally as the Chief Executive. For lack of a better term, I have used "unintended land mines" merely as a short-hand label, without any malice intended.

To the above named natural calamities should be added certain in-scrutable events, such as the new Chep Lap Kok airport fiasco. Built with state-of-the-art technology at a cost of HK$155 billion (approximately U.S.$20 billion), it was to replace the over-crowded and antiquated Kai Tak airport. On its first day of operation on July 6, 1998, however, nearly every-thing went wrong. Software bugs caused the computer system to blank out and created massive chaos. Four of the 38 air-bridges proved to be faulty and not working properly, and passengers had to wait on the tarmac for hours before they could disembark. Some had to wait as long as five hours for their luggage. An estimated 10,000 bags missed flights. The cargo services were even worse. The largest, most advanced Super Terminal 1 was paralyzed. The backlog was so bad that air cargo services at Chep Lap Kok had to be suspended for over eight days. Part of the air cargo traffic was diverted to the old Kai Tak airport, and part, with Beijing's endorsement, was rerouted to airports in Macau and Shenzhen, the closest Chinese town to the north (Weng 1998, pp.7–8).

As Byron Weng (1998, p.9) points out, most of the other mishaps could be laid to external causes, but the fiasco at the Chek Lap Kok airport was "totally Hong Kong's own doing." The only caveat that might be entered is that the computer system and most other software facilities, as well as the hardware equipment and infrastructure, were supplied or built by British firms contracted by Governor Chris Patten during his last years before the handover (cf. Paau 1998, p.65). But, rightly or wrongly, most people pinned the blame on "arrogant and incompetent officials." In public opinion polls, 68.5 percent of the respondents thought Chief Executive Tung was too slow in responding to the airport problem.

Some bureaucratic *faux pas* or idiosyncratic practices may have been initiated from very low levels, as Tung's avowed attempt to "depoliticize" postcolonial Hong Kong (see LAU Siu-kai's chapter on this) may have made this possible if it results in his hands-off policy on certain low-level decisions. But, whatever the reason, Tung could not escape the blame. One such incident was the bizarre move by the SAR's Immigration Department to affix a stamp on all travel documents held by Taiwan visitors warning them not to display emblems of "the authorities of the Taiwan region" and "not to engage in any activity that may embarrass the HKSAR government" (Ching, Lian, and Weng 1998, p.22). It turned out that the requirement came from the whims of lower rungs of the department, and was discontinued only after Tung's aides received complaints from Taiwan's representative office.[28] This incident may well be telling evidence of an entrenched bureaucracy often alleged to work at cross purposes with the Chief Executive.[29]

In his chapter, LAU Siu-kai notes that the Chief Executive's relationship with his principal officials (hailing from the top rungs of the bureaucracy) was an "uneasy one." Except for the Secretary of Justice—who was hand-picked by Tung Chee-hwa—all the SAR's principal officials were holdovers from the colonial regime. Some were even groomed by Governor Chris Patten to fill top administrative posts for the transition. They have stayed on under the ill-defined "no change" pledge and because Tung wanted to keep them for the sake of continuity. As Hong Kong has had no tradition of political appointments since the colonial days, all positions of ministerial rank were, and still are, filled by individuals promoted from the civil service. The end result is that it insulates the bureaucracy, including the top officials, from the reach of the Chief Executive's power or even supervision.[30] In one instance, Financial Secretary Sir Donald Tsang even committed U.S.$1 billion out of the HKSAR's coffers to the International Monetary Fund's U.S.$16 billion bailout package for Thailand, without prior consultation with the Chief Executive. The unusual event, attesting to the Financial Secretary's enormous independent power, cannot find support in the Basic Law, and would, arguably, be comprehensible only within the context of the political bequest by the last British Governor. I use the word "bequest" here in a peculiar sense, as I doubt if any top official serving under Patten during his tenure as Governor would have such enormous independent power. But, it was made possible by his deliberate endeavor to insulate the bureaucracy for the future.[31] The Chinese pledge to keep intact Hong Kong's inherited governmental system, with the exception of Patten's elected Legco, may also have helped to leave the SAR's bureaucracy very much to itself.

Before proceeding any further, let me inject a brief commentary on the British legacy in Hong Kong, in order that the discussion here will be seen in more balanced light. It should be noted that there is a drastic difference between British views of their legacy and those held by some independent commentators in Hong Kong, let alone Beijing's views. Purely seen from the constituencies in Britain, the legacy of the colonial administration in Hong Kong was chiefly "formed from the characteristics of the territory on the eve of retrocession" (Hook 1997, p.553). Leading local Hong Kong pundits, however, tend to take a longer view of history on the British legacy (e.g., Chan 1997, p.567ff.), some even over the entire colonial era of over 150 years (Liu 1997, p.583ff.), in their assessment.

In the discussion here, I tend to concentrate on the material effects of Governor Patten's reforms on the SAR government, in the way it functions or fails to function after its inception in July 1997. British contributions over the longer span of time in the past, such as the legacy of the rule of law, and a clean government, etc., are not at issue, much less questioned. I would add, nevertheless, that in this discussion no value judgment is intended. Presented here is a matter-of-fact assessment of the identifiable effects from the more immediate bequest of Governor Patten, solely for the purpose of explicating what has happened, and why, since the departure of the British rule. The kind of post-handover tug of war between Tung Chee-hwa's Executive Council (Exco) and the legislature, especially following the May 1998 elections, is a case in point.

The running tug of war between the Pattenesque insulated bureaucracy, as noted above, and the Chief Executive seems to have affected the government's efficiency in its responses to the spate of natural and man-made disasters ranging from the bird flu to the Chek Lap Kok airport fiasco. It seems to have accounted for the publicly perceived ineptitude of the SAR's bureaucracy and the high officials hailing from its ranks.[32] In the *Yazhou Zhoukan* survey (1998, p.44), respondents laid blame on many of these officials, who owed their positions of eminence to Governor Chris Patten in his campaign to build an institution that was presumably equipped to withstand the effects of future regime change. The perceived ineptitude, nevertheless, undermined the public's traditional confidence in Hong Kong's civil service. Chief Executive Tung Chee-hwa, who managed to receive high ratings on a par with the best British Governors in the past (*Yazhou Zhoukan*, p.45), had his hands tied because of the "no change for 50 years" frame of reference under which the former colony was returned to Chinese sovereignty. Here one finds a paradox of sorts in that a bureaucratic elite coterie picked by the outgoing Governor Christopher Patten on political grounds (i.e., their ability to "defend democracy") has proven to be inept in handling the many haphazard crisis situations that happened to plague Hong Kong in its first year as a Chinese SAR. To put it more graphically, Patten's well-intentioned reform, in this particular instance, has turned out to be a hidden landmine, if not an outright Trojan horse.

In her chapter, Beatrice Leung notes seven constraints on the Chief Executive's power and, in addition, the *mediocrity* of the "high administrators" in government, who are holdovers from the previous regime. This may explain the Hong Kong SAR's inability to cope promptly and efficiently with the sudden economic setbacks amidst a financial crisis coming out of the blue. But, it also points up an anomaly in retrospect: The belated attempt by the departing British to ensure checks and balances, and the elaborate design, sealed by the Sino-U.K. agreement and the Basic Law, to institute a labyrinth of delicate balance between the Chief Executive and the rest of the SAR government, and between the SAR and China proper under the "one country, two systems" model, has been responsible for the legacy of an entrenched bureaucracy in the SAR government and, equally, for the literal hands-off policy of Beijing that has left the Chief Executive to fend for himself. While the Hong Kong public is gratified with the consequential continuance of press freedom, human rights, the rule of law, openness of government, etc., it has lost confidence in the SAR's ability to cope with the ongoing economic downturn and other crisis situations (*Yazhou Zhoukan* 1998, 42f.). This is confirmed in the various chapters in Part I below on the SAR's domestic scene, which together illustrate and support the paradox syndrome enwrapping Hong Kong in its historic transition.

Again, in his chapter, LAU Siu-kai speaks of the failures of Tung Chee-hwa's leadership strategy, including his attempt to depoliticize the Hong Kong polity. The truth is that, with an entrenched bureaucracy and a

Legco that was artificially thrust into the forefront of the political stage, neither the Chief Executive nor Beijing can do much better vis-à-vis what they have inherited from the British following Hong Kong's reversion, lest they break their "no change" pledge. Much less can the Chief Executive hope to succeed by way of, in Lau's term, "depoliticizing" a system that had been deliberately, albeit with an allegedly lofty intent, politicized under Patten.

One indication of the obduracy of the holdover high officials from the colonial era is afforded by what transpired during a hearing held by the SAR's legislature (Legco) on 9 July 1998. "Our airport has become the laughing stock of the world. Do you still think you are completely immune from any responsibility for that?" asked one legislator of Dr. Hank Townsend, the Airport Authority chief executive. Altogether, 18 high-level executives or government officials were present at the hearing, including Dr. Townsend. But no one would admit to a share of the responsibility, nor did they see a need for anyone to be held responsible for the fiasco (Weng 1998, p.8).[33] The Chinese hands-off stance, it seems, has paradoxically contributed to providing a gratuitous blind shield for the holdover officials from being accountable either to the Chief Executive or to the Legco. Not until very late in the game did Tung realize that the public's disgruntlement with the civil service would ultimately translate into disillusionment with the Chief Executive himself. Belatedly, in his second state-of-the-SAR report, Tung singled out the 190,000-strong civil service as being targeted for a performance review. He demanded higher efficiency and a 5 percent reduced expenditure in three years.[34]

As James Tang (1998a, p.11) has aptly observed, preserving colonial political institutions while granting limited democracy has proved to be "costly" for the HKSAR. The insulation of the bureaucracy alluded to above, while well-intentioned, seems to have incurred the "cost" of inaccountability, as shown in the Chek Lap Kok case. Another instance of political "cost," I might add, is the tension-ridden executive-legislative relations in the SAR government after the 24 May 1998 election, which produced the territory's first elected legislature after the handover. The new Legco has replaced the previous provisional Legco, which was largely a lame duck during the initial transition period.

To fully appreciate this point, we have to recall that elections for the Legco were unknown until the last years of the long British colonial rule. The first ever direct elections were held only in 1991, or seven years after the signing of the 1984 Sino-U.K. agreement on the return of Hong Kong (Lo 1997, p.181). As part of a drive toward democratization in the territory, the move was hailed as a necessary guarantee of a democratic legacy for the future of Hong Kong after 1997. Earlier we noted Chang Hu's

(1996) comment that in their post-treaty wholesale change of the political system in Hong Kong, the British maneuvered to shift the gravity of a long-standing executive-led government to the legislature. If anyone had doubts before about the long-term effects of this structural change, they should be dispelled by now, considering what has happened to Tung Chee-hwa's SAR government after the handover. Despite the Basic Law, which sought to continue the SAR's governance structure after Hong Kong's historic executive-led model, the novel direct elections the British introduced for the Legco shortly before their departure, according to James Tang (1998, p.12), have left a Trojan horse, as it were, for the SAR in that it has proven impossible for the Chief Executive to try to maintain a semblance of executive predominance. The irony is that, at a time of financial crisis hitting the region after 1997 and the resultant economic meltdown, Hong Kong finds itself in dire need of a strong, executive-led government, something the former colony always had under British rule until its final years.[35] But, such is lacking when it is sorely needed, thanks to Patten's reform with all its good intentions.

Although the 24 May 1998 elections did not return any single majority party, the Democrat Party (with 19 seats, out of a total of 60) became the largest party in the Legco, and all the major parties across the pro-Beijing and pro-democracy ideological divide joined hands to force the SAR government to do their bidding, such as adopting an economic rescue plan, which resulted in a deficit budget for 1998–99. Despite what the pundits had predicted, the first elected SAR Legco turned out to be assertive, prompting one Executive Councilor (i.e., member of Tung's advisory body) to refer to all Legco members as the "opposition" (Tang 1998, 13). I should add that, despite the urgent needs arising from the unanticipated high unemployment rate created by the economic slowdown, Hong Kong since the colonial days never has had an unemployment insurance program. And, ironically, the government's unemployment insurance proposal did not even get off the drawing board, simply because the Legco's predominant coalition led by the Democrats opposed it on the ground that unemployment insurance (reflecting a reflex from colonial times) would only encourage the "lazybones."[36] And Tung's government could do little about it. Before the inauguration of the elected Legco after the May 1998 election, by contrast, the government was able to push through the previous less assertive (albeit admittedly less representative) Provisional Legislature the necessary legislation making operative, before the start of the year 2,000, a Mandatory Provident Fund for retirement.[37] This, incidentally, is the first ever comprehensive retirement scheme covering more than just government employees in Hong Kong. Under British rule, only the latter, presumably mostly British expatriates, had retirement benefits from their employment.

When viewed with hindsight, the original British endeavor to shift the center of gravity in government from the executive to the legislature has, in effect, resulted in emasculating the executive power of the SAR government to such an extent that, as shown in the scrapped unemployment insurance measure, the government cannot even have its way in helping the needy people caught in the clutch of a totally unanticipated recession, which by the spring of 1999 had produced an unprecedented 6 percent jobless rate, not to mention a 2.9 percent underemployment rate. Again, when this emasculation of executive power, however unfortunate it has turned out to be, has resulted from Patten's professed attempt to democratize Hong Kong for posterity, what a paradox-*cum*-irony it is.[38] An undeniable development since the May 1998 election was the less than normal working relationship between the Legco and the Chief Executive. Although Tung Chee-hwa was in part to blame for his scanty appearances before the Legco,[39] the problem may not be a one-way street. At issue is that the kind of confrontational relationship as it exists between Tung and the Legco should come as no surprise. This is so because of the two opposing pulls on the government: (a) the Pattenesque bequest, raising expectations in the Legco that it is the center of governmental gravity, and (b) the Basic Law, which purports to favor political predominance of the executive, as had always been the case before 1984.

Still another paradox is that, despite pre-handover predictions by many pundits and the mass media to the contrary, Hong Kong residents' confidence in the SAR's political future went up to a rate of 135 points (using April 1996 as the base of 100), even as widespread unemployment was rising in early 1998. Their confidence in the territory's economic prospects, by contrast, declined to 49.7 points.[40] HO Loksang notes two more minor paradoxes: (a) the record number of returnees from abroad to Hong Kong after the end of the transitional uncertainty, plus a low record number of emigrants, has only exacerbated the territory's job squeeze; and (b) Chief Executive Tung's policy of catering to the housing needs of the people-in-the-street, in contradistinction to the former British fixation with the elite and corporate interests, has led to the fiasco of the real estate slump triggered by his new TPS project, as Ho explains in his chapter below.

To the above seemingly unending list of paradoxes surrounding Hong Kong's experience regaining itself from the end of colonial rule, we should add another one, in the sense in which the word paradox is used in this book. Governor Patten, according to Dimbleby (1997, p.140), once recalled his trip to Beijing in which he discovered that the Chinese were "more obsessed with [anti-Beijing leader] Martin Lee, with Hong Kong becoming a focus for unrest in China, and with the consequent threat to their control over Hong Kong, than we had been aware." If that obsession and fear were true—which I have no doubt were—then the very fact that in the months since the handover Hong Kong has not degenerated into a citadel of anti-Beijing activism, even in spite of its hands-off policy, was undeniable evidence that Beijing's reported fears were largely unfounded. Thus, a paradox in itself. This, incidentally, may explain why the HKSAR's Secretary of Justice, Elsie Leung, disclosed in July 1998, or one year after the handover, that she would not be presenting draft legislation on subversion, secession, sedition, and treason, as required by the Basic Law (Art. 23), to the Legco for debate any time soon (Weng 1998, p.18). Speculations that Beijing would push for such legislation at the earliest possible chance had, again paradoxically, prompted widespread apprehensions before the 1 July 1997 takeover date.

In this connection, it bears noting that the future of the SAR's autonomy does not lie just with the guarantee of the Basic Law, or Beijing's hands-off policy, but also with the residents' restraint in not allowing Hong Kong to become an anti-communist bastion against China proper. Equally, as Ting Wai's chapter counsels, Hong Kongers owe it to themselves to take judicious cognizance of the cardinal principles governing the unitary system cherished by Beijing for the whole of China. This condition—which testifies to an implicit two-way traffic (of mutual abstention from meddling in each other's internal matters) in the "two systems" commitment in the "one country, two systems" schema, which guarantees Hong Kong's autonomy—is a prerequisite for the sustainability of Beijing's hands-off policy toward the SAR. It is in this light that some of the reported events pointing to the opposite direction to restraint on the part of Hong Kong's hard-pushing democratic leaders should be cause for concern.[41]

The list of similar paradoxes could go on. But, the above is enough to illustrate what I have tried to convey in this introduction to a volume that proclaims as its essential goal the unraveling of the myth of the super paradox surrounding Hong Kong during its transition from a British colony, after 155 long years, to a Special Administrative Region under Chinese sovereignty.

Although the facts and data used here were largely culled from the SAR's first one and a half years, the patterns that we see emerging, which we explore in our expositions, are, we believe, likely to continue into the twenty-first century, so long as the general contours of the system inherited from the pre-1997 era (with the post-treaty British changes) remain in place for the vaguely defined, mandatory 50-year duration. Hence, we hope that the background, and the paradigm of paradoxes presented here, can serve as a common framework for serious attempts to understand post-1997 Hong Kong under Chinese sovereignty. The book has provided not only the "what," but also the "why," and even how to find the "why."

For fear that the uninitiated would think that what they have heard from the casual reportage on Hong Kong's problems—including the natural calamities (e.g., the bird flu, the fish-killing toxic algae "red tide," the dengue fever epidemic, etc.) and the economic deterioration among other socio-economic woes—was brought about by the territory's return to China, let me reiterate that the Chinese government in Beijing has thus far stuck to a hands-off stance even to a fault. If they have erred, they have done so on the side of being too aloof, not the other way around. According to the same survey series conducted by the Hong Kong Policy Research Institute alluded to above, the Hong Kong people's increasing confidence in the SAR's political future, even amidst an economic recession, was for a large part derived from their satisfaction with Beijing's non-intervention policy toward the territory (*Yazhou Zoukan* 1998, p.43).

Skeptics may intuitively ask if an invisible hand, presumably from Beijing, is not behind most major decisions of the SAR government. While maintaining some skepticism is always healthy in search of truth, it is instructive to take note of an instance that took place in the SAR government's brief interventions in the course of combating the perceived threat of external hedge funds to the territory's capital and stock markets during the summer of 1998. Martin Lee, like other skeptics, suspected that Beijing must have had a hand in Hong Kong's momentary deviation from its long tradition against government intervention in the markets, and openly suggested so, while speaking on RTHK radio. But, after discovering no material evidence to substantiate his allegation, Lee swiftly and courageously retracted his off-the-cuff statement two days later.[42]

The chapters below, each from its own perspective, endeavor to show that in a wide range of areas, from political to social, from the rule of law to press freedom and human rights, Hong Kong has fared much better than was presaged by the media and the prophets of doom, whose premature doomsday prophesying had aroused worldwide alarm before 1997. The chapters on the domestic aspects of the SAR's experience, in Part I, probe into the wide gaps between prior downcast expectations and subsequent outcomes, which are normal and produce no surprises. The only exception on the domestic scene is the unanticipated economic slowdown, contrary to earlier, more confident, forecasts. I must hasten to add, in this connection, that the SAR's future seems to lie in whether its government and populace can continue to work successfully, in conjunction with Beijing, to maintain the delicate balance that need be in place between their "two systems," to make the "one country" idea in the equation work. A detractor, though, was the brief dispute, the first since handover day, between the SAR and Beijing over the SAR Court of Final Appeal's ruling in the *Chan Kam-nga* case, which briefly triggered a dispute over who has the ultimate authority to interpret the Basic

Law. Since I have more to say on the case in the concluding chapter, I shall not say more about it now.

In the external domain, on the other hand, the SAR's experience contradicted the earlier, somehow optimistic—in some cases even upbeat—forecasts by pundits and international law experts, for reasons dealt with in the chapters covering the SAR's foreign affairs, in Part II.

We are not out to paint a rosy picture of the SAR's survival during the territory's transition after return to Chinese sovereignty, but rather to set the record straight. It is necessary, and even urgent, to do so, in order to do justice to Hong Kong, precisely because the world was confounded—nay, misled—for so long by what we now know in retrospect were grossly misguided predictions.

This introductory chapter has relied on, and cited, a wide range of outside sources besides the chapters in this book. The purpose is not only to round out the true picture about Hong Kong, at times to update the facts, but, more important, also to demonstrate that the views expressed in the various chapters here were widely shared and supported by experts from a wide cross-section of fields. Particularly noteworthy were the positive official U.S., British, and European Union evaluations cited above. Implicitly, the contributors to this volume are aware that the wide gaps between the prior dismal predictions and the benign post-handover political reality in Hong Kong (despite the fortuitous economic reverses) may, paradoxically, make it hard for the general public in the West to accept what is the true reality as it has turned out. The reader may find that all the contributors, out of caution and concern for their own credibility and reputation, have often opted to understate their case. Nevertheless, we share the same conviction that the true story must be told, and right now.

Notes

1. Jesus to his disciples while sailing in a storm, Matthew 8:26 (King James version).
2. Most representative of this view was Red Flag Over Hongkong (Chatham, NJ: Chatham House Publishers, 1996).
3. These updated figures were based on "The U.S.-Hong Kong Relationship Today," a talk by U.S. Consul-General, Richard A. Boucher, at the Hong Kong Rotary Club, March 3, 1998.
4. "Taiwan Says Mainland Has Kept Its Promise Regarding Hong Kong's Autonomy," Ming Bao (Hong Kong), 12 March 1998, p.13, quoting sources from Taipei.
5. The Economist, 28 March 1998, p.25.
6. South China Morning Post (hereafter referred to as SCMP), 2 October 1998, p.4.

7. "SAR Corruption Lowest in 5 Years, Survey Shows," SCMP, 6 April 1998, at B-2.
8. See Hong Kong Police Review: A Statistical Supplement 1997, p.50. I am indebted to Dr. Wai-kin Che, of Lingnan University, for calling this to my attention.
9. United States-Hong Kong Policy Act Report, as of April 1, 1998; as required by Section 301 of the United States Hong Kong Policy Act of 1992, 22 U.S.C.5731 as amended. See p.1. Washington, DC: U.S. Department of State.
10. Technically, Article 3 (5) refers to "current social and economic systems." It is absolutely apparent, from the legislative history, the intent of the Joint Declaration, and the annexed documents to the Declaration, that the current political system is included under the "current social and economic systems" umbrella. If not, then the loophole would be even bigger.
11. Cf. Article 24, the Vienna Convention on the Law of Treaties (1969), text in 8 I.L.M. 679ff (1969).
12. The only discussion I know that calls attention to the British efforts at wholesale change of the Hong Kong political system after 1984 is in an article published in Taiwan, by Hu Chang 1996 (see References).
13. Strictly speaking, the so-called "unofficial members" had one seat more than the "official members" appointed by the Governor. But, the latter, voting as a bloc, almost always had majority control. On lack of democratization before the 1980s, see Lo Shiu-hing 1997, pp.33–100.
14. The Basis Law likewise calls for legislation (by the SAR's legislature) "to prohibit foreign political organizations or bodies from conducting political activities in the Region, and to prohibit political organizations or bodies of the Region from establishing ties with foreign political organizations or bodies" (Art. 23). This, it may be argued, is in keeping with the old Societies Ordinance that was "current" in 1984–85.
15. Article 18, the Vienna Convention on the Law of Treaties (1969), text in 8 I.L.M. 679ff (1969).
16. I shall not name the publication or its co-authors, for to do so would heap on it a dignity it does not deserve and, moreover, would only smear the name of the organization sponsoring its publication.
17. These similar domestic problems include an overburden of external debt, overcapacity, over-reliance on foreign investment, lack of sound government regulations, government-business cronyism, etc. Hong Kong, however, does not have its full share of these problems. For instance, its external debt is very low, its government regulations are sound, and government-business cronyism is almost unheard of, thanks to the efficacy of the Independent Commission Against Corruption. Cf. Huang 1998.
18. Asian analysts tend to emphasize external factors, such as speculation by hedge funds, while external commentators, including the IMF, seem to zero in on endogenous causes. For a more balanced review, see Asia Development Research Forum (ADRF) Hong Kong Meeting, 18–19 May 1998, Summary of Discussions. See also Montes 1998. In view of the reported declaration of war by

George Soros' Quantum Fund on Hong Kong's financial market, (SCMP 28 August 1998, p.1), it would be hard to maintain that the region's financial crisis was brought on solely by domestic woes, however.

19. Even Chris Patten, the last British Governor, disclaimed that the British departure was a cause of Hong Kong's economic downturn and said in an interview on CNN TV, New York, on 18 September 1998, that he was "lucky" that his Governorship coincided with the territory's good economic times.

20. Protesters broke through tight security to shower government officials and guests with anti-Beijing slogans at the Hong Kong Convention and Exhibition Hall, where the Chinese National Day celebrations were being held, on 1 October 1998. Nobody was arrested by police, and the protesters were allowed to lay a wreath under the flag pole in honor of the victims of the 4 June Tiananmen crackdown in 1989. *SCMP,* 2 October 1998, p.4.

21. In fact this point has sometimes been made by local analysts in Hong Kong. See Wang 1998, p.10.

22. Articles 1 and 2, the Basic Law of the HKSAR.

23. Rene Albrecht-Carrie (1973, p.155) characterized Austria-Hungary as "not a national state," for its lack of "unity in diversity."

24. Cf. Lee Kuan Yew, *The Singapore Story: Memoirs of Lee Kuan Yew* (New York: Simon & Schuster, 1998).

25. The total includes six international organizations (such as the International Typhoon Committee) in which Hong Kong is an Associate Member, and 19 others (such as the FAO) in which Hong Kong participates as part of the PRC delegation. See U.S. State Department's "Hong Kong Policy Act Report," dated 1 April 1 1998, to Congress, pp.33–5.

26. While, in comparison with the bureaucracy and the Legco, he was given somewhat higher marks for his performance, Tung received very low approval rating when judged alone. In a worst case, nearly 20 percent of the public gave Tung a zero mark for his first year administration, according to a poll conducted by the opposition Democratic Party; *SCMP,* 2 October 1998, p.4.

27. *SCMP,* 3 May 1998, p.6.

28. I learned this from Mr. Wei-lien Li, head of the Office for Hong Kong and Macau Affairs, in Taiwan's Executive Yuan (or Cabinet), 31 August 1998, in Taipei.

29. Evidence that the public may be aware of this unbudging bureaucracy that Chief Executive Tung Chee-hwa had to contend with, was the fading faith in the SAR's inherited civil service from the British colonial era. Even after the outbreak of the regional financial crisis resulting in a deepening economic downturn, a survey showed that while Mr. Tung enjoyed 60 percent approval rating in late January 1998 (as against Chris Pattern's 48 percent in February 1997), the civil service received only 51 percent support (as compared with the 73 percent satisfaction rate registered in 1995). See *SCMP,* 23 January 1998, p.6.

30. The British guarded their moves, including the rationale for promoting certain top civil servants to important government positions, with absolute secrecy. It

is known that they had shipped all confidential data, including personnel files concerning top civil servants, back to London before they left on 1 July 1997. See Paau, p.63.

31. *SCMP,* 12 August 1997, p.1; *Hong Kong Economic Daily,* 12 August 1997, p.A-2.

32. Although "civil service," "bureaucracy," and "high officials" were used in this discussion, I am referring to three separate groups hailing from the same background within the Hong Kong government context, which has no tradition of political appointees. "Bureaucracy" encompasses both the rank and file of the civil service and its higher echelons, from which the bulk of the "high officials," comparable to a ministerial rank elsewhere, were recruited.

33. Subsequent investigations into who was responsible for the fiasco were made possible only after initial media reports on the attitudes of the officials who appeared before the Legco's first hearing had generated a public uproar.

34. Xin Bao (*Hong Kong Economic Journal*), 8 October 1998, p.5; *SCMP,* 9 October 1998, p.7. Later, the SAR government announced, in a document known as *Civil Service into the 21st Century,* a proposal that, if approved, would replace, over a span of ten to 20 years, the current blanket life-tenure system of the civil service with one of contractual employment, subject to renewal based on performance. Only one-third of the civil service would be given permanent contract terms. *SCMP,* 9 March 1999, p.1. It remains to be seen if the new system is going to increase civil service efficiency.

35. There seems to be a consensus that at a time of economic meltdown, amidst the region's financial crisis of 1997–98, Hong Kong needs a strong executive-led government; but the Patten reform shifting the political gravity to the legislature bequeathed a much emasculated executive unequal to the challenge of the time. *SCMP,* 19 September 1998, p.15.

36. *Ta-kung Pao Daily,* 10 June 1998, p.1. In response to my inquiries, the Democratic Party, after reconsidering its position on the issue, replied weeks later, in carefully couched terms, that its Legco members were "basically" sympathetic to the needs of the unemployed. Under apparently mounting pressures months later, the Democrats announced they favored commissioning a full study into insuring workers against joblessness. SCMP, Oct. 2, 1998, p.4.

37. SCMP, 13 November 1997, p.1.

38. Despite the backfiring of his reform moves in the name of democratizing Hong Kong, as explained in this section, Mr. Patten does not seem to be aware of the paradox, let alone have second thoughts, if one could judge by what he said in his new book, *East and West,* in which he seems to describe all the world's (including Hong Kong's) problems as attributable to too many people willing to "kowtow" to Beijing; *SCMP,* Agenda Features, 23 August 1998, p.1.

39. This was the view of some editorial writers; see an editorial bearing the title "Poor Relations," in *SCMP,* 9 October 1998, p.20.

40. Data were from a serial survey project conducted by the Hong Kong Policy Research Institute, revealed in May. See report in *Ming Pao,* 1 May 1998, p.6.

41. In late spring 1999, a pro-democracy group branded subversive by Beijing was reportedly poised to make its first appearance before a United Nations hearing

on China's human rights record in Geneva. Szeto Wah, a member of the De-mocratic Party in the Legco, and head of the Hong Kong Alliance in Support of the Patriotic Democratic Movement in China, was to speak on human rights in China during the U.N. hearing, scheduled for 11–17 April. Two other Legco members from the same party, including Chairman Martin Lee and James To, were reported also headed for Geneva, to air concerns over the rule of law and the pace of democracy in the SAR. In addition, the same group, led by legislator Szeto Wah, was sponsoring a visit to Hong Kong by Wang Dan, one of the better known leaders of the student demonstrations at Tianan-men in 1989, who has been on medical parole in the United States since 1998. If Wang should be denied a visa by the SAR government, the Szeto Wah group would invite Wang's mother from China to participate in the rally for the tenth anniversary of the 4 June episode, to be held in Hong Kong. *SCMP,* 3 April 1999, p.2.

42. "Martin Lee Acknowledges Mistake; Withdraws Allegation [Regarding Beijing's Hand]," *Xin Bao* (the *Hong Kong Economic Journal*), 15 September 1998, p.7.

References

Albrecht-Carrie, Rene. 1973. *A Diplomatic History since the Congress of Vienna,* rev. ed. New York: Harper & Row.

Boucher, Richard A. 1998. "U.S. Policy and the Asian Economic Crisis," English text of an op-ed piece published in a local Chinese-language newspaper; courtesy of United States Information Service (U.S.IS), Hong Kong.

Chan, Ming K., ed. 1994. *Precarious Balance: Hong Kong Between China and Britain, 1842–1992.* Armonk, NY: M. E. Sharpe.

Chan, Ming K. 1997. "The Legacy of the British Administration of Hong Kong: A View from Hong Kong," *China Quarterly* no. 151 (September).

Chang, Hu. 1996. "The Mainland Chinese Takeover Preparations for 1997," *Mainland China Studies* 39 no. 10: 92–105.

Chang, David W.W., and Richard Chuang, 1998. *The Politics of Hong Kong's Reversion to China.* New York: St. Martin's Press.

Ching, Frank. 1997. "Misreading Hong Kong," *Foreign Affairs* 76, no 3: 53–66.

Ching, Frank, Yl-zlieng Lian, and Byron Weng. 1998. "Indicators," in *CSIS Hong Kong Update.* Washington, D. C.: Center for Strategic and International Studies.

Dimbleby, Jonathan. 1997. *The Last Governor: Chris Patten & the Handover of Hong Kong.* London: Little, Brown.

Enright, Michael L, Edith E. Scott, and David Dodwell. 1997. *The Hong Kong Advantage.* New York: Oxford University Press.

Hook, Brian. 1997. "British Views of the Legacy of the Colonial Administration of Hong Kong: A Preliminary Assessment," *China Quarterly,* no. 151 (September): 553–56.

Huang, Zhilian. 1998. "The Unexpected Opportunity for Hong Kong's New Genre of Enterprises in the Midst of the East Asia Financial Crisis," *Haixia pinlun* (Straits Review) (Taipei), no. 86 (February): 44–6.

Liu, Shuyong. 1997. "Hong Kong: A Survey of Its Political and Economic Development over the Past 150 Years," *China Quarterly*, no. 151 (September).

Lo, Shiu-hing. 1997. *The Politics of Democratization in Hong Kong*. New York: St. Martin's Press.

Miners, Norman. 1995. *The Government and Politics of Hong Kong*, fifth ed. Hong Kong: Oxford University Press.

Montes, Manuel F. 1998. *The Currency Crisis in Southeast Asia*, updated ed. Singapore: Institute of Southeast Asian Studies.

Mushkat, Roda. 1997. *One Country, Two International Personalities*. Hong Kong: Hong Kong University Press.

Paau, Danny. 1998. "Curious Maneuvers? Certain Moves of the British Government in Hong Kong Before Departure," in Danny Paau, ed. *Reunification with China: Hong Kong Academics Speak*. Hong Kong: Asian Research Service.

Quine, Willend V. 1965. *The Ways of Paradox*. New York: Random House.

Sung, Chen-chao. 1997. "The Economic and Political Implications of the Asia Pacific Regional Financial Crisis," *Haixia pinglun* (Straits Review) (Taipei), no. 82 (October): 12–15.

Tang, James. 1998. "Politics in Hong Kong: Democracy in Retreat?" in *Hong Kong: The Challenges of Change*. New York: Asian Society. Report issued on 10 September 1998.

Tang, James. 1998a. "Executive-Legislative Relations in Hong Kong," in *CSIS Hong Kong Update* (August). Washington, D.C.: Center for Strategic and International Studies. Report issued on 2 September 1998.

U.S. Department of State. 1998. *U.S.-Hong Kong Policy Act Report* (to Congress), as of 1 April 1998; as required by Section 301 of the United States Hong Kong Policy Act of 1992; 22 U.S.C 5731 as amended.

Wang, Ka-ying. 1998. *A Retrospective and Prospective View on "One Country, Two Systems" After One Year*. Hong Kong: The Hong Kong Center for Cross-Strait Relations.

Weng, Byron. 1998. "Weathering a Growing Discontent," in CSIS *Hong Kong Update* (August). Washington, D.C.: Center for Strategic and International Studies.

Yazhou Zhoukan. 1998. "*yu-ce die yanjing; pingfen jin jige* (Predictions [regarding post-handover Hong Kong] Proven Unfounded)," in the *Yazhou Zhoukan* (AslaWeek) journal (Hong Kong), issue for 6–12 July: 42–50.

PART I

The Domestic Scene

CHAPTER 1

Government and Political Change in the Hong Kong Special Administrative Region

LAU Siu-kai

The first year of the Hong Kong Special Administrative Region (SAR) is a tumultuous year packed with unexpected events and broken predictions. Since then, pre-handover fears of blatant Chinese interference in local matters and serious abridgment of human rights and freedoms have proven to be unfounded. However, in late October 1997, Hong Kong was attacked by the Asian financial turmoil, which, some argue, also acted as a catalyst in exacerbating the long-standing structural problems of its economy. If the inauguration of the SAR in July 1997 was greeted by the people with a mixture of hope and misgivings, by the end of its first year the prevalent mood in Hong Kong was one of despondency, despair, and fear. To the SAR government headed by Tung Chee-hwa—a new government with uncertain legitimacy—the first year of the SAR was one of both challenges and opportunities. After an auspicious start, however, the Tung administration has since lost much public support and, more ominously, the confidence and goodwill of many people. At the end of the first year of the SAR, the government found itself in a political predicament from which it was difficult, if not impossible, to escape.

It is undoubtedly true that external and accidental factors were playing havoc with the efforts of the Tung administration to build an effective and popular government. The way Tung Chee-hwa exercised his leadership and powers significantly explained why he was not able to take advantage of his initial popularity and the opportunities to craft a powerful governing coalition that would do his bidding. A conservative shipping tycoon before his elevation to the highest office in the SAR, Tung's instinct was apolitical, probably even anti-political. He was convinced that Hong Kong, after so many years of political turbulence, should become depoliticized. To depoliticize a highly politicized society is a decidedly difficult task. It demands above all political wisdom and skill. In light of Tung's overriding political goal, what was glaringly lacking in Tung's leadership was a coherent strategy to ensure that the goal would be achieved. The end result, quite unintended, was just the opposite. At the first anniversary of the SAR, in spite of the subsidence of the key political disputes that had plagued the previous one and a half decades of colonial Hong Kong, the Tung administration found itself besieged on all sides. Looking ahead, in view of the unfavorable social and economic circumstances and the rise of political challengers, the situation of the SAR government is likely to remain difficult unless a drastic change in Tung's leadership style and his leadership team is made in time.

Rise and Decline of the Popularity of the New Regime

In a formal sense, the SAR government came into existence only on 1 July 1997, when Hong Kong was handed over by Britain to China. Politically speaking, however, when Tung Chee-hwa was elected to be the SAR's first Chief Executive by the 400-strong Selection Committee on 11 December 1996, it can be said that the new regime was already born. Tung was chosen to lead the SAR not because he was a bold and charismatic political veteran—which he certainly was not—but because he was considered a "good guy," a moderate person, a compromise candidate who did not seem likely to threaten vested interests. Most importantly, it was widely believed that he had the blessings of Beijing.

Popularity polls on Tung have been carried out since his selection. The trajectory of Tung's popularity ratings over the span of one and a half years shows an initial high at the time of electoral victory, then an uneven downward trend until the handover in July 1997. Thereafter, Tung's popularity score rose until the onslaught of the financial turmoil in late October 1997. Afterwards an overall trend of sliding popularity took hold, which was reversed somewhat after March 1998, but not for long. Not surprisingly, the rise and fall of Tung's—and his government's—popularity had much to do with public evaluation of the Chief Executive's performance and the objective conditions in Hong Kong, particularly conditions that had immediate impact on the mood and interests of the people.

In a poll commissioned by the *Apple Daily*, and undertaken by the Hong Kong Institute of Asia-Pacific Studies at the Chinese University of Hong

Kong immediately after his electoral victory in December 1996, Tung received an average performance score of 71.3 (maximum score: 100; passing score: 50) from the respondents. This represents an all-time high. The score fell to 63.7 at the time of the handover in July 1997. Afterwards, it climbed to the post-handover high of 68 in September 1997. After that, it fell gradually and reached a low of 59.1 in January 1998, rebounding a little to 63.3 in February 1998 and 62.4 in March 1998.[1] Thereafter a basically downward trend tookhold, being 60.1 in April 1998, 56.2 in May 1998, 60.9 in June 1998, and 56.5 in July 1998.[2]

Findings from the polls commissioned by the *South China Morning Post* and administered by the Asian Commercial Research Ltd. tell the same story. In June 1997, immediately before the handover, 57 percent of respondents were satisfied with Tung's performance. The figure rose in the next few months to 78 percent in July, 82 percent in August, and 89 percent in October. Then it dropped precipitously to 76 percent in December.[3] By early August 1998, it fell to the historic low of only 56 percent.[4]

Tung's initial success in building his personal popularity was all the more remarkable considering the fact that as a political newcomer with hardly any political experience he even surpassed Chris Patten—the last colonial Governor of Hong Kong and a veteran politician with a populist flair—in popularity by a small margin. In my questionnaire survey of Hong Kong residents on the eve of its reversion to China (mostly in May and June 1997), Tung received a slightly higher level of public trust (26.4 percent) than Chris Patten (21.5 percent).[5]

The trend of public support for the SAR government, which formally came into being on 1 July 1997, basically mirrored that for Tung. Nevertheless, while Tung was slightly more popular than Patten, the SAR government was far below the colonial government in public acceptance. In my 1997 survey, it was found that on the eve of the handover the SAR government was trusted only by 31.7 percent of the respondents, whereas 52.8 percent expressed trust in the departing colonial government. Similar findings were obtained by the polls carried out by the Social Science Research Centre of the University of Hong Kong. Whereas the proportion of the respondents who trusted the colonial government in June 1997 had stood at 62.9 percent, the figures for the SAR government in July and August 1997 were found to be 50.7 percent and 48.3 percent, respectively.[6]

After its inauguration in July 1997, the SAR government initially saw a rise in public support, followed by a decline after the onset of the financial turmoil. The bi-monthly polls conducted by the Home Affairs Department clearly indicate fluctuations in public ratings of the performance of the government. The proportion of people who were satisfied with the overall performance of the SAR government was 50 percent in July 1997, rising to 53 percent in September. In November, it fell dramatically to 45 percent and stood at 43 percent in January 1998. In March 1998 it again rose to 48 percent.[7] Thereafter it again dropped precipitously to 32 percent in May and 25 percent in July.[8]

Somewhat different from changes in the public evaluation of the overall performance of the SAR, changes in public attitudes toward other aspects of the SAR government showed roughly a downward trend after July 1997. The trend was somewhat reversed temporarily in March 1998. Again, according to the polls of the Home Affairs Department, the proportions of respondents saying that the civil service was efficient were 50 percent, 45 percent, 45 percent, 38 percent, 44 percent, 34 percent, and 28 percent, respectively, in July 1997, September 1997, November 1997, January 1998, March 1998, May 1998 and July 1998. The corresponding figures for those who thought the civil servants' working attitude was good or very good were 59 percent, 55 percent, 56 percent, 55 percent, 55 percent, 51 percent, and 50 percent, respectively. The figures for those who were satisfied with the government's performance in disseminating information on its policies and actions to the general public were 62 percent, 57 percent, 56 percent, 54 percent, 60 percent, 49 percent, and 49 percent, respectively. With regard to those who believed that the government was concerned about public opinion on public affairs, the figures were 55 percent, 55 percent, 50 percent, 56 percent, 62 percent, 55 percent, and 59 percent, respectively. Lastly, 62 percent, 63 percent, 60 percent, 61 percent, 68 percent, 62 percent, and 62 percent of respondents respectively were of the opinion that the government took into account public opinion when deciding on policies and actions.[9]

Public Perception of Hong Kong's Political and Economic Situation

Changes in public attitudes toward Tung Chee-hwa and the SAR government pretty much reflected people's satisfaction or dissatisfaction with the political and economic situation of Hong Kong as well as their own circumstances. Upbeat feelings were translated into favorable views of the new regime, whereas pessimism was the cause of negative attitudes.

In the half-year between the election of Tung as the Chief Executive in December 1996 and the inauguration of the SAR in July 1997, Tung and his leadership team suffered from an uneven decline in popularity. During the half-year concerned, the most important work of Tung and his team was to lay the groundwork for the establishment of the SAR in close collaboration with the Preparatory Committee of the SAR—a body established by the Standing Committee of the National People's Congress of China. At first, Tung was warmly received by the people as a result of his decision to retain the senior civil servants of the outgoing colonial regime as his principal officials in February 1997, and his appointment of Andrew Li as the head of the judiciary. Later on, however, Tung suffered setbacks in his efforts to win the hearts and minds of the people, largely because of a number of important decisions he had made. The support he gave to the decision of the Preparatory Committee to abolish those amendments to Hong Kong's laws made in accordance with the 1991 Hong Kong Bill of Rights Ordinance, as well as to downgrade the legal status of the Ordinance itself, was the primary factor behind the fall in Tung's popularity. The proposal of the SAR government in April 1997 to introduce the concept of "national security" as a consideration in regulating the formation of societies and the holding of parades by the authorities had aroused public fears about the curtailment of political freedom. This, together with his controversial appointment of Executive Councilor Leung Chun-ying—the boss of a prosperous chartered surveyor firm—as his chief adviser on housing policy in March 1997, had dented his political popularity. Nevertheless, according to my 1997 survey, despite the pre-handover decline in the popularity of Tung and his government, Hong Kong people were still by and large fairly satisfied with Tung and his SAR government immediately before the handover. Twenty-four percent of the respondents rated their performance as good, 17.3 percent saw it as poor, and 37.4 percent regarded it as about average.

The continuous rise in popularity of Tung in the early months of the SAR was chiefly the result of a sense of relief felt by the people, who had been haunted for more than a decade by anxieties about the future of Hong Kong. In other words, the increase in public trust in Tung and the SAR government had little to do with the performance and leadership of the new rulers. It was largely the result of the fact that the widely held pre-1997 doomsday scenarios failed to come true, thanks to Beijing's hands-off policy. Most people agreed that the way Beijing handled Hong Kong affairs after the SAR government began to function after the handover was nothing short of impeccable. Shortly after the establishment of the SAR, the Central Government retired Lu Ping, the high profile Director of the Hong Kong and Macau Office, under the State Council, and Zhou Nan, Director of the Hong Kong Branch of the New China News Agency. Lu and Zhou had been heavily involved in the Sino-British negotiations over the future of Hong Kong back in the early 1980s, and were heavyweights in charge of Hong Kong matters after that. Over the years they had accumulated formidable political authority and stature. Their continuation in office, it was feared, would severely threaten Tung's freedom to act in the governance of the SAR. The replacement of Lu and Zhou by the low-key Liao Hui and Jiang Enzhu, coupled with the reduced functions of the two departments under their respective charge, was a master stroke of Beijing to convince Hong Kong and the international community of China's sincerity in respecting the territory's autonomy.

The way the People's Liberation Army (PLA) garrison conducted itself in Hong Kong was another factor that allayed the public's fears. Before the handover, most Hong Kong people harbored some paranoiac uneasiness about the PLA, which not infrequently aroused imageries of corruption, cruelty, special privilege, and unruliness. The virtual invisibility of the PLA garrison in the SAR since the handover went a long way toward easing those fears, so much so that within several months public acceptance of the PLA had appreciably increased. According to two consecutive polls by the Department of Journalism and Communication at the Chinese University of Hong Kong, conducted in August 1996 and September 1997, the proportion of people feeling uneasy about the PLA dropped from 30.3 percent to 10.9 percent in a year's time.[10] Similarly, in an October 1997 poll commissioned by *Apple Daily* and carried out by the Hong Kong Institute of Asia-Pacific Studies at the Chinese University of Hong Kong, 43.9 percent of respondents reported that they were satisfied with the performance of the PLA garrison in Hong Kong. This figure was higher than that reported in July 1997 (29.6 percent), but slightly lower than in September 1997 (45 percent).[11]

The prudence exercised by the Chinese leaders in Beijing and by Chinese officials in the SAR with respect to Hong Kong affairs was very impressive and took China's critics, both locally and abroad, by surprise. With the explicit purpose of fortifying Hong Kong's international status, Beijing in November 1997 even agreed to submit reports on human rights in Hong Kong to the United Nations. This was in spite of the fact that China then was not yet a party to either the International Covenant on Civil and Political Rights or the International Covenant on Economic, Social and Cultural Rights.

The overall record of Beijing in handling Hong Kong affairs since the handover was greatly appreciated by the Hong Kong people. After the handover, there was a conspicuous increase in public satisfaction with the way China handled Hong Kong matters. The polls conducted by the Hong Kong Transition Project—an inter-institutional research project led by the Hong Kong Baptist University—found a trend of rising satisfaction with the way China dealt with the SAR. In February 1995, it was found that as many as 62 percent of the people were dissatisfied. But the figure fell dramatically to 22 percent in January 1998. On the other hand, while only 17 percent were satisfied in February 1995, the proportion of people satisfied soared to 61 percent in January 1998.[12]

In fact, public satisfaction with the way Beijing handled Hong Kong affairs after the handover led to a gradual increase in public trust in the Chinese government itself. Obviously, there was still a long way to go before one could see a relationship of mutual trust between Beijing and the Hong Kong people. It was found that in June, 1997, only 28.5 percent of the people trusted the Chinese government, according to a poll of the Social Science Research Centre of the University of Hong Kong. The figure, however, rose to 31.1 percent in July and 32.7 percent in August the same year.[13] Generally speaking, there was a conspicuous difference in public trust in the Chinese government between the first and second half of 1997. Whereas in the first half of 1997, 29 percent of people said they were trustful and 42 percent were mistrustful of the Chinese government, the figures in the second half of the year changed to 32 percent and 30 percent respectively.[14]

During the first months of the SAR, all pollsters were able to discern an optimistic mood in Hong Kong. According to the polls conducted by the Home Affairs Department, the proportion of people who were satisfied with the situation in Hong Kong rose from 77 percent in July 1997 to 81 percent in September 1997. On the other hand, the pessimists in Hong Kong showed a significant drop in number. While 11 percent of respondents expected the situation in Hong Kong to get worse in the next year or so, the figure fell to a minuscule 6 percent in September 1997. Regarding the future, 72 percent of the respondents in July 1997 were confident that Hong Kong would continue to be prosperous and stable. In September, the figure jumped to an unprecedented high of 80 percent.[15]

The Social Science Research Centre of the University of Hong Kong had also found a trend of rising confidence in Hong Kong's future in the early months of the SAR. According to the polls conducted in 1997, the percentages of respondents expressing their confidence in Hong Kong's future showed a steady pattern of increase throughout the year, from 66.7 percent in April, 71.8 percent in May, 75.2 percent in June, 77.5 percent in July, 83.1 percent in August, to 84.7 percent in September.[16]

More specifically, people were more confident in the practicability of "one country, two systems" and of the motto "Hong Kong people ruling Hong Kong." A poll commissioned by *Apple Daily*, but executed by the Hong Kong Institute of Asia-Pacific Studies at the Chinese University of Hong Kong, showed that in July 1997 only 13 percent of respondents said that their confidence in these areas had declined. The figure just a month before had been 24.2 percent. On the other hand, 62.6 percent of the respondents in July 1997 reported a rise in confidence in these areas, as compared with 45.4 percent in June 1997. These findings clearly showed that there was an unmistakable rise in public confidence in the future of Hong Kong in its reincarnation as a Chinese SAR.[17] Furthermore, according to the polls commissioned by the *South China Morning Post* and carried out by Asia Commercial Research Ltd., confidence in the economy of Hong Kong increased from 80 percent in June 1997 to 87 percent in August 1997, shortly before the financial crisis hit the region. People's confidence that their political freedoms would be maintained rose from 54 percent to 69 percent.[18]

There was an abrupt change in the public mood after late October 1997, however. The preponderant factor turning an upbeat mood into a downcast one was the economic crisis triggered by the financial turmoil in the region. The immediate manifestations of the crisis were plummeting stock prices and tumbling property values, when the SAR government resorted to hiking the interest rates in its resolve to defend the peg of the Hong Kong dollar to the U.S. dollar. The after-effects of the financial crisis were multiple and devastating. Hong Kong plunged rapidly into an economic recession. Bankruptcies and unemployment were on the rise. Unlike economic hard times in the past, this time even the middle classes felt the pinch. Measured by its magnitude and impact, the economic crisis that began in late October 1997 was unprecedented in the post–World War II history of the territory.

To the extent the economic recession magnified the structural inequities in Hong Kong's economy, the worst possible implication was that it might be quite some years before the people could see a conspicuous economic revival. As if the economic crisis were not enough, after October 1997 Hong Kong was hard hit by a slew of mishaps whose convergence at more or less the same time was mind-boggling. They included the outbreak of the bizarre avian flu that, while claiming only a few lives, had aroused deep fears among the people because of its life-threatening potential. Apart from this, the fall-off of tourism, a series of bus crashes, a string of medical accidents caused by human negligence, and the poisonous "red tide" that killed many fish, a local staple food source, together cast a dark cloud over Hong Kong. The new airport fiasco, which struck Hong Kong at the time of the first anniversary of the HKSAR—an example of unmitigated human error—was the final straw which shattered public confidence in the government and left people intuitively pondering what had gone wrong after the end of colonial rule.

The grim public mood that took hold of Hong Kong after late 1997 contrasted starkly with the upbeat outlook in the early months of the SAR. The polls of the Home Affairs Department recorded a sudden drop in public satisfaction with the situation in Hong Kong, from 70 percent in November 1997 to 58 percent in January 1998. Whereas 25 percent of respondents in November 1997 expected the situation in Hong Kong to get worse in the next year or so, by January 1998 the proportion had jumped to a post-handover high of 38 percent. Similarly, while 72 percent in November 1997 had some or a great deal of confidence that Hong Kong would continue to be prosperous and stable, the figure fell off to 62 percent in January 1998.[19] In March 1998, as the stock market stabilized and the Hong Kong dollar stayed strong, public mood did brighten up a bit. Sixty-one percent of the respondents expressed satisfaction with the current situation in Hong Kong, 24 percent expected the situation to get worse, and 67 percent were confident that Hong Kong would remain prosperous and stable.[20] Thereafter, the economy deteriorated further, bringing down the public mood with it. As a result, the confidence rate was only 59 percent in May and July, 1998.[21]

A December 1997 poll, conducted by the Social Science Research Centre at the University of Hong Kong, found a plurality of respondents—42 percent—expecting the overall situation in Hong Kong to be worse next year.[22] The same pollster, in a January 1998 survey, even found a majority—62.2 percent—seeing the overall condition of Hong Kong then as worse off than that before the handover.[23] Public confidence in the future of Hong Kong got a good beating. The proportion of respondents expressing confidence in Hong Kong's future declined to 74.4 percent in October 1997. It was 72 percent in November 1997, 68 percent in December 1997, and 50 percent in January 1998.[24]

Since the onslaught of the financial turmoil, people have become more concerned about the social and economic problems of Hong Kong. According to polls conducted early on, in October 1997, by the Social Science Research Centre of the University of Hong Kong, 53.5 percent of respondents had expressed concern about social issues, 33 percent about economic issues, and 5.3 percent about political issues. In December 1997, however, economic issues became the chief concern of a majority of the people (55.2 percent). People who were concerned about social issues dropped to 28.2 percent, and those concerned about political issues, further down to 4 percent. In October 1997, 48.5 percent were satisfied, and 27.3 percent dissatisfied, with the economic conditions; but the figures in December 1997 changed to 19.8 percent satisfied and 50.2 percent dissatisfied.[25]

In the midst of unprecedented and unforeseen economic difficulties, economic and political confidence in Hong Kong waned. In a November 1997 poll commissioned by *Apple Daily* and carried out by the Hong Kong Institute of Asia-Pacific Studies of the Chinese University of Hong Kong, it was found that 29.3 percent of the respondents expected the political situation to worsen. Even more people—40 percent—expected the economic situation to worsen.[26] In a June 1998 poll by the same organization, while the proportion of respondents (12.2 percent) despairing of Hong Kong's political situation had dropped substantially, the proportion having a dim view of its economic situation remained more or less the same (39.7 percent).[27]

In a poll commissioned by the *South China Morning Post* and carried out by Asian Commercial Research, it was found that confidence in the economy fell from 82 percent in August to 60 percent at the beginning of December 1997. People's confidence in their own financial circumstances during the coming year was down from 79 percent in August to 66 percent in early December 1997.[28] At the end of March 1998, the same pollster found that the proportion of respondents confident in the Hong Kong economy dropped further to 51 percent, and those confident in their own financial circumstances dropped to 63 percent.[29] More alarmingly, by August 1998 less than half (37 percent) of the Hong Kong people had confidence in the economy over the next 12 months.[30]

The economic confidence index prepared for the *South China Morning Post* and *Ming Pao Daily News* by the pollster AC Nielsen.SRH in January 1998 "plummeted 21 points to match the 73-point low during the slump of winter 1995. Political confidence had been similarly shaken, with the index falling to 87, close to levels of winter 1995 and after the Tiananmen Square massacre." The report added: "Less than one percent of the 1,016 respondents interviewed by telephone . . . said the SAR government performed very well over the past six months, and only 21 percent thought it had done better than average." Furthermore, it continued, "sixty-five percent of respondents said they had confidence in the future, compared with 84 percent in October [1997] and 79 percent in July 1997. There was less optimism when it came to the economy and personal finances. Just nine percent of those surveyed expected their personal financial situation to improve—compared with 22 percent in October 1997—while 29 percent expected it to get worse. In October 1997 only 16 percent thought their situation would worsen. A mere 11 percent expected an upturn in Hong Kong's economic situation in the coming year, while 50 percent foresaw deterioration. In October 1997, 26 percent expected improvement and 14 percent predicted worsening."[31]

In such a context of a gloomy political and economic mood, there was a corresponding erosion in public support for the new SAR regime. Nevertheless, the Tung administration could take comfort in the fact that the people did not blame the new regime for the plight of Hong Kong. But in a time of economic hardship and other mishaps, it was impossible for the government to emerge unscathed. What did hurt Tung and his government most was that they were seen as incompetent in handling crises. When people begged for leadership, it was not provided by the Tung administration. By the end of Tung's first year in office, public disappointment with him reached a dangerous level. In a poll conducted by the Social Science Research Centre of the University of Hong Kong in June 1998, a plurality of people (43.5 percent) were of the view that Tung's performance was inferior to that of Patten, with only 13.7 percent taking the opposite view. At the same time, there was an increase in public nostalgia for colonial rule. A plurality of people (39.9 percent) thought that Hong Kong would fare better if it continued to be under colonial rule, with only 14.8 percent thinking otherwise.[32] In yet another poll by the Social Science Research Centre of the

University of Hong Kong, 40 percent of respondents were found to object to Tung seeking another term of office. Of the people questioned, "34 percent supported a second term, but mainly because there were not other suitable candidates."[33] To explain the change of political fortune of the new regime, we need to analyze the governing strategy of Chief Executive Tung, why it enjoyed initial success and why it failed to sustain its popular support and prestige under crisis conditions.

The Governing Strategy of Tung Chee-hwa

The governing strategy crafted by Tung Chee-hwa faithfully reflected his political beliefs, his style of leadership, his perception of the political and economic situation of Hong Kong, and his assessment of the opportunities and constraints faced by the new regime. The primary objective of Tung was to promote Hong Kong's economic development and social stability in a context of depoliticization. Right from the beginning, the new regime was under the dark shadow of public nostalgia for colonial rule and the narrow elitist base of the regime. As a political conservative in the traditional Chinese sense, Tung was much taken aback by the heavily politicized and conflict-ridden environment of Hong Kong in the years since the signing of the Sino-British Joint Declaration, which sealed the territory's political future. Tung was fully aware, however ironically, of the lingering public fondness for the departed colonial regime and persisting public fears of changes in the status quo. Still, he was determined to blaze a new political trail by resetting the public agenda, placing his administration on a new political support base, and reshaping the relationship between government and people.

In its last years, the colonial regime resorted to democratic and occasionally even demagogic appeals to maintain public support in an environment of declining governmental authority. Under the five-year rule of Chris Patten, the departing regime even deliberately manipulated the anti-Communist sentiments of the Hong Kong people to advance their political interests. Public nostalgia for colonial rule, though diminishing over the years, still remained palpable on the eve of the handover. The British were particularly remembered for their democratic reforms and human rights legislation, all of which took place only after, not before, the 1984 Sino-British Declaration on the return of Hong Kong to China. Given Tung's political conservatism, the constraints of the Basic Law, the SAR's mini-constitution, and the skepticism of Beijing, his administration could never outdo the colonial regime on the democratic front. (James Hsiung speaks of this as a "revolution of rising expectations," created by the departing British, which the successor SAR government can hardly cope with.) Realistically, the new regime has to seek political legitimacy and public trust through other means.[34]

Since Tung's electoral victory in December 1996, the new regime has adopted a four-pronged approach in quest of public support, drawing upon its strengths and the opportunities that beckoned. The principal strategy was to convince the Hong Kong people that the new regime was given four-square support by Beijing. A relationship of mutual trust between the Central Government and the SAR government was depicted as not only capable of averting Chinese interference in local affairs, but also as crucial in procuring China's support in resolving Hong Kong's problems. Since China had played a critical role in Tung's ascendance to the top post of the SAR, he has the unswerving support of China. A telling example of this were the complementary actions taken by China in support of the SAR's controversial policy of trying to stem the tide of children born on the Mainland, of Hong Kong parents, seeking entry into the territory immediately after 1 July 1997. In the absence of backing from Beijing, the SAR would be beset with serious social problems in its early months.[35] Another example was China's decision not to devalue the renminbi when the Hong Kong dollar was under attack by international speculators, in an avowed effort to shore up its stability.[36] And China had to incur tremendous economic costs by that decision. In addition to cooperating actively with the Tung administration before and after the handover, Chinese leaders went out of their way to openly applaud Tung for the purpose of bolstering his prestige in the SAR. In return for China's staunch support, Tung had taken every care to reassure the Central Government that Hong Kong would not be allowed to pose any political threat to the Communist regime in Beijing.[37] In the first year of the SAR, Tung was fortunate in that he had not been challenged by any large-scale anti-Communist actions in Hong Kong, thanks to the self-restraint of the people, including those who had not been friendly to Beijing. In any case, the solid backing given by China to Tung did, over time, convince the people that Tung was his own man in the SAR. This was definitely of great help in his search for public support.

Another tactic of Tung to seek public trust was to appeal strenuously to Chinese tradition. Without any doubt, Tung was an ardent believer in traditional virtues. The political philosophy advocated by Tung was manifestly traditionalist in tone, though Tung was not able to articulate a coherent political doctrine. In his speeches, Tung resorted liberally to values such as collectivism, the family, harmony, peace, filial piety, respect for the elderly, benevolence, obligation to the community, modesty, and integrity.[38] Tung's ideal government was a Confucian one practicing paternalistic, benign, and active rule. Unlike the colonial regime, Tung, while extolling the free market, was no blind believer in laissez-faire. In Tung's mind, the government had a constructive role to play in promoting economic development. Furthermore, Tung's ideal society was one devoid of conflicts, particularly those political in nature. In that society, people were respectful and deferential to authorities, especially political authorities. To all appearances, Tung's governing philosophy should be anachronistic and at odds with a society which was highly modernized and Westernized. This however was not the case. Past studies have repeatedly indicated that traditional beliefs remain strong in the ethos of the Hong Kong Chinese.[39]

My 1997 survey also shows that on the eve of the handover, Tung's way of thinking had broad public appeal. In the survey, a large majority of respondents—69.7 percent—agreed with the view that "a good government should treat the people as though they are its own children." Most (68.2 percent) respondents were of the view that the government should enact laws to punish people who did not take care of their parents. Moreover, many more (75.6 percent) respondents agreed than disagreed (12.4 percent) with the view that a good government was one that taught the people how to conduct themselves. To the respondents, traditional Chinese values were preferable to Western values. A plurality (32.1 percent) of respondents considered traditional Chinese values such as loyalty, filial piety, benevolence, and righteousness suited Hong Kong better than Western values of freedom, democracy and human rights. In addition, a majority of respondents—59.1 percent—placed social order above individual freedom. And a plurality— 44.7 percent—also said that public interest took priority over human rights. Overall, as the majority of Hong Kong people shared Tung's political beliefs, his initial success in winning a decent level of public endorsement was not at all surprising.

The third aspect of Tung's governing strategy was to draw upon his own favorable personal image as perceived by the people. As a political novice, Tung was unknown to the people until immediately before his electoral victory. Tung had the benefit of not having his reputation tarnished and compromised by the political bickerings in the transitional period before the handover, which had destroyed the career of many an aspiring politician. While as a politician Tung was definitely dwarfed by the populist and astute Chris Patten, Tung's personal image of sincerity, integrity, modesty, amiability and diligence had, in no small measure, allayed public fears of an arbitrary and repressive SAR government. His personal image furthermore went a long way toward convincing the people that he would run a benevolent government at a time when Tung still had no policy achievements.

The last prong of Tung's governing strategy was his unequivocal promise of government performance in the economic, social, and livelihood spheres. Many long-standing economic, social, and livelihood problems, which had not been seriously dealt with under colonial rule and were getting even worse during the pre-handover transition,[40] furnished the Tung administration with an opportunity to win public support. As a Chinese-style politician, Tung envisaged a government more interventionist on social-welfare and livelihood issues than its colonial predecessor, though he was prudent enough to pay homage to the free market. In his inaugural speech on 1 July 1997, Tung declared that the foremost task of his government was "to enhance Hong Kong's economic vitality and sustain its economic growth."[41] He vowed to improve Hong Kong's education, with particular emphasis on the elementary and secondary levels. He promised to resolve the housing problem by increasing the overall housing supply at a target of not less than 85,000 flats a year, with the aim of achieving a home ownership rate of 70 percent within ten years. The problem of the elderly was also a major focus of the new government. Tung pledged to "develop a comprehensive policy to take care of the various needs of our senior citizens and provide them with a sense of security, a sense of belonging and a sense of worthiness."[42] Tung's preoccupation with the long-term economic future of Hong Kong was further elaborated in his policy address to the Provisional Legislature in October 1997. Acknowledging the fact that Hong Kong could no longer depend upon low-wage manufacturing and services to propel its economic development, Tung declared that the SAR government would encourage enterprises to develop into higher value-added activities. Toward this goal, Tung would set up and chair a commission on strategic development.[43] All of Tung's promises were undoubtedly credible at the time they were made, for the economy was still vibrant and the SAR government had inherited from the colonial regime an enormous amount of fiscal reserves and no external debt.

Tung's emphasis on economic and social issues was basically in accord with public preferences. In my 1997 survey, the respondents were asked about the top priority of a good government. Among those with definite answers, 34.5 percent chose social stability, 33.4 percent people's livelihood, 20.5 percent economic development, 5.4 percent human rights and freedom, and 1.7 percent democratic development. By the same token, among issues of economic prosperity, social stability, personal freedom, and democratic development, 52.6 percent ranked social stability as most important, 23.3 percent economic prosperity, 10.4 percent democratic government, and 7.7 percent personal freedom.

Without any doubt, the Tung administration had an auspicious start. The steadfast support of the Central Government was crucial to the build-up of its authority. Tung's personal image initially endeared him to a people who had very limited knowledge of the ability and history of their new chief. The political beliefs championed by Tung struck a responsive chord in society, and his policy agenda was in general accord with that of the public. In a nutshell, political paternalism as advocated by Tung was well received by a society that until very recently was reeling from years of political turmoil and was in desperate need of a respite from political upheavals. The initial success of Tung's governing strategy was reflected not only in his rising popularity in the early months of his rule, but also in the generally tranquil political ambience in the SAR. All of a sudden, the inception of the SAR coincided with the eclipse of politics in Hong Kong.[44]

The Unraveling of Tung's Governing Strategy

The governing strategy of Tung Chee-hwa represented in essence a project of depoliticization. By deliberately focusing on social, economic, and livelihood issues, Tung sought to draw people's attention away from political issues such as democratization and toward practical bread-and-butter concerns. By putting forward a public agenda drastically different from that of his predecessor and shadow competitor, Chris Patten, Tung in effect attempted to place his government on a political base at variance with that of the colonial regime. According to the findings of my 1997 survey, Patten's supporters and Tung's supporters differed sharply in political values. People who were Westernized and democratically oriented were more inclined to support Patten. Tung, in contrast, drew support largely from people who embraced traditional values and held strong nationalistic sentiments. There were also marked differences in the socio-demographic characteristics of the two groups. Generally speaking, Tung's supporters were older in age, less educated, and lower in income. These people were more susceptible to the traditionalistic appeals of Tung. Needless to say, they also harbored great expectations that Tung would cater to their material needs and normative aspirations.

Just like that of any governing strategy, the long-term success of Tung's governing strategy hinged upon a set of favorable conditions. Foremost among these would be a prosperous economy and an institutional structure that could implement the policies originating from Tung's paternalistic goals. The early success of Tung's governing strategy owed in no small measure to public confidence in his determination and ability to keep his paternalistic promises. The congenial economic environment in the first months of the SAR had boosted public confidence. Nevertheless, for his governing strategy to succeed in building for him and his government a powerful and durable political support base, Tung's *personal paternalism* had to be turned into *institutional* or *organizational paternalism*. In other words, in order to retain and strengthen public support over time, Tung's governing philosophy had to become successful practice on a long-term institutional and organizational basis. Personal paternalism, if not institutionalized or organized, would sooner or later become a political scourge for Tung, because if he could not deliver on his promises embedded in personal paternalism, he would be increasingly seen by the people as not only incompetent, but also hypocritical. As a result, he would suffer from declining popularity and authority. In the early months of the SAR, a set of favorable conditions did exist to allow Tung to convert his personal paternalism into a more lasting institutional paternalism. Over these months, the economy of Hong Kong was still in very good shape. The Provisional Legislature, dominated by business and professional interests, was willing to do the bidding of Tung and his government. Tung had at his disposal vast powers to appoint his supporters to various public offices. He had also a free hand to devise public policies in accordance with his preferences, while his opponents—the democrats—found themselves in the political wilderness and had no toehold in the establishment. Last but not least, Tung was riding high in his popularity, and was in a favorable position to shape the public will. In retrospect, however, it turned out that these early months had largely not been put to constructive use.

The failure of Tung's governing strategy was due to the fact that, aside from its success in enlisting China's support for the SAR government, it primarily was made up of a set of promises to the people. In order for the promises, with respect to economic development and social well-being, to be realized, a fast-growing economy was indispensable. Only rapid and sustained economic growth would provide the necessary public revenue for Tung to satisfy his key constituencies—the business community and the grass-roots people—and to ward off challenge from the liberal forces in Hong Kong. The economic crisis starting in late October 1997 had upset the apple cart. Being unprepared psychologically for the sudden change of the territory's economic fortunes, it was inevitable that the people would blame Tung and his government for their plight, even though they were fully convinced of the external origin of the crisis. Nevertheless, the Tung administration suffered tremendously from the crisis in terms of popularity, because it was widely perceived as neither competent nor caring. Public officials were criticized as aloof from the people and incapable of handling crisis situations. Mishaps other than the economic crisis—particularly the avian flu and the new airport fiasco—further reinforced public dissatisfaction with the performance and demeanor of Tung and his officials.

Admittedly, given the international origin of the economic crisis, which was beyond Hong Kong's control, it was impossible for the SAR government to come up with effective remedies to cure Hong Kong's economic woes. However, if Tung's governing strategy was sound, he could at least minimize the damage suffered by his government and even turn the crisis into an opportunity to bolster his personal popularity and the prestige of his government. In other words, if Tung could convince the jittery people that the SAR government was capable of minimizing the adverse impact of the crisis and make them believe that his government was caring and benevolent, it was possible that public support for him would not decline so steeply.

At the time when Hong Kong was hard hit by the economic crisis, Tung's paternalism remained basically his own personal convictions. It did not have the necessary institutional structure to translate it into sound public policies that would, in turn, be successfully implemented. At the same time, Tung had failed to organize a governing coalition around him, a coalition that shared his political beliefs and commanded respect and trust from the people. In short, an institutional structure to support his personal convictions was lacking. The absence of an institutional component in Tung's governing strategy reflected also the lack of a workable political strategy on his part. The reasons for this are legion.

Before his elevation to the top post of the SAR, Tung Chee-hwa had all along been a shipping magnate running a family business founded by his father—a businessman cast in the Confucian mold. Tung's background had produced in him a visceral distaste for politics. Not only did he find political conflicts repugnant, but he abhorred politicking of all sorts. His disdain for politics and his disgust at political strife before the handover instilled in him an urge to depoliticize the SAR, once it was installed. However, Tung did not realize that a deliberate strategy of depoliticization by any government required either the suppression of political conflict by coercive means or the dilution of politics by a wise political strategy. The option of using coercion to do away with politics was out of the question, for that would stigmatize Hong Kong in the international community and deprive it of the much-needed support from the West. The strategy of depoliticization by political means would demand an active political strategy skillfully devised and executed by the government. It would also call for firm and powerful leadership from Tung himself. However, instead of adopting an active political strategy targeted at depoliticization, Tung instead sought to depoliticize by passively ignoring and avoiding politics. The low political profile adopted by Tung after assuming office was undoubtedly a reflection of his anti-political bent, his lack of charisma, his elitist instinct, and his lack of political decisiveness and populist skills. In contrast to the Patten government, Tung's was very reluctant to reach out to the people. Government policies were not adequately explained to the public, and the public was only minimally consulted before major policy decisions were made. This passive, or more appropriately escapist, political strategy wobbled along in the first several months of the SAR, when political tranquillity prevailed. Its flaws, however, became evident after Hong Kong found itself in an economic crisis, when widespread discontent surfaced.

Tung's lack of an active and aggressive political strategy reflected his hitherto inability to consolidate a coherent leadership core. Being a newcomer in Hong Kong's political scene, Tung had no political allies who could give him loyal and capable political backing or help him build a supportive political network in society. His Executive Council (ExCo), presumably his "inner cabinet," was basically a form of political compromise or balancing act. He chose his Executive Councilors from different political backgrounds, who could hardly form a cohesive team. What was worse was that most of the ExCo members had strong connections with big business. They shared with Tung the same disdain for politics, and were by and large detached from the people. Politically, they were not in the combative mood, not the type willing to take on the Legco in defense of Tung's policy; nor did they have the political ability to win mass support for Tung. As a result, the ExCo members were unable to win public trust. More unfortunately, their popularity was severely damaged by a slew of political scandals that showed their lack of sensitivity to the public mood.[45]

Tung's relationship with his principal officials was an uneasy one. Except for the Secretary of Justice, who was handpicked by Tung, all the principal officials were holdovers from the colonial regime. Unlike in the decolonization in other former British colonies, where the colonial government played an influential role in grooming future political leaders, the colonial government in Hong Kong—reacting to its fights with China—had to transform senior civil servants into politicians in the hope that they would dominate the post-1997 scene. In fact, the most prominent of them were deliberately groomed to fill top administrative posts in the last moment of colonial rule by Chris Patten. Tung's political values and policy agenda were not heartily shared by his principal officials, who were still immersed in laissez-faire thinking, anti-Communism, and a parochial "Hong Kong first" mind-set. They regarded themselves as the rightful successors to the colonial rulers, hence the legitimate custodian of Hong Kong's vital interests. Tung was not highly regarded by his principal officials, who were particularly irked by his hands-on approach, personal arbitrariness, and irresolution. At the same time, the principal officials were haunted by a palpable sense of insecurity. They were much plagued by the fear that sooner or later Tung would replace them with more dependable political allies. In any case, notwithstanding his uneasy relationship with his principal officials, Tung had no choice but to continue to rely on their service, for the sake of maintaining the morale of the civil servants, who are still the mainstay of his administration.

The relationship between the ExCo and the principal officials was even more strained than that between Tung and the latter. Unlike under colonial rule, in which civil servants exercised strong influence on who would become ExCo members, most of the ExCo members appointed by Tung held different political views and policy ideas from those of the civil servants. Moreover, a portion of ExCo members held some of the principal officials in contempt, which was fully reciprocated. Originally Tung intended to have the ExCo members lead and oversee the principal officials by giving some of them a prominent role in policy-making and political leadership. This, however, led to the eruption of latent and open friction between them. Eventually, the ExCo proved unable to perform the function expected of them by Tung, who had perforce to depend more on his principal officials. Still, the relationship between the two groups was difficult at best.

The inability of Tung, the ExCo members, and the principal officials to work as a team has severely marred the functioning of the new government. The lackadaisical performance of the Tung administration during crisis situations was a major reason for the decline of Tung's popularity. By the end of Tung's first year in office, none of the major policy goals spelled out by Tung as his top priorities had been achieved. Instead, in the areas of housing supply, land sales, public health and hygiene, and the linguistic medium of instruction in secondary schools, there was a lack of clear policy direction or singleminded implementation of policies. The Tung administration was even widely blamed for aggravating Hong Kong's already difficult conditions. The public became convinced that the principal woes of the new government were that it lacked decisive leadership and that it was out of touch with the public.

The failure of Tung to build a strong leadership core naturally also resulted in the corollary failure to establish a governing coalition in society. Despite his enormous powers to make political appointments and formulate public policies, Tung has not been able to convert them into a political alliance in support of his administration. The kinds of political appointments he has made fail to draw a large enough pool of politically influential figures into his fold. As a matter of fact, what he has done simply fails to produce a broad-based political coalition; instead, a large number of disgruntled political actors has converged on his administration.

Not only were the liberals excluded from Tung's favor, many pro-China figures, likewise, were dismayed by the cold shoulder given them. Consequently, when Tung was under attack, few came to his rescue. At the same time, the inconsistency in his policies has left people in nearly all social sectors disappointed or aggrieved. This, coupled with Tung's ineptitude in resolving Hong Kong's economic problems, has severely weakened his government's appeal to the business community and the grass-roots people alike, both of whom were initially favorably disposed toward the new regime.[46] By the end of his first year in office, Tung found himself a lonely political leader, besieged on all sides.

Conclusion

By now it is patently clear that Tung's governing strategy has failed. A needed change of strategy has, however, yet to be made. Beginning in the second half of 1998, Tung seemed to have some awareness of the flaws in his governing strategy. He has adopted a slightly higher political profile. His officials were more forthcoming in explaining public policies and soliciting public support. Though the Asian financial turmoil has temporarily subsided, after the turn of 1999, Hong Kong is still mired in continuing economic recession. The difficulties faced by the Tung administration remain inexorable. With the continuous unfolding of the after-effects of the financial crisis and the public's lingering fears and discontent, the SAR government will continue to encounter difficulties in governing. Already, the appearance of a more hostile legislature after the May 1998 election, where the liberals and grass-roots-oriented politicians are represented along with those representing functional constituencies, poses a formidable challenge to Tung and his government. To survive in this more turbulent environment, let alone exercise effective governance, the Chief Executive needs a drastic change in his governing strategy and style. Unless and until a broad-based governing coalition led by a cohesive leadership core is forged, and a more aggressive and proactive political strategy launched, the SAR government headed by Tung will not be able to reverse the trend of its declining popularity.

The bottom line of all this is that the Hong Kong SAR's problems, as this chapter has endeavored to demonstrate, are essentially its own, including a perceived inept Chief Executive, and not, as the prophets of doom had predicted, the arbitrary meddling by the Chinese sovereign after the territory's return from British colonial rule. The solutions are anticipated by the source of the problems: they have to come from within the SAR.

Notes

This paper is based on a research project entitled "Indicators of Social Development: Hong Kong 1997." The project was generously funded by the Research Grants Council of the Universities Grants Committee. The project is a joint endeavor of the Hong Kong Institute of Asia-Pacific Studies at the Chinese University of Hong Kong, the Centre of Asian Studies at the University of Hong Kong, and the Department of Applied Social Studies at the Hong Kong Polytechnic University. I am grateful to Ms. Wan Po-san, Research Officer of the Institute, for rendering assistance to the project in many ways. Special thanks are due to my two dedicated research assistants, Mr. Shum Kwok-cheung and Mr. Yiu Chuen-lai, for their help in administering the questionnaire survey and in data preparation.

1. *Apple Daily,* 6 April 1998, p.A12.
2. Ibid., 27 July 1998, p.A2.
3. *South China Morning Post,* 8 December 1997, p.1.
4. Ibid., 9 August 1998, p.1.
5. A total of 701 face-to-face interviews were successfully completed, yielding a response rate of 49.7 percent.
6. *Hong Kong Economic Journal,* 29 August 1997, p.8.
7. Home Affairs Department, *Report on a Telephone Opinion Poll in March 1998* (Hong Kong: Home Affairs Department, 1998), p.25.
8. Idem., *Report on a Telephone Opinion Poll in July 1998* (Hong Kong: Home Affairs Department, 1998), p.25.
9. Ibid., pp.26–8; and Home Affairs Department, *Report on a Telephone Opinion Poll in July 1998,* pp.26–8.
10. *Ming Pao Daily News,* 4 February 1998, p.A7.
11. *Apple Daily,* 4 November 1997, p.A14.
12. *Hong Kong Economic Times,* 23 January 1998, p.A18.
13. Ibid., 29 August 1997, p.8.
14. Ibid., 24 December 1997, p.A18.
15. Home Affairs Department, *Report on a Telephone Poll in January,* Appendix V, pp.1–2.
16. *Ming Pao Daily News,* 1 October 1997, p.A1.
17. *Apple Daily,* 31 July 1997, p.A6.
18. *South China Morning Post,* 3 August 1997, p.1.
19. Home Affairs Department, *Report on a Telephone Opinion Poll in January,* Appendix V, pp.1–2.
20. Idem., *Report on a Telephone Opinion Poll in March 1998,* pp.31–32.
21. Idem., *Report on a Telephone Opinion Poll in July 1998,* p.33.
22. *Apple Daily,* 23 December 1997, p.A17.
23. Ibid., 20 January 1998, p.A17.
24. *Apple Daily,* 26 November 1997, p.A15, and *Hong Kong Economic Times,* 28 February 1998, p.A18.
25. *Oriental Daily News,* 31 January 1998, p.A13, and *Apple Daily, 31* January 1998, p.A7.
26. *Apple Daily,* 1 December 1997, p.A17.
27. Ibid., 27 June 1998, p.A16.
28. *South China Morning Post,* 7 December 1997, p.1.
29. Ibid., 5 April 1998, p.1.
30. Ibid., 18 August 1998, p.1.
31. Ibid., 19 January 1998, p.1.
32. *Ming Pao Daily News,* 20 June 1998, p.A6.
33. *South China Morning Post,* 7 August 1998, p.8.
34. Lau Siu-kai, "The Search for Political Legitimacy by the Hong Kong Special Administrative Region Government" (unpublished manuscript 1998).
35. On 9 July 1997, a controversial bill forcing Mainland-born children seeking permanent Hong Kong residency to obtain prior endorsement of their right of

abode was rushed into law by the Provisional Legislative Council. Under the Immigration (Amendment) No. 5 Ordinance, Mainland children entitled to residency since 1 July 1997 under the Basic Law would have to apply for a certificate of entitlement before their status was confirmed. Application must be made outside of Hong Kong. The law was retroactive to July 1, and threw into doubt the fate of more than 2,000 Mainland children already in the SAR claiming right of abode. The ordinance deliberately aimed to admit these children into Hong Kong in an orderly manner. The legality of the ordinance was strongly challenged by the legal community in Hong Kong. Immediately after the enactment of the ordinance, China worked together with Hong Kong to confirm the status of the child applicants and to deter "illegal" entry into Hong Kong by strengthening border controls.

36. When China's Premier Zhu Rongji spoke to the leaders of the Chinese community in London on 4 April 1998, he confirmed that the most important consideration for not devaluing the renminbi was Hong Kong. He said that Hong Kong was part of China and if China devalued the renminbi it would have a huge impact on Hong Kong. See *South China Morning Post,* 5 April 1998, p.1.

37. Among other things, Tung had inserted the concept "national security" into Hong Kong's laws as a justification for the authorities to prohibit the formation of societies and holding parades. He also tore down the "Republic of China" flags hoisted by pro-Taiwan elements on 10 October 1997, and ordered the police to physically isolate the anti-Communist demonstrators during their demonstrations.

38. See, for example, Tung Tee Haw, *Building a 21st Century Hong Kong Together* (Hong Kong: Tug Tee Hwa, 22 October 1996); idem., *A Future of Excellence and Prosperity for All.* Speech by the Chief Executive the Honourable Tung Chee-hwa at the Ceremony to celebrate the Establishment of the Hong Kong Special Administrative Region of the People's Republic of China, 1 July 1997 (Hong Kong: Printing Department 1997); and idem., *Building Hong Kong for a New Era.* Address by the Chief Executive The Honourable Tung Chee-hwa at the Provisional Legislative Council meeting on 8 October 1997 (Hong Kong: Printing Department, 1997).

39. Lau Siu-kai and Kuan Hsin-chi, *The Ethos of the Hong Kong Chinese* (Hong Kong: Chinese University Press, 1988).

40. Lau Siu-kai, "The Fraying of the Socio-economic Fabric of Hong Kong," *The Pacific Review,* vol.10, no. 3 (1997), pp.426–41.

41. Tung, *A Future of Excellence and Prosperity for All,* p.13.

42. Ibid., p.16.

43. Tung, *Building Hong Kong for a New Era,* pp.5–6.

44. Lau Siu-kai, "The Eclipse of Politics in the Hong Kong Special Administrative Region," *Asian Affairs,* vol. 25, no.1 (spring 1998), pp.38–46.

45. According to a telephone poll by the *Ming Pao Daily News,* 33 percent of respondents saw the Executive Councilors of the SAR as performing more poorly than their counterparts in colonial Hong Kong. Only about a quar-

ter—26.7 percent—were satisfied with the work of Tung's Executive Councilors, whereas 29.2 percent weren't satisfied *(Ming Pao Daily News, 10* September 1997, p.A6). The popularity of the Executive Council plunged to a very low point after it was reported that some ExCo members had attempted to interfere in the work of the Electoral Affairs Commission in the drawing of electoral boundaries, and that Tung had decided to hold a birthday party for ExCo member Chung Sze-yuen—a controversial politician. In the polls commissioned by *Apple Daily* and conducted by the Hong Kong Institute of Asia-Pacific Studies, at the Chinese University of Hong Kong, it was found that the proportion of respondents who were satisfied with the performance of the Executive Council was 18.3 percent in July 1997, 20.5 percent in August 1997, 14.9 percent in October 1997 and 10 percent in July 1998 *(Apple Daily,* 29 October 1997, p.A6 and 27 July 1998, p.A2). Similarly, the polls of the Social Science Research Centre at the University of Hong Kong showed that only 23.9 percent of respondents were satisfied with the performance of the ExCo members in December 1997. The figure dropped to 22.1 percent in January 1998 *(Oriental Daily News,* 31 December 1997, p.A19, and 30 January 1998, p.A15).

46. On 18 February 1998, the SAR government handed out a modest yet wide-ranging package of tax cuts to comfort a jittery community and boost a sagging economy. Costing $13.6 billion in the next financial year and nearly $100 billion up to 2001–2, the package featured relief on the cost of mortgages, increased personal tax allowances and half a percentage point cuts in profits tax and rates. It meant 99 percent of taxpayers would pay less. The first budget of the SAR government was well received by the community, particularly by the middle classes. It helped boost the popularity of Tung a little bit. Still, the basic political predicament of Tung remained.

CHAPTER 2

Hong Kong in the Midst of a Currency Crisis

HO Loksang

Introduction

In an article presented at a conference in February 1997—well ahead of the currency crisis and the recession in Hong Kong that came in its wake, I wrote that "there are causes for concern. . . . These causes for caution have nothing to do with the leaving of the British or the imposition of ' Red Flag Over Hong Kong.' They have to do with internal dynamics."(Ho, 1997) I further warned that if the government was not careful and if it did not act in time, the problems could be serious."(Ho, p.89). A year later, a cover story in *Business Week* (16 February 1998) carried the message: " Hong Kong: Don't be Fooled by the Markets. The Economy has Serious Problems."

The central theme in the *Business Week* article was that Hong Kong had become less competitive. It cited more expensive secretaries, more highly priced real estate, and more costly port services. These are indeed problems. But as late as the first half of 1997 Hong Kong was still growing rapidly. Its publicly listed companies were still highly profitable. The government was still running large surpluses. Even more ironically, a book entitled *The Hong Kong Advantage,* jointly authored by Michael J. Enright, Edith E. Scott, and David Dodwell, was published in that year by Oxford University Press. Despite high costs, Hong Kong was thriving on the basis of its ability to tap the cheap labor in South China, to serve as a middleman facilitating trade, syndicated loans, and stock flotations, and to offer excellent infrastructure and management skills to those multinational companies that have chosen to set up regional headquarters in Hong Kong.

Was Hong Kong's success merely a mirage or a bubble? My earlier analysis suggests otherwise. My earlier analysis, in February 1997, raised four delicate problems that require careful handling by the new SAR government. They were a fiscal problem, a monetary problem, a social problem, and a political problem. Unfortunately, the new SAR government went overboard in trying to address certain issues, particularly housing, which had shown up as a top concern in opinion polls, and did not take heed.

The Emerging Problems and the Genesis of the Crisis

I pointed out in my earlier paper that Hong Kong has always relied on land rent for a huge chunk of its fiscal revenue. Land rent is captured by way of land auctions, land premiums collected upon lease modifications and upon lease expiry, various taxes on land-related transactions, rates, stamp duties on land transactions, profits tax on land developers and on banks, which derive profits from mortgage loans, etc. Because Hong Kong's property prices were at a high level and further increases could not be taken for granted, I warned that Hong Kong government's revenues could dwindle quickly if there is a dramatic downturn in property prices.

I warned then that the ongoing asset price inflation would worsen any problem posed by a strong U.S. dollar, given the link to the U.S. dollar. I warned that because prices and wages tend to be rigid downwards, a strong U.S. dollar could start off a major recession and that if investors tried to sell their Hong Kong assets and convert them into foreign currency a problem of major dimensions could happen. I warned that income inequality was being compounded by inflation and was causing social tensions to heighten. Finally, on the political front, I warned about possible confrontations. I suggested that should political confrontations take place and get out of hand China could get involved. The worry was real that political pressures on the government could sway its policy in a way that proved detrimental to Hong Kong's long-term interest.

As things developed, the delicate policy problems were not handled carefully as had been hoped, and the internal dynamics proved disastrous for the Hong Kong economy. There was no political confrontation as had been feared, and the Central Government truly honored its pledge of no intervention into Hong Kong affairs. There was no surge in corruption as some people had feared. Rather than taking over Hong Kong's rich fiscal and foreign exchange reserves, China pledged to defend the Hong Kong dollar any time upon request. By all accounts the Central Government was well behaved and highly supportive of the SAR government while giving it the free hand to maneuver.

In response to public outrage over excessively high housing cost, the Chief Executive made three pledges and actually went about putting them into reality. In his Policy Address (akin to the U.S. President's State of the Union Address) made in 1997, Mr. Tung said, "I set my Administration three main targets: to build at least 85,000 flats a year in the public and private sectors; to achieve a home ownership rate of 70 percent in ten years; and to reduce the average waiting time for public rental housing to three years." Since that time his administration announced an unprecedented five-year Land Disposal Programme (31 March 1998) with great details, specifically to give teeth to the housing production target announced. His administration also announced in December 1997 a Tenant Purchase Scheme (TPS) allowing sitting tenants in selected public rental housing blocks to purchase their own flats at deep discounts, giving credibility to the 70 percent home-ownership ratio target. Finally his administration is committed to an accelerated public housing construction program that lends credence to the pledge of shortening the queue for public housing. By all accounts Mr. Tung had a clear vision. He wanted to dramatically reduce the cost of housing in Hong Kong and to boost home-ownership. Unfortunately, his administration was not aware of the implications of the economic dynamics involved.

The public housing sales program was intended to increase home-ownership, while the 85,000 production target was intended to reduce housing cost. In the context of public outcry against excessive increases in home prices both initiatives seem eminently appropriate. Unfortunately, the Tenant Purchase Scheme (TPS) had two serious flaws, and these flaws were so serious that the result was a total collapse of the housing market. The first flaw was to give the richer tenants, who had all along been told that they would have to pay double rent or even market rent, the right to buy their units at deep discounts just like the other tenants. The second flaw was that public housing units sold were allowed to trade just like other private housing units without requiring that future buyers must meet certain eligibility criteria. While these two conditions certainly helped sell the units they also made Home Ownership Scheme (HOS) flats, and other forms of lower-end private housing, unattractive. Unable to sell their own units, the owners of these flats could not climb the housing ladder to higher level flats. Higher level flat-owners, in turn, found that their potential buyers had disappeared. They, in turn, could not climb the housing ladder. The delicate ecological balance was suddenly disturbed and the entire housing market was paralyzed. The problem was further compounded by the ambitious 85,000-a-year production target. It served to warn potential buyers that housing is no longer a good investment. Suddenly the housing market, which had already been badly injured by the South East Asian currency crisis, collapsed. Table 2.1 shows that the decline in residential property prices since the last quarter of 1997 has been quite dramatic. Given that the residential sector is unlike the office premises sector in that there is no noticeable excess supply, the sharp declines in property prices would be quite puzzling without referring to the TPS.

With the rapid decline in asset values many households found that their equity dwindled rapidly. Net assets readily turned into net liabilities. Everybody tightened their belts. Rather than helping families to own their own homes, the new housing policy ruined many dreams to own homes as homebuyers were forced to give up their deposits when banks found it necessary to refuse to lend the required amount, and as some home-owners were forced to give up their homes altogether. While the interest to buy homes fell, construction activities also slowed down rapidly and applications to convert non-residential land into residential land dropped sharply. This is in sharp contrast to the situation before the change in policy.[1]

Table 2.1 Percentage Change of Price Movements of Private Property during Period (1989=100)

Period	Residential	Office Premises	Retail Premises	Residential	Office Premises	Retail Premises
1991	153	97	143			
1992	215	133	200	40.5	37.1	39.9
1993	237	159	244	10.2	19.5	22.0
1994	293	222	285	23.6	39.6	16.8
1995	272	188	277	-7.2	-15.3	-2.8
1996	298	184	287	9.6	-2.1	3.6
1997	420	206	382	40.9	12.0	33.1
1998	299	129	274	-28.8	-37.4	-28.3
1995: Q4	264	176	260			
96: Q1	277	182	281	4.9	3.4	8.1
96: Q2	289	183	283	4.3	0.5	0.7
96: Q3	298	178	283	3.1	-2.7	0.0
96: Q4	330	198	288	10.7	11.2	1.8
97: Q1	395	214	340	19.7	8.1	18.1
97: Q2	429	215	371	8.6	0.5	9.1
97: Q3	433	196	409	0.9	-8.8	10.2
97: Q4	422	193	396	-2.5	-1.5	-3.2
98: Q1	354	159	342	-16.1	-17.6	-13.6
98: Q2	321	144	294	-9.3	-9.4	-14.0
98: Q3	263	110	238	-18.1	-23.6	-19.0
98: Q4	253	105	209	-3.8	-4.5	-12.2

Source: Hong Kong Census and Statistics Department: Hong Kong Monthly Digest of Statistics, various issues.

Note: Percentage for quartes are for quarter-on-quarter and those years are for year-on-year.

One could easily blame the currency turmoil and the formation of the asset price "bubble" for these troubles but there is no doubt that a gigantic policy error has turned an otherwise relatively minor difficulty into a major disaster. While the property market in 1996 through early 1997 was overheated and was due for a correction, the 200 to 300 basis point increase in the mortgage lending rate since the beginning of the financial turmoil cannot explain the 50-plus percent price decline in property prices from their recent peak. What prices are "reasonable" cannot be determined simply on the basis of a comparison between the price of an average flat against the income of a university graduate, as many commentators do. First, the prices of the average flats were "affordable" to the extent that real people were paying such prices for them, their buying power being buoyed by the high prices that their previously owned flats sought. Second, the average salary of university graduates has lost its earlier meaning and significance because there are so many of them today, given the rapid expansion of the tertiary education sector in recent years. In short, the property market became a dangerous bubble only when peoples' confidence collapsed, and to the extent that purchasing power was channeled away.

For decades Hong Kong households had put money into their homes as a depository for their life savings. Indeed, public housing tenants are known to be big savers and they have participated actively in home purchases.[2] With public housing available at very deep discounts (they are available to the sitting tenants at as low as 12 percent of the market price), private sector housing has suddenly become unattractive as an investment vehicle, and homes lost their function as a depository for savings.

As the retail sector depends very much on the new purchases people make when they change houses and on the perception of economic security, the collapse of the housing market tore the retail sector into shreds. The index for the value of retail sales hit a record low of 84.5 in February 1998— well below the base year index of 100 for 10/94 to 9/95. The volume index then was a pitiful 77.5, from which level there seemed to be a slight recovery in the year, but to the same painfully low level the index returned in November (See Table 2.2). It was reported that some 300 restaurants closed down business in the first quarter of 1998.

Hong Kong's External Economic Relations

Actually, despite the unfavorable labor cost and high land cost that have been afflicting Hong Kong, the SAR continues to be rated quite highly in terms of its international competitiveness. The International Institute for Management Development (IMD) put Hong Kong's 1998 rating at number 3, just behind the U.S. and Singapore, identical to that of 1997. The rating was based on scores in eight categories of performance, namely, domestic economy, internationalization, government policy, financial institutions, infrastructure, management, science and technology, and quality of the population. Even a better rating, at number 2 globally, was given by the World Economic Forum as recently as 17 August 1997. As an economy, Hong Kong remains as open as before, fully committed to free trade. The Heritage Foundation has, since 1996,

Table 2.2 Index of Retail Sales (Monthly Average of 10/94–9/95=100)

			Value Index (% change year-on-year)		Volume Index (% change year-on-year)
1995		101.1	+4.7	99.7	-1.4
1996		107.5	+6.3	101.3	+1.6
1997		112.7	+4.9	102.4	+1.1
1996	October	111.1	+7.9	102.4	+3.3
	November	106.0	+7.5	97.2	+2.9
	December	126.9	+8.4	116.5	+3.0
1997	January	128.3	+15.8	119.8	+11.4
	February	103.0	-3.2	96.1	-6.8
	March	107.5	+5.5	99.2	+0.9
	April	106.6	+10.0	96.7	+5.8
	May	113.0	+10.1	101.7	+5.6
	June	109.2	+8.4	98.1	+4.0
	July	117.8	+7.4	106.1	+2.2
	August	118.7	+7.7	107.7	+2.9
	September	112.4	+5.7	102.5	+2.5
	October	113.8	+2.4	101.4	-0.9
	November	105.6	-0.4	94.0	-3.3
	December	116.7	-8.0	105.3	-9.7
1998	January	114.6	-10.6	105.6	-11.8
	February	84.5	-17.9	77.5	-19.4
	March	94.2	-12.3	86.3	-13.0
	April	91.5	-14.2	81.6	-15.6
	May	95.6	-15.4	84.8	-16.6
	June	91.2	-16.5	82.1	-16.4
	July	98.6	-16.3	88.7	-16.4
	August	94.5	-20.4	85.3	-20.8
	September	88.2	-21.5	80.2	-21.7
	October	91.9	-19.2	83.1	-18.1
	November#	84.8	-19.7	77.9	-17.2

Source: Census and Statistics Department, Hong Kong SAR Government

given Hong Kong top rating in the Index of Economic Freedom, ahead of Singapore. As a financial center, Hong Kong is committed to prudential regulation and transparent dealings. In a survey conducted in 1997 by the Industry Department (i.e., Survey of Regional Representation by Overseas Companies in Hong Kong), 91 percent of the responding companies considered that, compared with the previous year, the overall attractiveness of Hong Kong as a regional headquarters or regional office had improved in 1997 or was more or less the same. The majority of them also held the same view on banking and financial facilities, infrastructure, tax regime, geographical location, and access to information. Although cost of office/factory space and staff cost were generally regarded as unfavorable, over the years the number of new regional headquarters set up had increased, from an average of 21 in the 1980–84 period to an average of 61 in the 1990–96 period.

Another survey carried out by the Industry Department, targeted at the manufacturing sector, found the majority of the 400 companies surveyed were optimistic about 1997 and the following year (Survey of External Investment in Hong Kong's Manufacturing Industries 1997): 237 companies expected business for 1998 to be better or equal to that in 1997.

It must be noted that competitiveness comprises many factors. If low land cost and low labor cost were sufficient to make a country competitive, then Indonesia would be very competitive indeed, particularly after the sharp depreciation of the rupiah. Recent developments told us that there is much more to competitiveness than low land cost and low labor cost. Indeed, in the wake of the riots in Indonesia, the rupiah fell sharply but garment orders that would have gone to Indonesia found their way to Hong Kong (*Hong Kong Economic Times*, 20 May 1998)! As a result, the prices for Hong Kong's garment quotas rose sharply. Hong Kong's political and social stability, evidently, have contributed to the SAR's competitiveness.

Dr. Victor Fung, Chairman of the Hong Kong Trade Development Council, recently noted that Hong Kong has now emerged as a coordinating center for various transnational economic activities. An example of such activity is offshore trade originating from Hong Kong but involving production outside Hong Kong that is ultimately sold to another country. The TDC estimated that such offshore trade amounted to U.S.$ 130 billion in 1997, some 5 percent of the GDP. Dr. Fung estimates that some 40 percent of the 300,000 small and medium enterprises in Hong Kong are involved in transnational business operations in one form or another. He points out, quite correctly in my view, that Hong Kong's competitiveness is not based on duplicating what other people can do, but on excelling in areas in which others do not have the expertise.

Success in such untraditional activity has earned Hong Kong the title of "the leading example of the Virtual Economy" (Sheng 1997). The virtual economy is defined, according to Professor Rosecrance of University of

Table 2.3 External Trade (HK$ million)

	Imports		Domestic Exports		Re-Exports	
	HK$ million	Year-on-year % change	HK$ million	Year-on-year % change	HK$ million	Year-on-year % change
1993	1,072,597	+12.3	223,027	-4.7	823,224	+19.2
1994	1,250,709	+16.6	222,092	-0.4	947,921	+15.1
1995	1,491,121	+19.2	231,657	+4.3	1,112,470	+17.4
1996	1,535,582	+3.0	212,160	-8.4	1,185,758	+6.6
1997	1,615,090	+5.2	211,410	-0.4	1,244,539	+5.0
1998	1,429,092	-11.5	188,454	-10.9	1,159,195	-6.9
1996 October	138,106	+5.2	19,432	-6.5	110,472	+8.2
November	131,054	+4.9	18,014	-4.4	100,396	+7.0
December	131,409	+2.9	18,217	-9.6	98,446	+3.5
1997 January	131,765	+1.1	17,274	-9.1	102,100	+0.9
February	102,273	+6.6	12,594	-11.3	76,465	+0.3
March	128,104	+6.0	14,799	+2.9	90,726	+10.2
April	139,099	+4.6	16,463	-4.2	104,525	+5.7
May	134,527	+3.5	16,834	-4.5	100,169	-0.1
June	131,720	+6.4	17,904	+5.9	101,138	+10.1
July	150,983	+9.0	20,278	-2.5	116,354	+4.7
August	139,766	+5.8	19,528	+7.4	109,964	+3.1
September	133,197	+1.9	19,032	+4.2	107,830	+0.0
October	152,931	+10.7	20,039	+3.1	122,919	+11.3

(continues)

Table 2.3 (continued)

		Imports		Domestic Exports		Re-Exports	
		HK$ million	Year-on-year % change	HK$ million	Year-on-year % change	HK$ million	Year-on-year % change
1997	November	133,018	+1.5	17,992	-0.1	160,687	+6.3
	December	137,822	+4.9	18,722	+2.8	105,716	+7.4
1998	January	113,194	-14.1	15,232	-11.3	99,110	-2.9
	February	106,753	+4.4	11,557	-8.2	76,402	-0.1
	March	123,590	-3.5	14,489	-2.1	94,096	+3.7
	April	133,190	-4.2	15,467	-6.0	100,518	-3.8
	May	124,454	-7.5	16,226	-3.6	97,030	-3.1
	June	123,258	-6.4	17,685	-1.2	98,998	-2.1
	July	123,119	-18.5	18,217	-10.2	100,865	-13.3
	August	119,873	-14.2	17,519	-10.3	101,621	-7.6
	September	115,675	-13.2	16,088	-15.5	98,131	-9.0
	October	118,711	-22.4	15,638	-22.0	102,358	-16.7
	November	113,809	-14.4	15,144	-15.8	97,943	-8.2
	December	114,260	-17.1	15,206	-18.8	92,287	-12.7

Source: Census and Statistics Department, HKSAR Government

California at Los Angeles, who coined the term, as one that "pushes toward a downsizing and relocation of its production capabilities. It performs the headquarters functions and produces its goods within the territory of another state" (Sheng 1997, p.28).

Table 2.3 shows that Hong Kong's external trade in the first quarter of 1998 was not seriously affected by the financial crisis. Although exports grew only slightly on a year-on-year basis, the merchandise trade deficit actually declined. Overall, the external sector, apart from tourism, was not, at the time, a particularly negative influence on the Hong Kong economy.

Of course, Hong Kong cannot be immune from the currency crisis of the region. Especially hard hit was the tourist industry. Tourists from Southeast Asia fell to a trickle as the purchasing power of the nationals in ASEAN, Japan, and Korea were eroded by the deep depreciation of their currencies. In the first quarter of 1998 Japanese and Korean tourists fell some 60 percent from a year ago. Total tourist arrivals fell 24 percent from a year ago (*Ming Pao*, 4 April 1998).

The Hong Kong dollar was attacked by speculators and, in defense of the dollar, Hong Kong's interest rates were raised. The best lending rate rose in stages from 8.75 percent in September 1997, finally reaching 10.25 percent in January 1998, with commensurate consequences on consumer loan rates and mortgage rates. It was widely believed that international speculators had sold short in the stock futures market, and had profited from stock price declines resulting from interest rate hikes consequent upon the Hong Kong dollar's decline in the forward market. It was with this hard-learned lesson that the SAR government made a decisive, bold, and surprising move: in August 1998, the government spent some $120 billion mopping up shares dumped by major players, bringing the Hang Seng Index from around 6,600 to around 7,800 in a couple of weeks. As a result, the SAR government has become a major shareholder for a number of blue-chip companies, including the Hong Kong Bank, Sun Hung Kei Property, and Cheung Kong Holdings. There was wide support in the local community, but the international community were quite critical. Over time, however, the criticism appeared to give way to applause, particularly after the Federal Reserve Board decided to bail out Long Term Capital when this major hedge fund suffered huge losses. Still, in order to dispel public concern that there is conflict of interest involved when the Hong Kong government is both shareholder and regulator, an independently managed company in the name of Exchange Fund Investment was established in October 1998. Chaired by a former Chief Justice, Mr. Yang Ti-liang, the board comprised legislators, academics, and professionals.

Up until the end of 1997, Hong Kong appeared to have fared quite well compared to other economies in the region. As a gesture to show that it is a good neighbor, Hong Kong even offered help to Thailand and Indonesia as part of an international effort to salvage those badly hit economies. But the economy started to suffer "internal injury" starting in December 1997.

Developments in the Local Economy and Policy

As Table 2.4 shows, the number of transactions in existing homes suddenly plummeted in December, coinciding with the announcement of the Tenant Purchase Scheme on 8 December. The sharp decline in existing home transactions was only temporarily interrupted by the stimulative budget and an interest rate drop in March. By destroying the ability of Home Ownership Scheme housing owners to move up to higher quality housing, the TPS is really the culprit behind the dramatic decline in property prices and the emergence of "negative asset" for hundreds of thousands of home-owners. According to *Ming Pao,* "Within four months from May 1998 the number of negative asset homes had gone up from 36,500 to 97,000 by September."

Another development, also quite damaging, was that in the wake of the currency crisis, some investment bankers and security houses had got into serious trouble. Most notable was the fall of Peregrine, the leading local investment bank, which had pioneered the flotation of H-shares from China in the Hong Kong Stock Exchange. Peregrine fell because of the "melt-down" of the Indonesian economy leading to heavy losses in its bond market involvement. Another failure of a financial institution worthy of mention is the case of C.A. Pacific, following which hundreds of clients suffered heavy losses not knowing that their stockholdings had been pledged as collateral to a subsidiary and were not recoverable. Yet another case was the failure of Forluxe Securities. These latter two incidents point to the need for greater protection of clients' assets. At the same time they highlight the difficulties that many brokers face as a result of a dramatic downturn in stock prices and trading volumes following the financial crisis and the plunge in property prices.

As the property market collapsed, some law firms also got into trouble. Cases of lawyers giving up their practices began to surface, and layoffs of legal secretaries and clerical helpers followed. Within the construction sector, unemployment jumped as workers were laid off while housing construction projects were put on hold. Indeed, according to the Construction Workers' Association some 55,300 construction workers worked less than 10 days a month in February. Close to 75 percent of construction workers were unemployed or underemployed (*Ming Pao,* 22 April 1998). Official unemployment figures also indicated that the construction sector is among the worst hit of all industries in the current slump.

While the plunge in property prices has such dire consequences throughout the economy, affecting the retail sector, the banking and financial sector, the real estate brokerage sector, the construction industry, and the legal and accounting professions, many observers were hopeful that the short-term pain would in the end benefit Hong Kong by making it more competitive. There is little doubt that, other things being equal, if land and labor costs were lower Hong Kong would be more competitive. Unfortunately, other things are not equal. As the property market collapsed, bad debt and drying credit would cripple many firms and ruin many lives. With the disappearance of land as an important source of revenue, Hong Kong's taxes would have to be increased, or the government would have to cut back its services.

What urgently needs to be done is to take concrete measures to revive peoples' confidence in the property market. As I argued elsewhere (*South China Morning Post,* 18 May 1998), the property market is actually an amazingly efficient way to diffuse and create more income and wealth. In order for the benefits of the virtual economy to be shared by more people, those who have the money should be willing to put it in property and should be prepared to move up the housing ladder when they can afford it. Because the government collects land premiums, the benefits of prosperity is captured by society at large through the government. When people move from smaller homes to bigger ones they make their older homes available for the less well-off, and benefit a whole range of industries from finance and real estate to decorators and developers, and from furniture and appliance retailers to lawyers and movers. As these various sectors enjoy strong growth in employment, spending from these sectors, in turn, boosts spending in other sectors.

Aware of the importance of rebuilding confidence, the SAR government has taken a number of concrete steps. Apart from a stimulative budget announced in February 1998, which included an unprecedented mortgage interest deductibility allowance, the government announced a moratorium on land sales from 22 June 1998 through March 1999. In February 1999 the government announced a cautious and relatively small-scale land sale program aimed at retaining the fragile confidence of the Hong Kong people. This was on top of other initiatives such as boosting the quotas for cheap or interest-free loans for home-buyers made available through either the Housing Authority or the Housing Society, and a loan insurance mechanism that facilitated the offer of mortgage loans, providing 85 percent financing for the appraised values of homes. However, short of revamping the Tenant Purchase Scheme, which is really the crux of the problem, all these efforts will prove ineffective.

A serious obstacle to the recovery of the Hong Kong economy is the tightness of credit. It is well-known that high real interest rates would dampen economic activity. With inflation clearly trending down to negative territory toward the end of the year while lending rates are holding up at nearly double-digit levels, borrowers are squeezed. What is even more damaging is that many banks, out of caution, are not willing to lend at all. In an

Table 2.4 Transactions in Existing Homes

	1997 Oct.	1997 Nov.	1997 Dec.	1998 Jan.	1998 Feb.	1998 Mar.	1998 Apr.	1998 May	1998 Jun.	1998 July	1998 Aug.	1998 Sept.	1998 Oct.
Number	9069	9546	4198	4016	3217	6063	5102	4816	3750	3690	3848	3564	2993
Value	36.0	34.5	15.3	14.9	11.0	19.8	17.8	15.4	11.5	9.6	9.4	9.0	7.0

Source: Ming Pao, 10 November 1998

Table 2.5 Civil Service Pay Increases since 1988

	1988	1989	1990	1991	1992	1993	1994	1995	1996	1997	1998
Senior	9.56	13.43	15.00	10.43	11.17	9.76	9.47	9.98	7.68	6.90	6.03
Middle	9.93	14.81	15.00	10.43	11.60	10.66	9.89	10.14	7.67	6.81	5.79
Junior	9.62	14.81	15.00	10.43	11.60	10.66	9.89	10.14	7.67	6.81	5.79

Source: Wenhui bao, May 12, 1998

attempt to forestall bank failures, the Hong Kong Monetary Authority proposed to raise the permitted capital adequacy ratio to 10–12 percent from the current 8 percent (*Hong Kong Economic Journal,* 1 May 1998). Even though most banks have a comfortable margin of capital adequacy, the decline in property values could easily eliminate this margin of comfort. The proposed rule could start another round of credit tightening, further damaging the ability of the economy to recover.

Apart from increasing the availability of credit and lowering interest rates, Hong Kong would seem to need to curtail the salary increases of our civil service and non-civil service public offices. The Chief Executive of the Hong Kong Monetary Authority, for example, currently earns some $8 million a year. Others earning similarly astronomical salaries are the heads of the Airport Authority, the Hospital Authority, the Kowloon Canton Railway Corporation, and the Mass Transit Railway Corporation. Compared to the salaries of these high officials, the salary of the SAR Chief Executive is very modest indeed.

Table 2.5 shows the salary increases of civil servants since 1988. Long years of salary increases have now generally put civil service pay way out of line with private sector pay. Although civil service pay increases are supposed to make reference to salary increases in the private sector, the private sector salary increase surveys are not representative. They also fail to take account of the frequent occurrences of layoffs and streamlining which lead to de facto salary cuts for many workers in the private sector who have to accept lower pay to stay employed. Such high rates of salary increases widen the gap between the haves and the have-nots. As a result, income inequality was worsened, and the business environment in Hong Kong was adversely affected.

While many commentators have argued that the linked exchange rate is to blame for Hong Kong's economic woes (see the editorials of *Hong Kong Economic Journal,* 21 and 27 April 1998) the damage of the linked exchange rate to Hong Kong in this particular connection is likely to have been exaggerated. Although the Hong Kong dollar's exchange value had risen along with the U.S. dollar under the linked exchange rate system, the trade-weighted effective exchange rate index had risen only moderately. While it stood at 129.3 in July 1997, it rose to a high of 138.6 in January 1998, and softened before peaking at 139.3 in August 1998. Since then it softened markedly, reaching 131.8 by December 1998, which is not much higher than that before the onset of the crisis. In any case, in the past the moderate increases in the effective exchange rate index could have been absorbed without serious problems. Movements in the effective exchange rate index suggest that Hong Kong's problems in 1998 originate mainly from within. The TPS is a big problem. Another problem, perhaps equally damaging, is the failure of the Hong Kong Monetary Authority to safeguard peoples' confidence in the link, rendering it necessary for Hong Kong to suffer very high, recession-causing interest rates.[3]

Conclusions

Post-1997 Hong Kong promises bountiful opportunities. In search of such opportunities, a record number of returning emigrants was recorded in 1997, while a recent record low of emigrants left the territory. The significant number of net immigrants is clearly one important reason for the unemployment rate to shoot up.

The Hong Kong economy is known to be vibrant and flexible. Over the years, Hong Kong has gone through all kinds of ups and downs, including the shock of the Cultural Revolution on the Mainland in 1967. During the 1970s the two oil price shocks hit Hong Kong rather badly, and Hong Kong had seen its stock market index drop from 1700 to less than 200 in the 1973–75 period. Yet Hong Kong people have never felt so depressed and so alarmed.

The problem is clearly nothing of the sort as portrayed by *Fortune* magazine, which carried a cover story on the "Death of Hong Kong." *Fortune* foretold a Hong Kong beset by rising corruption and curtailment of civic freedoms. It foretold a Hong Kong that was ruined by intervention from the Mainland. Yet nothing like this happened. China left Hong Kong entirely on its own. As an example, the SAR allowed the commemoration of the 4 June incident in Victoria Park. The PLA has been extremely well-behaved. While top Chinese leaders and retired Party veterans did come to Hong Kong, none of them ever interfered in Hong Kong's administration. China honored its commitment and kept its word.

In contrast to what many people feared, the Tung Chee-hwa administration was extremely sensitive to calls from the grass-root level to help them.

Table 2.6 Immigration and Emigration 1994–1997

Year	Number of Emigrants	Number of Returning Immigrants
1997	30,900	127,000
1996	40,300	63,900
1995	43,100	50,100
1994	61,600	20,400

Source: Wen Wei Pao, 26 March 1998, citing Secretary of Security; *Ming Pao,* 19 May 1998.

His administration announced unprecedented tax cuts that benefited everyone but gave only a symbolic tax relief to the business sector in the form of a 0.5 percentage point reduction in the profit tax rate. His massive housing construction program was entirely an initiative in response to calls from Hong Kong people to address the housing problem. The Tenant Purchase Scheme was an initiative to put teeth behind the pledge to raise the home-ownership ratio to 70 percent in ten years. All in all, Tung Chee-hwa was far more responsive to the needs of the man in the street than to those of the chief executive officer or the chairman in the boardroom. Yet his policy, tragically, is a flop in spite of the best of intentions. Eager to achieve, he staggered because he attempted to run faster than he could.

Notes

1. The Director of Lands Department noted that land transactions in 1997, through land auctions, land exchange, land grants, and lease modifications would allow the construction of 45,000 units of housing, up 60 percent from the 96 level of 28,300 units. (*Hong Kong Economic Journal,* 23 January 1998).
2. According to a survey conducted from October 1992 through March 1993 public housing tenants accounted for 24 percent of all home purchases during the period. See the Interim Report on Long Term Housing Strategy, October 1993.
3. The Chief Executive of the Hong Kong Monetary Authority had been criticized for relying exclusively on high interest rates to defend the Hong Kong dollar. After much soul searching he amended his position and announced seven technical measures to strengthen the currency board system on 5 September 1998. The measures boil down to an increase in availability of liquidity when the Hong Kong dollar is under stress.

References

Fung, Victor. 1998. "Enhancing Hong Kong's Uniqueness and Competitiveness," keynote speech delivered to the Central Policy Unit forum on 22 April 1998.

Ho, Lok Sang. 1997. "Hong Kong in the 21st Century," in *New Asia,*vol.4, no. 2, (summer): 88–98. This was presented in the International Symposium on the East Asia Region, Okinawa, Okinawa Economic Association. 22–25 February 1997.

Sheng, Andrew. 1997. "Hong Kong and Japan in East Asian Finance," in *Money and Banking in Hong Kong,* vol. 2: 27–40. Hong Kong: Hong Kong Monetary Authority.

Hong Kong Economic Journal, also known in Chinese as Xin Bao, is a local daily newspaper in Hong Kong.

South China Morning Post (SCMP) is an English-language daily in Hong Kong.

CHAPTER 3

The Securities and Futures Markets: From 1 July 1997, The Year under Review

Anthony Neoh

A Momentous Day

On 1 July 1997, China resumed the exercise of its sovereignty over Hong Kong, and the former British colony became a Special Administrative Region (SAR) of China. Historically and constitutionally, Hong Kong entered a new era. However, as history is no more than a continuum of past, present, and future events, Hong Kong could not therefore escape from its own history and that of the rest of the world. That is why this review has to start from a global context.

Hong Kong in the Global Context

From day one, the Hong Kong SAR found itself in a continuing process of globalization unprecedented in world history. This continuing globalization of the markets requires each market to continuously rethink its systems of regulation. Regulation has become a much more dynamic process requiring every society not only to think of their own needs but also the needs of other societies. The globalization of the markets has been the result of two important developments. First, the breakdown in barriers to capital flows. Second, continuing and rapid advances in technology. Electronic international payments systems make it possible for international fund transfers to be made instantaneously, and multi-currency payments to be completed within a matter of hours. A worldwide web of custodians and depositories of financial instruments enable delivery to be made against payment within three days of a transaction, a universal standard first recommended by an influential industry group, known as the Group of Thirty. Trading derivatives in any scale, to any degree of complexity, would have been impossible without the software and computer processing power available today. The consequence of all this is that losses in one market thousands of miles away can mean financial ruin for venerable financial institutions within a very short time, as we have seen in the case of Barings Brothers. These developments make every market think hard about old assumptions, but more particularly about what is common between them. With increasing globalization, there is going to be more and more which is common between markets. Hong Kong found itself, on 1 July 1997, deep in the middle of this ferment of thought.

Hong Kong as an International Financial Center

It is because of this preoccupation that Hong Kong has become the freest and most open international financial center in the world. Many world-renowned studies (such as those conducted by the World Economic Forum, the Hoover Institute, the Heritage Foundation, and successive National Treatment Studies by the U.S. Government) have so concluded.

Capital freely flowed (and continues to flow) in and out of Hong Kong. The Securities and Futures Commission (SFC) was, and is, very much part of a growing interdependent global financial system. Hong Kong's foreign exchange market deals on a 24-hour basis around the world. It averages U.S. $91 billion a day in spot foreign exchange dealings alone (the derivative markets in currencies adding another U.S. $56 billion a day), making it the fifth largest foreign exchange center in the world. But like all international financial centers, Hong Kong and those who deal with us, have to take the mutual risks involved in these (and indeed other) financial transactions, namely, the credit risk, settlement risk, market risk, legal risk, and regulatory risk, to name but a few. Hong Kong was (and is) a place that was (and is) kept constantly aware of the fact that it is an important element in a global financial village. Yet, by the same token, Hong Kong competes as each market reckons with the free choice of investors.

Take Hong Kong's listed companies for example. In 1984, the Jardines group, after nearly 140 years in Hong Kong, decided to change its corporate domicile to Bermuda. The Hong Kong Stock Exchange, then listing only domestic corporations, decided to change its rules to allow Jardines to continue to list. Since then, Hong Kong's Stock Exchange has played host to an increasing number of foreign domiciled companies. In fact, nearly 60 percent of the 567 firms currently listed in the Hong Kong Stock Exchange (see www.sehk.com.hk) are foreign domiciled companies.

Turning Adversity into Virtue

Faced with a flight of companies to Bermuda, Hong Kong turned potential adversity into a virtue by making its listing rules sufficiently flexible to accommodate foreign domiciled companies. Jardines has since left the Hong Kong stock market, but the result to date of Hong Kong's response to potential adversity is a stock market with a market capitalization of U.S.$350 billion, the largest stock market in Asia outside of Japan, and the eighth largest in the world.

The Stock Exchange overhauled its listing rules to allow overseas domiciled companies to be listed, provided they fulfilled the required standards of shareholder protection and disclosure. These standards have consistently been improved so that many Hong Kong public securities issues are now globally offered. There are a growing number of cross listings between the Hong Kong Stock Exchange and the New York Stock Exchange, notable examples being some of the major blue chip companies and a number of Chinese state-owned enterprises, which would not have been listed in Hong Kong and New York but for the fact that they were able to comply concurrently with the standards of both jurisdictions.

The Application of International Standards

As part of the global market, Hong Kong had no choice but to apply international standards in all fields of financial regulation. On the banking side, Hong Kong was one of the first Asian markets to adopt the Basle capital rules and in 1998 also became the first Asian market to announce application of the Basle market risk capital rules when they came into effect in 1997. The importance of Hong Kong as an international banking center was recognized by the Bank of International Settlements (BIS) by its recent invitation to the Hong Kong Monetary Authority, along with eight other Asian central banks, including the People's Bank in Mainland China, to join a hitherto select club of OECD central bankers. In 1998, Hong Kong was chosen as the site of the Asian Regional Office of the BIS.

On the securities and futures front, Hong Kong has, since at least 1987, applied the standards developed by the International Organization of Securities Commissions (IOSCO), the principal international forum for the development of internationally recognized standards of regulation. IOSCO has in recent years made important strides, in which Hong Kong has actively contributed. The fact that the Hong Kong SFC chaired the

Technical Committee, the key committee in IOSCO charged with the development of international standards, underscores recognition of our commitment to this important task. This commitment is, at the same time, bolstered by bilateral regulatory agreements between Hong Kong and all the major markets of the world. The SFC has, at this moment, a total of 28 such agreements, including two with the securities and futures regulator on the Chinese Mainland. In the United States, the Hong Kong SFC has strong ties with the SEC and the CFTC, cemented by a regulatory agreement in 1996.

Market Development in Hong Kong

International investors would generally choose those markets that are liquid, efficient, transparent, and replete with all manner of product for investment and risk transfer. It was here that Hong Kong spent most effort and resources in the past seven years by the establishment of central clearing, a central depository for shares, automated trading systems in the stock and futures exchanges, and the launching of exchange-traded derivatives. At the same time, our financial prudential rules take account of the use of state-of-the-art financial risk models in use in international firms.

On the investment side, there were 46 mutual funds authorized for public sale in Hong Kong in 1978. The total was increased in 1985 to 161. In 1995, there were 1,183 domiciled in over 20 countries. In 1996, the number domiciled in as many countries went up to 1,300. There are now nearly 100 registered fund management firms in Hong Kong, employing some 600 registered investment advisers. Hong Kong is now the largest fund management center in Asia outside of Japan, and arguably, the largest international fund management center in Asia.[1]

Hong Kong also played (and continues to play) host to major financial institutions on the "sell side" of the business, from the United States, the United Kingdom, the European Continent (i.e., Europe other than U.K.), Japan, Taiwan, the Mainland of China, and other points in Asia.

On the banking front, there are 154 international banks operating in Hong Kong. In the securities and futures area, there are over 1,600 firms and over 16,000 individuals registered. Name almost any major international firm from any jurisdiction, and more likely than not, they are among our registered firms.[2]

Facing Competition

But where did all this place Hong Kong in the competition stakes? A 1996 report by the Nomura Research Institute (NRI) in Japan set out the requirements of a competitive international financial center as follows:

- efficient markets
- capacity for innovation
- good range of human resources
- wide use of the English language
- political stability, including an effective legal system
- good market reputation
- good business infrastructure, including low taxation and a reasonable cost structure
- good quality of life

The World Economic Forum's survey in the last three years preceding 1 July 1997, found the United States, Singapore, and Hong Kong at the top of the list. That meant that Hong Kong was found not only to have all the indicia that the NRI has indicated is required for a competitive international financial center, but has actually been listed, at least by this survey, as being at the forefront of competitiveness in the world. It is, however, axiomatic that since Hong Kong lives in a fast-changing world, where capital knows no borders and where choice reigns supreme, it could retain its competitiveness without being attentive to the needs of investors and the raisers of capital. That is why the SFC shall need to constantly listen and innovate, keep its antennae tuned, its senses keen, and its reactions quick.

As of 1 July 1997, a good case could be made that the hitherto openness that pervaded Hong Kong's society was truly at the root of its success. That openness may be seen not only at the level of trade and flow of capital. It is part of our everyday life. It enabled Hong Kong to look at change not as threats but as opportunities, and it filled Hong Kong with the air of dynamism that few would fail to experience soon after landing at Kai Tak Airport, itself soon to be replaced by a grand new airport for the twenty-first century. This dynamism was arguably the result of a sense of security in which the citizenry, and those who invest in or through it, felt that property, life, and liberty would be protected; that decisions that affected the commonweal were publicly discussed, and criticism of these decisions, no matter how harsh, may be freely expressed; that public servants were publicly accountable for their actions and may be amenable to challenge in the courts where necessary; that the government was, by and large, unobtrusive and efficient; that if any legal right was threatened there was recourse to an independent legal profession capable of getting justice from an independent judiciary. Some may wonder what is so special about this, asking rhetorically, "what is the big deal?" After all, this is what all democracies would take as a given. But, as events that unfolded after 1 July 1997 would show, this was indeed a big deal for all Asian economies.

A collective sense of security is not built overnight, nor can that be guaranteed by the most complete of written constitutions. That the SFC has built this collective sense of security in our legal and governmental institutions has been the sum total of the culture that has been built over 150 years of evolution of a community given a very special chance in history for East to meet the West and for the West to meet the East, to learn and live with each other's developing cultures and thereby enable both to prosper. The question was, as Hong Kong took a new road on 1 July 1997, would this culture continue to evolve? To answer this question, let us first look at the legal institutions after 1997.

The Legal Institutions Beyond July 1997

The Joint Declaration between the United Kingdom and the People's Republic of China and the Basic Law, Hong Kong's future constitution, explicitly built on this evolving culture. In 160 articles, the Basic Law set out the future institutions of the Hong Kong Special Administrative Region. They are familiar provisions because, by and large they seek to codify our way of life. In particular, they recognize Hong Kong's position as an international financial center, and mandate the free flow of capital, an independent monetary and taxation system, and a regulatory environment conducive to international transactions. As the markets of the world globalize, there are essential conditions for Hong Kong to grow in the future.

The provisions in the Basic Law that have been contentious are those that involve the relationship between the SAR and Central Government. It is perhaps ironic that provisions that are far less over-reaching in their scope than the residual prerogative powers of the British Crown, both in Hong Kong and the United Kingdom, should become a bone of contention. But that is probably not surprising. There is contention because these are untested ideas, whereas British constitutional law and practice has been tested in the past 150 years. The earnest debate of untested ideas enables each society to understand the issues with better clarity, and thereby fashion the best solutions. On 1 July 1997, Hong Kong could pin its hopes on three strands of history.

Firstly, the Basic Law was enacted in 1990, after four years of drafting and extensive public consultation and debate. Like all constitutions, it will be as good as the people who apply and interpret it. However, it is a constitution that stands to benefit from the accumulated ideas of two hundred years of development of constitutional law around the world. I am confident that with an open-minded legal profession and judiciary, this is a constitution that will enable our institutions to grow and meet the challenges of the future. In particular, because Article 39 of the Basic Law states that no law of the SAR may contravene the rights and freedoms set out in the International Covenant on Civil and Political Rights as applied to Hong Kong, those who live, work, invest in and through Hong Kong would be assured of the protections set out in this international document and the other provisions of the Basic Law.

Secondly, since the opening up of China after the Cultural Revolution, the legal and governmental institutions of Hong Kong have been widely admired by officials in China, and among those responsible for Hong Kong affairs there was good understanding of what made Hong Kong tick.

Thirdly, with the rapid economic growth of China and the region, China itself has become deeply conscious of the need for a legal system that can sustain economic growth. Hong Kong has been a useful exemplar in the process of building a legal system for a market economy, many in Hong Kong having been involved in advising Beijing on legislation or regulations of one form or another. As a big country, it cannot react with great speed, but no one can deny that China has made great strides when one takes account of the growing body of Chinese law and the flowering of private, independent lawyering.

The granting of the right of an accused person to independent and confidential legal counsel and representation upon arrest, the adoption of the presumption of innocence, and the placing of the burden of proof on the prosecution in the new Criminal Procedure Law promulgated in 1996 for implementation on 1 January 1997, was yet another example of the continuing development of the Chinese legal system. There was a fundamental revamping of the Chinese taxation system in 1994, and the rebuilding of the banking system to deal with a developing market economy is actively being pursued. These reforms would need time to be implemented, but any objective person would permit himself a degree of cautious optimism.

These were the conditions in which Hong Kong found itself on 1 July 1997. On that day of celebration, another drama was unfolding. In fact, the curtain was about to rise on the drama now known as the Asian financial crisis. On 2 July 1997, the Central Bank of Thailand announced that it could no longer defend the Thai baht and accordingly, it would allow the baht to float. Little was it known at the time that 99 percent of the U.S. $42 billion of the foreign exchange reserves of Thailand had already been committed in the forward markets, leaving Thailand with an effective reserves of U.S. $2 billion.[3] Hong Kong was still at the crest of a boom in property and stock prices. That happy position was not to last in this regional crisis.

A Year of Rapid Change from Boom to Bust

In fact, the year under review saw the end of a phase of unprecedented boom and the beginning of a decline. In the boom phase of the market, turnover was stood on its head when, contrary to its previous trend, the volume of trading in non Hang Seng stocks exceeded by far the trading volume in Hang Seng index stocks. All this was accompanied by rapidly rising trading volumes and share price movements bearing no relationship to the assets, profitability, or prospects of many companies outside of the Hang Seng Index. These were the market conditions which would, and did, put Hong Kong's markets to their most severe test since 1987. As can be seen from Figure 3.1, the Hang Seng Index reached its highest point in August 1997 and thereafter slipped inexorably. There were two drastic dips, one in October 1997, and another in January 1998. The Red Chips (Hong Kong domiciled and PRC majority-owned listed companies) Index and the H shares (Mainland domiciled and PRC majority-owned listed companies) Index performed similarly but with somewhat more violent force. However, the work of years of foundation-laying paid off. In the words of the Chief Executive of the Futures Exchange, Mr. Randy Gilmore: "The structures of the market were built for a hurricane. A hurricane came and the structures have held up."

The Securities and Futures Commission (SFC), in its report to the Financial Secretary made in January 1988, noted that market systems, particularly our trading, clearing, and settlement systems operated efficiently and effectively under the strain. Good risk management was in effect at all times. Despite the unusual share price movements observed, a fair and orderly market was generally maintained, margin lending levels had been massively reduced in brokerages and their unregulated finance company affiliates. A whole raft of policy and rule changes were suggested by the SFC and most of these appeared in the financial secretary's review published in May 1998.

The markets in the year under review claimed two major victims: Peregrine Investment Holdings Ltd. and the CA Pacific Group companies. In both cases, their failure could be attributed to the rapidity with which market conditions changed and while this did cast a long, psychological shadow over the market as a whole, the SFC's actions arrested any spreading to the rest of the market that these failures might have caused.

Peregrine Investments Holdings Ltd. was a publicly listed investment holding company that had 11 licensed entities providing a wide spectrum of financial services to the markets. As soon as a hoped for rescue by overseas investors fell through and it became clear that the Group would experience liquidity problems, the SFC issued restriction notices to protect the assets of the clients of all 11 licensed entities. As a result, all clients of those licensed entities have had all their investments or money returned.

CA Pacific Securities Ltd. and CA Pacific Finance Ltd. formed part of a group of financial services companies under a listed holding company. CA Pacific Securities was a securities dealer licensed by the SFC. It was also a member of the Stock Exchange of Hong Kong. CA Pacific Finance was a registered moneylender and, as such, did not require to be regulated by the SFC. The fact that it came to the attention of the SFC at all stemmed from activities started by the SFC in May 1997, consequent to an emerging and worrying trend then beginning to form in the market.

Meeting the Challenges of Change: Actions in Four Concurrent Directions

As can be seen from Figure 3.2, the level of trading in non-Hang Seng Index component stocks began to grow out of normal limits. Whereas trading in Hang Seng Index stocks typically take up 70 percent of the daily overall turnover, these stocks began to lose ground to non-Hang Seng Index stocks from about May 1997, when they fell to about 40 percent of the daily overall trading volume. In July, trading volume in Hang Seng Index stocks fell to 22 percent of the turnover. While this was happening, turnover began to increase significantly. At the same time, the prices of many non-Hang Seng Index stocks, particularly Red Chips, began to show irrational movements, indicating that there was a high level of speculation, driven possibly by rumor, market manipulation, or insider trading, or all three. There was no doubt that the market was overheating. As if to prove the point, a speech by the then Chairman of the SFC made in July 1997, that if we were not careful, Hong Kong's markets could turn into the "Wild West of the East," was greeted with derision. To meet this emerging situation, the SFC took action in four concurrent directions:

- Action against market malpractice
- Investor education
- Inspections of brokerages and affiliated finance companies
- Development of policy response to increased margin financing

Fair and Orderly Markets

The SFC's immediate priority was to ensure that the markets were fair and orderly. Thus, on 21 May 1997 a joint announcement was made by the SFC and the Stock Exchange expressing concern over the fact that there were unusual price movements and volume in trading in certain shares which appeared to bear no relationship to the assets, profitability, or prospects of the companies in question, and warning the market that, if necessary, shares may be suspended from trading until the SFC and the Stock Exchange could be satisfied that there was a fair and orderly market in the shares. As a result, there were 69 cases of suspension due to unusual price movements; 56 of these were requested by the companies, 7 were directed by the Stock Exchange and 6 were directed by the SFC. [4] Save in the most exceptional cases, where the directors have refused to make adequate disclosure to the market or where the company is in serious financial difficulties, trading would resume once the company makes adequate disclosure. In many cases, disclosures have been very detailed, indicating that there was indeed much information that the investing public would have been deprived of if no action had been taken. During the year, suspensions resulted in 9 investigations into possible market manipulation and 13 investigations into possible insider trading. These were on-going at the end of the year under review.

Informing Investors

In the area of investor education, the SFC issued brochures through the media and the Consumer Council, the most pertinent of which were "Choosing your broker or financial adviser" and "Monitoring your Investments." The SFC's web site had a dedicated Investor Education Section with an electronic facility for complaints. During the year, the SFC's enquiry hotline received an unprecedented number of complaints. Two popular television series had earlier in the year been aired under the auspices of RTHK (Radio & TV Hong Kong) on the pitfalls of unwise investment and on knowing your broker and financial adviser. As the markets continued to reflect high exuberance on the part of investors, these messages appeared to have fallen on quite a few deaf ears. This prompted the SFC to make a "Mrs. Sheep" TV commercial, warning investors to be careful in making their investments.

Margin Trading

The third prong of the SFC's actions was focused on margin trading and directed at brokerages and their affiliated finance companies. While margin trading has always been a feature of the Hong Kong markets, such trading had until May 1997 been at a relatively low level because of the low market turnover and because the pattern of trading was preponderantly in Hang Seng Index stocks. As the market practice was to lend no more than 50 percent of the value of the stock deposited as collateral, margin lending, prior to May 1997, was not a high risk activity.

As market conditions in May 1997 appeared to be trending toward significant change, the SFC undertook a survey of margin lending and, together with the Stock Exchange, inspected 103 stockbrokers and 56 affiliated finance companies. As the SFC had no regulatory powers over finance companies, much resistance was encountered in the course of these inspections. But by sheer perseverance, the SFC managed to persuade the brokerages most exposed to margin lending to reduce the level of credit extended to their clients. This tightening did help the broking industry, as a whole, to weather the sharp market decline in October 1997 and the poor credit environment that ensued. As the credit environment worsened, the SFC formed a task force in January 1998 to conduct further inspections and, as a result, total margin levels were reduced by a further $1.8 billion. The SFC succeeded in persuading all but one of the inspected broking firms and their associated finance companies to increase capital by $1.6 billion.[5]

In the course of these inspections, some companies in the CA Pacific Group were discovered to be encountering severe liquidity problems, due in large part to a connected property-related loan, which has since become the subject of a criminal charge. These liquidity problems meant that the securities firms might at any time go into default with Hong Kong Securities Clearing and were creating a huge potential for certain clients to be preferred against the rest. The Group was called upon to inject capital. In the event, CA Pacific failed to do so. To ensure that the interests of all investors are protected, the SFC presented a petition to wind-up CA Pacific Securities, and the Group itself presented a petition to wind up CA Pacific Finance, and provisional liquidators were appointed to preserve the assets and oversee their orderly disposition. It should be noted that the SFC's decisive action against CA Pacific not only served to ensure that the legal rights of their clients were protected, but it also served to demonstrate determination on the part of the SFC. That helped in preventing contagion to the rest of the market, as other firms responded to SFC concerns and found ways to inject capital and reduce their margin lending.

A Policy Response

In November 1997, the SFC developed a policy response to the increased margin activities of unregulated finance companies, and in early December the SFC submitted a report on margin lending activities, recommending that a working party be set up to consider the various policy options put forward in its report. The Financial Secretary acted with speed in forming the Working Party, chaired by the Deputy Secretary for Financial Services. By the end of the year under review, the work of the Working Party proposed legislative amendments bringing margin lending under tighter regulation. By the end of 1998, these suggestions were in the hands of the Law Draftsman.

Regulation Meets the Challenges of Change

Regulation is a continuing process of meeting the challenges of change. Thus, the four sets of actions that the SFC undertook represented the SFC's response to changing market conditions. Among the SFC's many activities, the primary accent was on strengthening the regulatory system. The Global Offering Mechanism as well as the Takeover Code were updated. There was a review of the Share Repurchase Code when the issue of treasury shares was considered. The use of plain language in all corporate finance activities, including initial public offerings, was implemented. The Fund Manager Code of Conduct was issued and a third edition of the Code on Unit Trust and Mutual Funds published. The SFC have made substantial strides in tackling, both within the SFC and in the market, the "Millennium Bug" computer problem. The SFC helped establish the Hong Kong Securities Institute. The SFC's international activities, through its Chairman, who also chaired the Technical Committee of IOSCO, brought Hong Kong important recognition among its colleagues worldwide as an important, trusted, and professional partner in the regulation of increasingly globalized markets.

An effective financial regulatory agency must continue to hone its skills and professionalism to meet the challenges of change in the markets. Arguably, the SFC lived up to its motto that it existed to serve the needs of the markets and the community, without fear or favor. The community could take the view that the dedication that the entire staff of the SFC has brought to their work had made a real difference in steadying the markets in the face of the strong winds that buffeted it for the year after 1 July 1997.

Postscript

On 17 August 1998, the Hong Kong government began a series of market intervention activities that attracted worldwide attention. Although these activities went beyond the period under review, their importance would merit at least a postscript. On that day, the Financial Secretary instructed the Hong Kong Monetary Authority (HKMA) to make use of the Exchange Fund to intervene in the stock and futures exchanges with a view to defending the Hong Kong currency. The best estimates in the markets at that time were that there were about U.S. $30 billion of notional value in forward contracts between the Hong Kong dollar and U.S. dollar. Many Hong Kong business entities and banks were hedging their currency exposure. There were estimates in the markets that some U.S. $9 billion of these forward contracts were held by hedge funds on the short side. Those on the long side of these contracts tended to be international players wishing to make a premium on the contracts or having Hong Kong dollar-funding requirements. At the same time, interbank interest rates have, since October 1997, been in double digits. In part, this rise in interest rates had been caused by the tight liquidity in the markets ensuing from the continuing retrenchment of foreign currency loans by foreign banks and the consequential refinancing in HK dollars by borrowers. Furthermore, only a handful of the 185 licensed banks in Hong Kong had a large HK dollar deposit base. In the absence of a discount window, the interbank market for HK dollar funds remained tight. The government believed that in shorting HK dollars (selling HK dollars for future delivery, and meanwhile borrowing the relevant amount of HK dollars), hedge funds were creating a situation whereby interest rates would climb and the stock market, including futures rates, would fall, thereby creating a "sure bet" if they were to sell short shares or index futures. It must be remembered that in all markets, "it takes two to tango": those who see a downside are matched by those who see an upside for a financial contract to be transacted. Thus, short sellers in the forward markets are always matched by people taking an opposite view.

What was in fact the position in the stock and futures markets? The reader is referred to Figures 3.3 to 3.16, where the following facts will emerge:

- By the beginning of August, the turnover in the stock market was less than HK $5 billion a day (Fig. 3.2).
- However, futures trading continued to be at a relatively high level, although by the beginning of August the daily level of trades went down to 20,000 to 30,000 contracts a day (Figs. 3.7 and 3.8).
- The premium (discount) between the stock and futures markets fluctuated greatly from October 1997 onwards but by August 1998, it had settled down to within 1 percent (Figs. 3.5 and 3.6).
- The fluctuating premium and discount created a large arbitrage trade (whereby buyers in the stock markets sold futures contracts short or vice versa), resulting in the Open Interest in the futures markets increasing from 80,000 to 100,000 contracts (Figs. 3.9 and 3.10).
- However, the market by the beginning of August was already deleveraging, namely, the stock market turnover was beginning to show upside signs and futures trading was showing slowing signs. This was consistent with the unwinding of arbitrage trades and, also, buying interest in the stock market (Figs. 3.11 to 3.14). Many fund managers and international brokers were in fact ready to start buying in the market by the middle of August.
- Short selling in the stock market was at a normal, low level before the market intervention but increased dramatically after the intervention (Figs. 3.15 and 3.16).

There are no published figures as to concentrations in the futures markets, nor were there published figures over the holdings of hedge funds. However, market estimates at the time indicated that most of the open interest in the futures market was accounted for in arbitrage trade. No firm in fact had more than 30 percent of the open interest. It was against this background that the Financial Secretary ordered the market intervention by the HKMA. From 17 August 1998 until the end of August, the HKMA gradually bought August month futures contracts as well as Hang Seng Constituent stocks in the rough proportions as they figure in the index. By the end of August, the HKMA had amassed index constituent shares to the value of HK $120 billion and about 50,000 August Hang Seng Index futures contracts. Because of the bidding up of the Index, the government's futures contracts expired in profit, estimated to be some HK $ 2 billion. If the people who were short the August futures contracts were hedge funds, they had only lost HK $2 billion (about U.S. $250 million).[6] That, of course, was not the case since the market concentrations were in arbitrage trades and not naked short sales of the August futures contract. If hedge funds were made to lose money in August, they would have accounted for a small portion of the U.S. $250 million profit made by the government. Fund managers and other institutional investors, who according to market rumors were about to buy before the intervention, were in fact sellers, and no doubt some became short sellers, as shown by the dramatic increase in shortselling toward the end of the intervention period. By the end of August the government was shorting September futures contracts. As the government was holding a large portfolio of shares at the time, selling the September futures short could be regarded as a hedge.[7]

Immediately after the end of August 1998, the Government announced a package whereby the HKMA would run a discount window for all authorized institutions and would be prepared to buy or sell HK or U.S. dollars from such institutions at the rate of HK$7.75 to the U.S. dollar. The package also contained tightening of regulations on the securities and futures markets. Interest rates immediately eased. In late September, the government announced that it would hold its portfolio of shares through a limited company and would make disclosures required under the regulatory regime. At the date of writing of this article, the portfolio remains with the government. The portfolio accounted for about 20 percent of the Hang Seng shares available for dealing in the market pool. Thus, substantial liquidity had been taken out of the markets.

In concluding, the writer cannot put it better than John Wadsworth, Chairman of Morgan Stanley Asia Ltd, who said in a seminar dealing with the market intervention that history will judge whether the intervention was brilliant. The Asian financial crisis has been an immense crucible to test the conventional wisdoms and regulatory structures of the region. It is beyond argument that each country has emerged from this period wiser in the ways of the markets and better equipped to deal with the fortunes of the future.

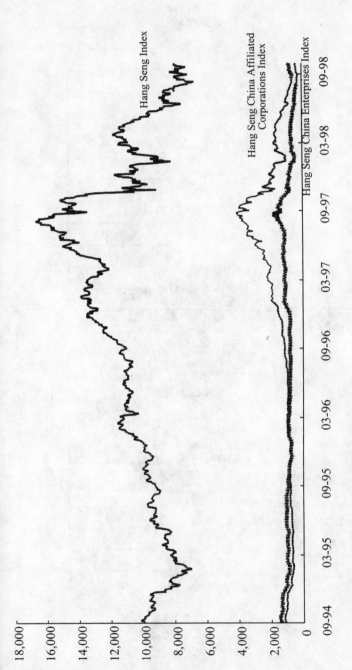

Figure 3.1 Hang Seng, Red Chips, and H Shares Index

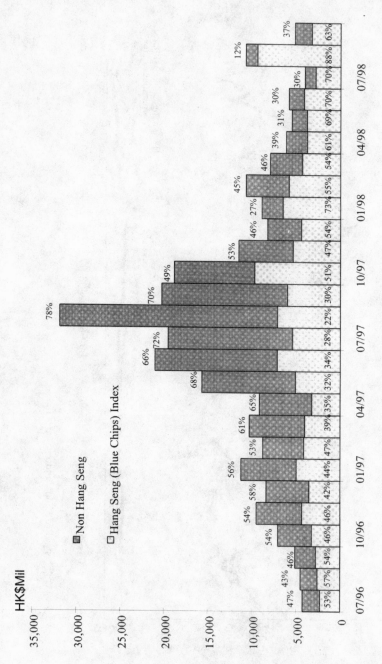

Figure 3.2   Average Daliy Turnover of Hang Seng Index Shares vs. Non-Hang Seng Index Shares

(HK$Bil)

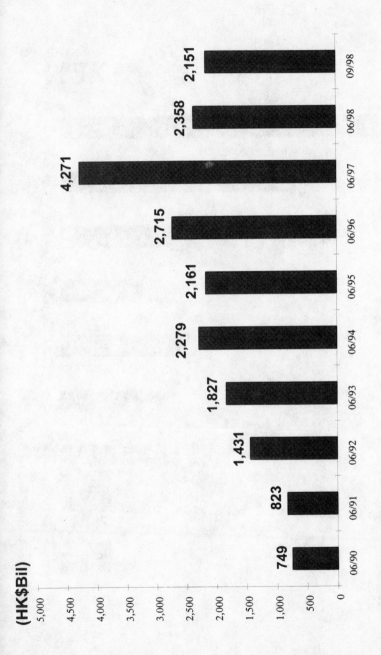

Figure 3.3   Stock Market Capitalization

92

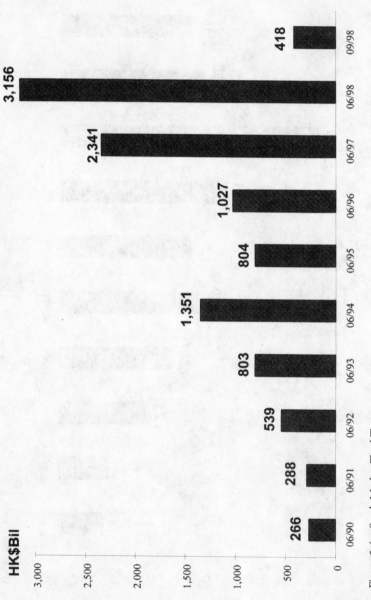

Figure 3.4   Stock Market Total Turnover

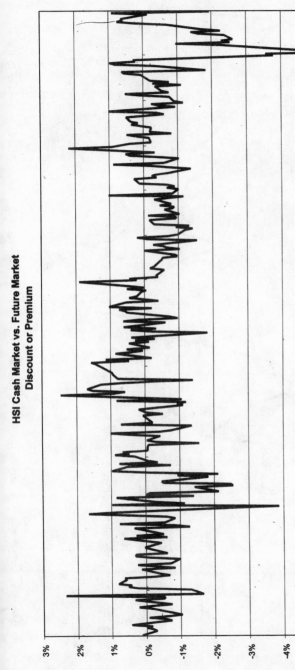

Figure 3.5   Discount (Premium) of the Futures Closing Price to the Stock Market Closing Price (September 1997 to September 1998)

94

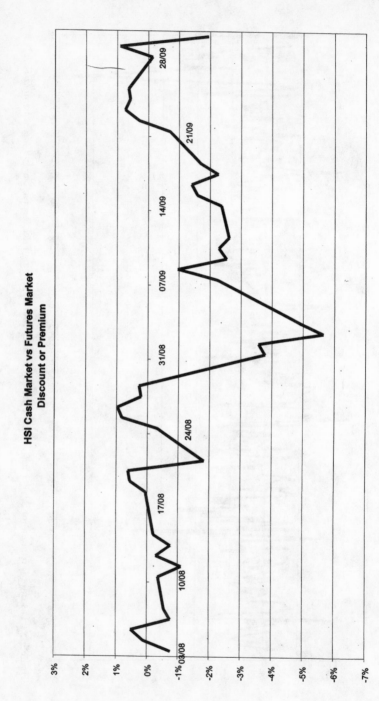

Figure 3.6 Discount (Premium) of the Futures Closing Price to the Stock Market Closing Price (August 1998 to September 1998)

95

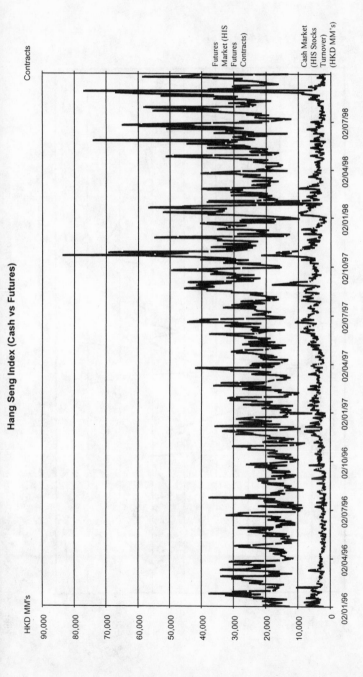

Figure 3.7 Stock Market vs. Futures Market (January 1996 to Spetember 1998)

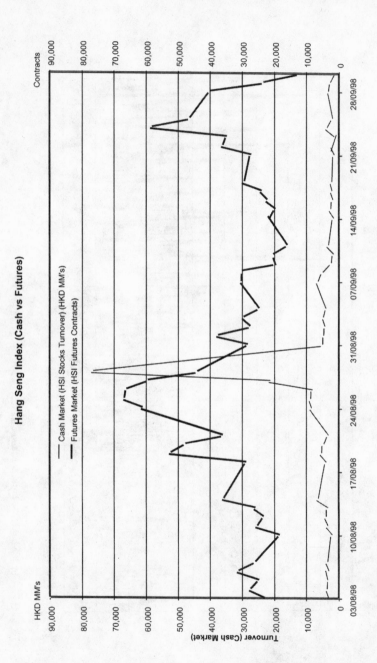

Figure 3.8   Stock Market vs. Futures Market Turnover (August to September 1998)

97

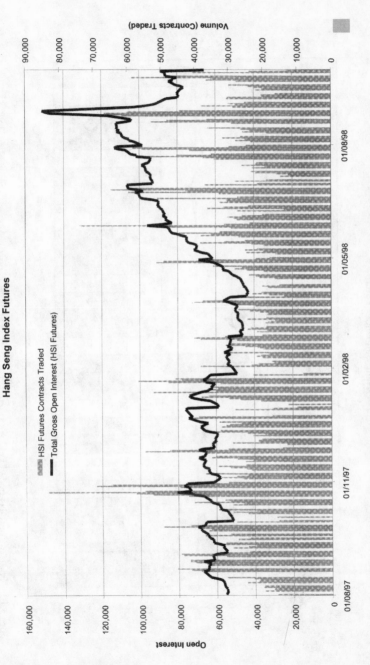

Figure 3.9 Futures Contracts traded vs. Open Interest (i.e. contracts not yet expired), August 1997 to September 1998

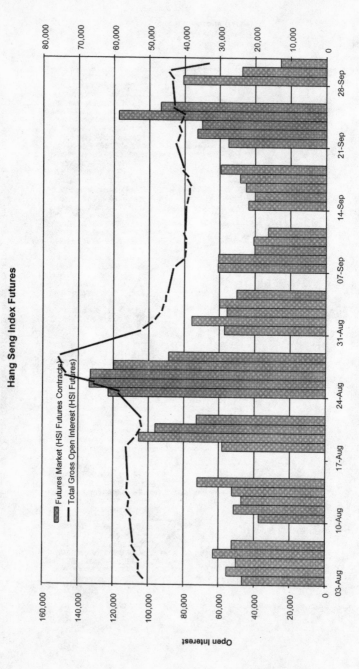

Figure 3.10 Futures Contracts traded vs. Open Interest (i.e. contracts not yet expired), August to September 1998

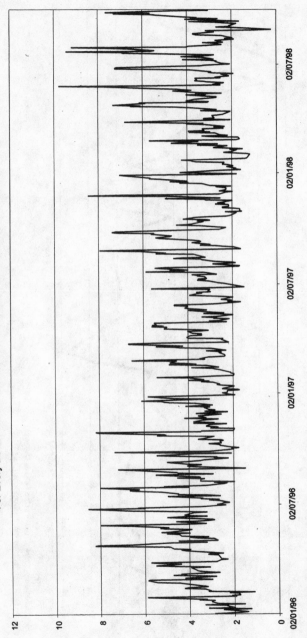

Ratio of Cash (HSI Constituents) vs Futures Markets Turnover
For Every Dollar Traded in the Cash Market, N Dollars Are Traded in the Futures Market

NB: Calculation of Dollars Traded in the Futures Market:
Futures Spot Month Close x $50 x Contracts Traded

Figure 3.11   Ratio of Stock Market Turnover to Futures Market Turnover (Trading Leverage ratio), January 1996 to September 1998

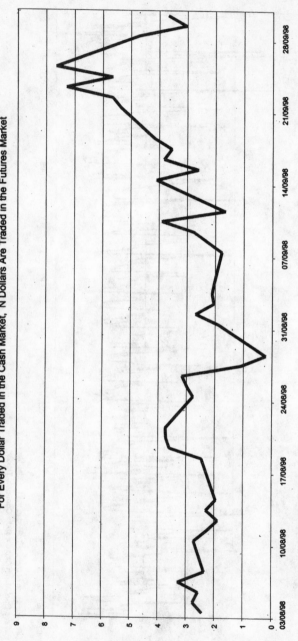

Ratio of Cash (HSI Constituents) vs Futures Markets Turnover
For Every Dollar Traded in the Cash Market, N Dollars Are Traded in the Futures Market

NB: Calculation of Dollars Traded in the Futures Market:
Futures Spot Month Close x $50 x Contracts Traded

Figure 3.12   Ratio of Stock MArket Turnover to Futes MArket Turnover (Trading Leverage ratio), August to September 1998

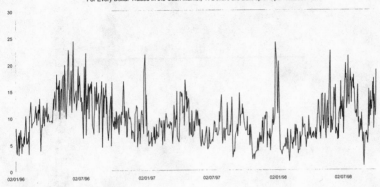

NB: Calculation of Dollars Traded in the Futures Market:
Futures Spot Month Close x $50 x Open Interest

Figure 3.13   Ratio of Stock Market Turnover to Futures Market Open Interest
(Residual Leverage Ratio), January 1998 to September 1998

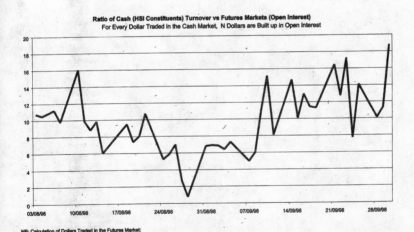

NB: Calculation of Dollars Traded in the Futures Market:

Figure 3.14   Ratio of Stock Market Turnover to Futures Market Open Interest
(Residual Leverage Ratio), January 1998 to September 1998

Figure 3.15   Shortselling of HIS shares January 1997 to September 1998

103

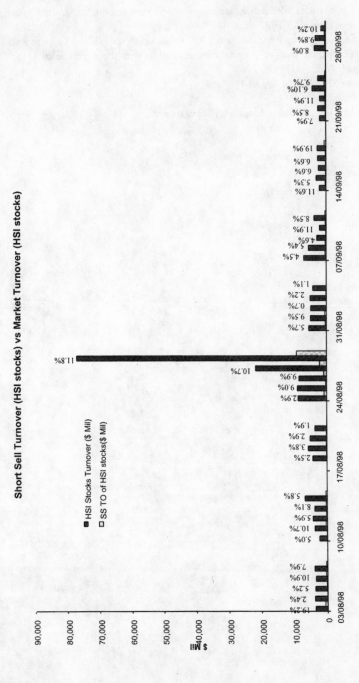

Figure 3.16 Shortselling of HIS shares August to September 1998

Notes

1. For pertinent information, see Securities and Futures Commission (SFC) Annual Report for 1997–98.
2. See Hong Kong Annual Report 1998, published by the HKSAR Government.
3. This part is based on personal communications with the International Monetary Fund (IMF) officials.
4. See SFC Annual Report 1997–98.
5. Ibid.
6. This is a calculation based on $50 per point.
7. Personal communication.

CHAPTER 4

Paradoxes of Hong Kong's Reversion: The Legal Dimension[1]

Daniel R. Fung, QC, SC, JP[2]

Introduction

During a speech made in Washington D.C. in May 1997, I made a rash prediction, namely, that after the imminent restoration of Hong Kong to Chinese sovereignty, the problems encountered would not be politico-legal but rather predominantly economic in nature. That particular prognosis appears to have hit the mark, since the Thai baht devalued on 2 July 1997, one day after Hong Kong's reversion to China. That event triggered the onset of the Asian financial crisis, the most serious economic recession the world has encountered since 1929, the ramifications of which have yet to be fully played out. Although both Hong Kong and the Chinese Mainland fared considerably better than their neighbors in East and Southeast Asia,[3] neither has been wholly spared from the onslaught.[4] Not surprisingly, the focus of the Hong Kong as well as the Chinese government's energies since July 1997 has been overwhelmingly economic.

Democratic Elections of the First HKSAR Legislature

The elections of May 1998, returning the first legislature of the Hong Kong Special Administrative Region (SAR) of the People's Republic of China (PRC), being the first set of democratic legislative elections held on Chinese soil[5] in 5,000 years of Chinese history, constitutes an interesting microcosm of how the transition has developed for Hong Kong. First, let us remind ourselves of the original expert prediction that, given such a complex and complicated system of elections (involving functional as well as geographical constituencies)[6] and limited democratization of Hong Kong's political system, Hong Kong would be lucky to get a voter turnout of 28 percent or 30 percent, roughly the same voter turnout as for the U.S. Presidential election. In fact, the voter turnout exceeded the most Panglossian of expectations, being an overwhelming 53.3 percent in the geographical constituencies and 63.5 percent in the functional constituencies.[7] Compared with the 1995 elections held during Chris Patten's administration (being the last set of elections under British rule), which had a 35 percent voter turnout,[8] as well as the 1991 elections (when Hong Kong enjoyed for the first time direct elections under geographical constituencies) which had a 39 percent voter turnout,[9] the 1998 elections constitute eloquent testimony on the part of the people of Hong Kong to participate in a democratic process, no matter how limited.

Of even greater significance are the results of the elections. The weight of punditry before 1997 was that, once China took over, neither Martin Lee nor his Democratic Party would be permitted to operate on Chinese soil. We were solemnly told that they would wither on the vine. The reality, however, has turned that prognosis on its head. Together with allies, the Democrats now control 18 seats in the legislature, constituting a one-third working minority. Again, the experts predicted before 1997 that, should the Democrats enjoy an electoral landslide, China, being preternaturally intolerant, would switch off the lights in Hong Kong. In fact, China has not missed a heartbeat since the results were declared. The reality is that, once Hong Kong reverted to being Chinese territory, Beijing became extraordinarily relaxed about how Hong Kong should run its own affairs. Indeed, the Central Government's attitude is best characterized as "Don't bother us; we've got too many problems on our hands but, should you need assistance, we'll extend financial and moral sustenance."[10]

Are the results of the 1998 elections a slap in the face for the Hong Kong SAR government? Is it a rejection of how the government has performed or delivered (or failed to deliver) over the first 11 months of its inaugural administration? Probably not. The sexy media headlines have tended to portray the Democrats as wearing white hats with the SAR government wearing the black hat. The reality, however, is infinitely more complex. What Hong Kong possesses today under the Basic Law, being its post-reversion constitution, is a quasi-presidential system. There are separate and unsynchronized elections for the Chief Executive who heads the executive arm of government as well as the legislature. This is not a parliamentary system. Just as in the United States where a sitting Democratic President does not guarantee a Democrat-dominated Congress, with the reverse being almost invariably the case (such that a sitting Democratic President virtually guarantees a Republican-dominated Congress and vice versa), the moral of Hong Kong's (as well as the United States') story is that the electorate loves a system of checks and balances. The governing principle appears to be that, in a presidential or quasi-presidential system, legislative elections tend to return an opposition and not a government, that job being left to presidential elections.

The Legal and Judicial Environment

I will return to the implications of the elections later. I propose to move to a related issue concerning the legal and judicial environment, complementing political developments to which I have adverted. Again, the expert prognosis before 1997 was that Hong Kong has had a good run for its money, and that whilst Hong Kong enjoyed a robust and liberal legal and independent judicial system and partook of the generous patrimony of the common law under British rule, all that would disappear down some juridical black hole after China resumed her sovereignty over Hong Kong or, at best, Hong Kong would metamorphose into a kind of legal no-man's-land. Let us, however, examine what has actually happened since China took over Hong Kong from Britain.

An illuminating microcosm is the decision of the *Hong Kong Special Administrative Region v. Ng Kung-siu*,[11] being a famous flag-burning case decided on 17 May 1998 in the magistrates' court. Paradoxically, flag-burning is as topical an issue in Hong Kong as it is in another common law jurisdiction, the United States. Indeed, I was asked by the U.S. State Department in April 1997 during one of my visits to Washington D.C. what would happen if someone were to burn the PRC national flag in Hong Kong after China took over. Would this individual be extradited to stand trial in Beijing? Would he or she be charged with some genre of anti-state or counter-revolutionary offence? Would this unfortunate person have to undergo reeducation through labor? I recollect my answer to have been that the matter would be handled by Hong Kong courts in accordance with Hong Kong law[12] as a matter falling within Hong Kong's autonomy.

That answer coming, however, on the eve of Hong Kong's reversion to Chinese sovereignty could not have been wholly persuasive. Indeed, I confess that I myself, not being possessed of a crystal ball, was only half-convinced at that juncture that my ostensibly confident prediction would necessarily coincide with the actual outcome. However, we now have a real life denouement of the posited scenario. On 1 October 1997, during the first national day celebrations held in Hong Kong after reversion to Chinese sovereignty, a group of about 12 protestors noisily demonstrating outside Hong Kong's Convention and Exhibition Center, just as Chinese leaders from the Central Government were alighting from their vehicles, pointedly waved a burnt and defaced PRC national flag in symbolic protest against China's takeover of Hong Kong. The event was extensively covered and telecast by the international as well as domestic media and there was, predictably enough, a certain degree of playing to the public gallery. The police were on hand, but did not move in with their truncheons. They merely took down the names and identity card numbers of the two principal demonstrators and released them on police bail. There was no violence.

The police took advice from the Department of Justice, which advised prosecuting in the magistrates' court, being the lowest rung on Hong Kong's judicial ladder. The Justice Department took a relaxed view and regarded the incident as a minor one that did not necessitate the choice of any higher judicial forum for the trial. The hearing took place on 24 April 1998 and the two defendants appeared with defense counsel, an eminent English barrister[13] practicing in Hong Kong specializing in human rights law and an activist who was former chairman of the English Bar's Human Rights Committee as well as the founding chairman of Hong Kong's Human Rights Monitor.

Not surprisingly, defense counsel argued that Hong Kong legislation that prohibited the burning or defacing of the national flag was unconstitutional in that it violated the right of free expression protected under Chapter III of the Basic Law (being that part of Hong Kong's post-reversion constitution entrenching human rights and civil liberties), Hong Kong's separate Bill of Rights as well as Article 39 of the Basic Law, which incorporated by reference into Hong Kong's domestic law, and the provisions of the International Covenant on Civil and Political Rights (ICCPR), one of which protects the right of free expression. The Justice Department, being the government's principal prosecutorial authority, drew to the court's attention, as it was duty-bound to do, the leading decision on flag-burning throughout the common law world, namely, the U.S. Supreme Court decision in *Texas v Johnson*.[14] In that case, decided in 1989, in a country that values free speech as a higher value than any other jurisdiction elsewhere in the world, the U.S. Supreme Court nevertheless came to a five-four split, the majority recognizing flag-burning as a legitimate form of free expression (which generated post-decision calls for constitutional amendment in the United States). The Hong Kong magistrate, after hearing full argument by both sides, adopted Chief Justice Rehnquist's minority decision and, following the approach pioneered by the Supreme Court of Canada in 1986 in the leading case of *R. v. Oakes,* justified the prohibition on flag burning as a reasonable, rational, and proportionate restriction on the right of free expression.

So what happened to the two defendants? They were bound over to keep the peace for one year, which is the lightest form of punishment that could be imposed by a Hong Kong court. The upshot of this sentencing decision was that, if the defendants did not offend again within the year, they would be neither fined, imprisoned, nor given any other form of punishment. The result is that the only people unhappy with this particular outcome were the two defendants and defense counsel, since judicial moderation (as well as the fact that the prosecution did not go overboard in demanding harsh penalties) is no facilitator of political martyrdom. Defense counsel vowed to appeal and indeed lodged an appeal on 1 June 1998.[15] At the time of revising this paper, the Hong Kong Court of Appeal had just delivered judgment, on 24 March 1999, overturning the magistrate's decision, striking down Hong Kong legislation prohibiting the defacement and burning of the national flag as unconstitutional vis-à-vis free expression provisions of the Basic Law and the ICCPR, and acquitting the two defendants.[16] A further appeal by the SAR government to the Court of Final Appeal is now a distinct possibility.

The moral of this particular story is victory for due process. The emergent message is that Hong Kong's legal system has not merely survived the transition intact, it has matured and strengthened. By acquiring a written constitution on reversion to Chinese sovereignty, Hong Kong has developed substantive judicial review as a fundamental legal culture, resulting in a Chinese judiciary acting for the first time in Chinese history as the vigorous arbiter of the constitutionality of executive as well as legislative acts, in much the same way as Chief Justice Marshall in 1803 arrogated to the U.S. Supreme Court in *Marbury v. Madison*[17] the power to determine the legality of acts of Congress.

On reversion to Chinese sovereignty, Hong Kong did not adopt an inward-looking jurisprudence, upholding peculiarly Hong Kong or Asian values. On the contrary, by virtue of the severance on 1 July 1997 of Hong Kong's constitutional umbilicus from the United Kingdom, Hong Kong was freed from an earlier culture of relying for judicial inspiration on predominantly English case law. Paradoxically, on reversion to Chinese sovereignty, Hong Kong, springboarding from Article 8 of the Basic Law,[18] which preserves Hong Kong as a common law jurisdiction, and Article 84 of the Basic Law,[19] which expressly permits the citation of authorities from other common law jurisdictions, rapidly became the most cosmopolitan in the common law world in taking the best the rest of the common law world has to offer. Thus, Hong Kong courts today regularly cite and follow, by way of persuasive precedent, decisions of the House of Lords, the Privy Council, the High Court of Australia, the New Zealand Court of Appeal, the Indian Supreme Court, the South African Supreme Court, the Supreme Court of Canada and last, but by no means least, decisions of the U.S. Supreme Court and U.S. federal District Court.

Indeed, Hong Kong has gone way beyond the common law world in seeking judicial inspiration. Since the enactment on 8 June 1991 of a Bill of

Rights replicating almost verbatim the text of the ICCPR, a process that was preserved on 1 July 1997 by the incorporation as part of Hong Kong domestic law of the provisions of the ICCPR under Article 39 of the Basic Law.[20] Hong Kong has cited and followed decisions of the European Court of Human Rights at Strasbourg (interpreting the similar provisions of the European Human Rights Convention) as well as those of the United Nations Human Rights Committee on individual references under the Optional Protocol. This jurisprudential development continues today.

This phenomenon neatly illustrates the obverse of Gunnar Myrdal's dictum. That great Swedish Nobel laureate described the United States in the 1950s as the most parochial country on earth since it was the most self-sufficient. By such criterion, Hong Kong being by definition one of the least self-sufficient of economies, constantly looking overseas for trade and investment, dependent on China for its daily food and drink, just cannot afford to be inward-looking.

The same attitude pervades the law as much as it does other aspects of Hong Kong life. Today Hong Kong's common law system is arguably the most cosmopolitan and most liberal of all common law jurisdictions throughout the world.[21] In terms of openness to adoption or adaptation of overseas jurisprudence, Hong Kong occupies the front rank. In Los Angeles, recently, a constitutional expert observed that, with increasing globalization, United States courts may one day become similarly enlightened. Indeed, the U.S. Supreme Court in 1997, for the very first time in American history, cited a non-American authority, being a decision of the Supreme Court of Canada. Hong Kong, by contrast, has always relied on overseas jurisprudence. Consider the judicial system and one will readily appreciate that something extraordinary has happened since the reversion to Chinese sovereignty. Contrary to the expert prediction that somehow Hong Kong would progressively deteriorate into a parochial non-entity at the edge of the Chinese polity, we have today a judiciary that is undoubtedly more vigorous and more cosmopolitan than at any time when Hong Kong was a British possession.

Let us survey, first, the structure of the judiciary. There is no nationality requirement for occupying any judicial position in Hong Kong, save only for two positions, namely, those of the Chief Justice and the Chief Judge of the High Court. These two positions must be filled by Hong Kong Chinese persons with no foreign right of abode.[22] However, every other judicial post may be filled by persons of any nationality provided that they are suitably qualified common lawyers.[23] Indeed, Article 92 of the Basic Law mandates the choice of judges on the basis of judicial and professional qualities and expressly permits recruitment from other common law jurisdictions. This leads to two consequences. First, judges from the Chinese Mainland, not being common lawyers, enjoy no right of entry into Hong Kong's judicial system. Secondly, over 50 percent of Hong Kong's judiciary today are not Chinese. This has pluses and minuses. However, in terms of preservation of the existing system, in terms of continuity, and in terms of locking Hong Kong into the international grid, it constitutes arguably an advantage. In terms of judicial appointments since 1 July 1997, more non-Chinese than Chinese judges have been appointed to the High Court, which underscores a political phenomenon well-illustrated in 1972 by U.S. President Richard Nixon who, with impeccable credentials as an anti-Communist crusader, possessed the necessary domestic political clout to stage an international historic coup in normalizing relations with China. Likewise, it took an SAR judiciary, confident of its status and its relations with the Chinese Mainland to appoint the people for the job on the basis of ability, judicial temperament, and professional competence, irrespective of nationality considerations irrelevant to the Basic Law.

Turning to the Hong Kong Court of Final Appeal, one finds a unique structure. No final court in any major jurisdiction enjoys the power of co-opting visiting judges from other jurisdictions. The matter evolved with some difficulty. Up until 30 June 1997, Hong Kong had no court of last resort of its own. Its court of last resort was the Privy Council, 8000 miles away, with not a single Hong Kong person sitting on its bench.[24] During the Sino-British negotiations over the future of Hong Kong, from 1982 to 1984, one of the major stumbling blocks confronting both Britain and China concerned what to do with the Privy Council after 1997. The second half of the negotiating period (namely, from mid-1983 to September 1984) focused on the restoration of Chinese sovereignty over Hong Kong and the need to preserve the latter's pre-existing systems insofar as they worked and could be adapted to the future constitutional reality. The doctrine of mirror-imaging, which permeated the second half of these negotiations, argued for replacement of the Privy Council by some sort of supreme court in Beijing, but such an outcome would have been disastrous for Hong Kong's common law system.

Ultimately, both sides to the negotiating table are to be congratulated for arriving at an imaginative, creative, and lateral solution. This involved replacement of the Privy Council not by a court in Beijing, but by one physically located in Hong Kong, staffed by Hong Kong common law judges, and with the power, additionally, to invite overseas jurists to participate in its deliberative processes. Today there are five seats in the Court of Final Appeal, filled by the Chief Justice leading three other permanent appointees, with the fifth seat to be filled ad hoc, on an ambulatory basis by visitors on a case-by-case basis.[25] Since Hong Kong's reversion to Chinese sovereignty, the Chief Justice has made six appointments to the overseas panel. They rank among the finest the common law world has to offer. From the UK are appointed Lord Hoffman and Lord Nicholls, arguably the two strongest Chancery judges sitting in the House of Lords today. From Australia are appointed Sir Anthony Mason, arguably her strongest Chief Justice since World War II (now retired), as well as Sir Daryl Dawson, a former judge of High Court of Australia. From New Zealand are appointed a former President of their Court of Appeal, Lord Cooke, who also sits on the Privy Council and is one of New Zealand's finest jurists since World War II, as well as another judge of the New Zealand Court of Appeal, Sir Edward Somers. No other major jurisdiction has a system whereby non-national, non-resident aliens enjoy an input into the judicial process at the very highest level.

Human Rights Protection

I turn now to another related aspect of the transition, that is, human rights protection. Again, the expert prognosis was that whereas Hong Kong under British tutelage was lucky to be conferred an internationalized human rights culture, once China took over all that would be a distant memory. Human rights could not survive, there would be considerable backsliding and civil society would disappear. However, the reality is that today Hong Kong is arguably better served in terms of human rights protection under its domestic regime, as well as in terms of international monitoring, than at any stage when it was a British possession.

The primary explanation for this paradox is the rise of constitutionalism in Hong Kong since reversion to Chinese sovereignty. For 156 years when Hong Kong was a British colony,[26] it had no written constitution. Hong Kong merely followed the British system in which the UK remains the only major sovereign state in the world without a written constitution. Today, however, under the terms of the Basic Law which conform with the Joint Declaration, Hong Kong now has a constitutionally entrenched bill of rights within Chapter III. This comprises 19 articles, 18 of which deal with specific protections for discrete forms of human rights and civil liberties. The remaining article, Article 39, is an extraordinarily creative one that incorporates, by specific reference, the provisions of the two most successful multilateral human rights treaties ever sponsored by the United Nations, namely, the ICCPR[27] and the International Covenant on Social, Economic, and Cultural Rights (ICESCR).[28] The individual provisions of these two covenants, adopted as domestic standards, have been enforceable in all Hong Kong courts since the reversion to Chinese sovereignty.

Of the approximately 138 states parties to the ICCPR and the ICESCR, the United Nations indicates that none has actually made the step taken by Hong Kong to adopt by reference and render domestically enforceable the provisions of these two international covenants. Since Hong Kong has taken this step as a common law jurisdiction, these international legal provisions do not constitute for Hong Kong merely a series of platitudes but are, on the contrary, living law interpreted by Hong Kong courts and used as a benchmark for measuring the constitutionality of legislative, as well as executive, acts.

This has inaugurated a sea change in Hong Kong jurisprudence since 1 July 1997. For the first time, Hong Kong possesses a judicial mechanism for checking and balancing overweening acts on the part of the other two branches of government. The right of abode cases[29] making their way through the courts over the last 12 months provide a good microcosm of this seminal development. The background to this litigation lies in the truism that, given massive disparity between Mainland China's and Hong Kong's respective standards of living, there has been since 1949, and remains today, considerable pressure for immigration from the Chinese Mainland to Hong Kong. This pressure has always been held in check by a vigorously policed border backed by law—in Hong Kong's case, the provisions of the Immigration Ordinance, which stringently regulates immigration from China. Beijing appreciates as much as Hong Kong does that any crack in the floodgates would overwhelm Hong Kong's delicate social and economic balance and destroy Hong Kong and it therefore polices the Shenzhen-Hong Kong border as vigilantly as does the Hong Kong SAR Government.

A new dimension developed in 1979 with the launch of Deng Xiaoping's Four Modernizations Movement, which opened up China to foreign trade and investment. Hong Kong immediately positioned itself as China's biggest single foreign investor and trading partner, with thousands of businessmen going into China and thousands of container vehicles crossing the border daily. Human nature being what it is, these businessmen and truck drivers acquire wives, either bigamously or for the first time, and families with offspring (legitimate as well as illegitimate) began to mushroom. The pressure for wives and children coming to Hong Kong for the purposes of family reunion began to mount.

Before 1997, none of this presented a real problem since controlled immigration was regulated by the Hong Kong Immigration Ordinance that, on the eve of the reversion, permitted the inward flow of 150 immigrants a day from the Chinese Mainland totaling some 55,000 per annum with consequent but manageable demands on public housing, transport, education, and public health.

All that changed with Hong Kong's acquisition of a written constitution on 1 July 1997. Article 24 of the Basic Law constitutionally confers the right of abode on six categories of individuals. One of those six categories is the person born of at least one parent who possesses a right of abode in Hong

Kong.[30] Claims by Chinese residents to right of abode in the Hong Kong SAR, coupled with judicial challenges (financed by Hong Kong government-provided legal aid) to the restrictive regime of the old Immigration Ordinance, arose immediately after the reversion.

The first question calling for judicial determination on such challenges was whether the claimant, pending processing of his claim, should remain in his place of current residence, namely, Mainland China or be permitted immediately to come to Hong Kong on assertion of claim. The Hong Kong government took the view consistent with international practice that, since there may be a world of difference between an assertion of a right of abode and proof of such right, claimants should remain in their place of origin pending processing of their claims. The claimants took issue with that proposition and the Hong Kong courts have ruled in the government's favor.[31]

On other issues, however, the courts have ruled against the SAR government. For example, are illegitimate offspring of a Hong Kong parent entitled to the benefit of Article 24 as much as legitimate offspring? The Hong Kong government has maintained that the *travaux preparatoires* underlying the drafting of the Basic Law make plain that the true intent of Article 24 was to benefit only legitimate offspring. However, the courts have ruled against the government on this issue, observing that there being no inherent textual ambiguity in Article 24, the words should be given their plain meaning. In other words, all those born (whether legitimately or otherwise) of at least one parent with a Hong Kong right of abode enjoy a right of abode in Hong Kong.[32]

A further issue canvassed by the courts is whether there could arise any retrospective (viz. post birth) acquisition of a right of abode. In other words, if at the time of the claimant's birth the entire family (being both parents and child) were Mainland residents, but at some subsequent stage one parent (usually the father) made his way to Hong Kong (either legally or illegally, usually illegally) and stayed the requisite number of years, can such parent's subsequent acquisition of a right of abode in Hong Kong retrospectively feed his offspring's claim? This issue has given rise to a see-saw battle, with some courts ruling in the government's favor and others against.

Both the foregoing issues were decided by the Court of Final Appeal on 29 January 1999 against the SAR government.[33] The judgment ruled that illegitimate offspring can reap the benefit of Article 24 as much as legitimate ones and that a claim to right of abode in Hong Kong under the parentage category may be acquired subsequent to one's birth. This decision has given rise to considerable controversy, not least because it has resulted overnight in a 5 percent increase in Hong Kong's population with consequent extraordinary strains on public housing, transport, education, and medical care.

Moreover, this 5 percent represents merely claimants physically in Hong Kong since before 1 July 1997 whom, though legally deportable as overstayers under the earlier court decision holding that claimants should, pending the processing of their claims, remain in their place of origin, the SAR government chose, for humanitarian and domestic political reasons, not to deport. As for numbers of claimants presently on the Chinese Mainland, estimates range between one to two million, representing another 15 to 30 percent increase in Hong Kong's population.

Of even greater controversy is a gratuitous *obiter dictum* made by the Court of Final Appeal on a point extraneous to its decision on the case that Hong Kong courts may strike down Mainland Chinese laws for inconsistency with Hong Kong's Basic Law. Such an issue was not, in fact, pertinent to the case, since the matter before the court concerned whether Hong Kong legislation (being certain provisions of the Immigration Ordinance) was consistent with the Basic Law. No one seriously questions the ability of the Hong Kong SAR courts to strike down, in the manner articulated by Chief Justice Marshall in *Marbury v. Madison,* SAR executive as well as regional legislative acts (viz. Hong Kong ordinances and subsidiary legislation) for non-conformity with the Basic Law. However, the Hong Kong Court of Appeal had earlier drawn the jurisdictional line clearly in ruling on 29 July 1997 in *HKSAR v. David Ma* (being, arguably, Hong Kong's *Marbury v. Madison*) that SAR courts, being regional courts, had no power to strike down national legislation. The Court of Final Appeal, in specifically overruling *David Ma,* had purported to extend Hong Kong's jurisdictional frontier into the Chinese Mainland.

Whether such purported extension was intended or inadvertent, at least two structural problems arise from the Court of Final Appeal's decision. First, given the status of Hong Kong courts as Chinese regional courts, and that of the Court of Final Appeal as Hong Kong's, but not China's, supreme court nor yet, indeed, China's constitutional court, it is difficult to see how SAR courts could legitimately arrogate to themselves the right to act as arbiters of the constitutionality of Chinese national legislation.

Secondly, given the status of the Basic Law as a national law of the PRC promulgated on 4 April 1990, as well as the constitution of the Hong Kong SAR, but not that of the Chinese Mainland, it is likewise difficult to see how the Basic Law could be considered the proper criterion for measuring the legality of other pieces of national Chinese legislation of co-equal status.

Controversy aside, this chapter is not concerned with whether the court's ruling was right or wrong.[34] On the other hand, the decision constitutes graphic illustration of the proposition that, contrary to expert prognosis, the Hong Kong judiciary not only did not metamorphose into a shrinking violet after reversion to Chinese sovereignty, but presented rather the spectacle

• Daniel R. Fung

of a robust constitutional arbiter, even to the extent, indeed, of raising the specter of bellicose regionalism.

On the international front, the expert prediction before 1997 was that as China had yet to accede to either the ICCPR or the ICESCR,[35] Hong Kong, after reversion to Chinese sovereignty, would no longer benefit from international monitoring thereunder. So far as concerned the four relatively minor multilateral human rights conventions applicable to Hong Kong,[36] the expert prognosis was again that the reporting process would be unrecognizable from that enjoyed by Hong Kong under British rule in which the Hong Kong government played an active role in reporting autonomously on Hong Kong's situation.

The reality, however, could not have been more different. First of all, China acceded to both the major multilateral human rights treaties shortly after Hong Kong's reversion to Chinese sovereignty: the ICESCR on 27 October 1997, at the time of the Clinton-Jiang summit, and the ICCPR in October 1998, six months before Premier Zhu Rongji's visit to the United States.

As regards the relatively minor covenants, there is another paradox. Until less than nine months before the reversion of Hong Kong to Chinese sovereignty, Hong Kong did not fall under the umbrella of the Convention on the Elimination of All Forms of Discrimination Against Women (CEDAW),[37] as Britain did not extend the application of that convention to Hong Kong until October 1996. Conversely, China had been a signatory since 1980 and had been reporting under the CEDAW for sometime. Accordingly, Hong Kong had never submitted itself to international scrutiny under the CEDAW throughout its period as a British colony. Hong Kong presented its initial report under the CEDAW in September 1998 and attended and participated in the UN hearings in New York on 2 February 1999 in a manner as autonomous as that enjoyed by Hong Kong as a British colony reporting under the other human rights treaties.

The Hong Kong SAR government delegation attended the UN hearing under the leadership of the Chinese ambassador to the UN flying the PRC national flag. However, questions from the CEDAW committee were fielded by the Hong Kong team. The hearing was regarded generally as a success and is expected to form the broad template or blueprint for future reporting on Hong Kong under the other human rights treaties.

Reality on the Ground

Given the structural protections outlined above, has the reality on the ground actually kept pace with the structures that Hong Kong has erected? A few critical figures tell an interesting story. From 1 July 1997 to 31 March 1998, Hong Kong had 1,260 public demonstrations,[38] a figure higher than during any nine-month period when Hong Kong was a British colony. No application to hold public demonstration has been rejected by the Hong Kong SAR government since 1 July 1997.

A total of 626 societies were formed in the Hong Kong SAR over the same period of time. Presently, over 9,000 non-governmental organizations (NGOs) are registered in and operate in or from Hong Kong. These include Amnesty International, Justice, Lawyers' Committee for Human Rights (a New York-based NGO that is extremely active in supporting Vietnamese boat people litigation and the right of abode challenges), the right-wing, conservative American think-tank,Heritage Foundation (which has established in Hong Kong its only branch outside of its headquarters in Washington DC), Asia Society (which again has in Hong Kong its only branch outside of its headquarters in New York) and the International Chamber of Commerce (ICC) (which likewise has in Hong Kong its only branch outside of its headquarters in Paris). All of these NGOs operate without let or hindrance from the SAR government and no single application for registration as a society has ever been rejected by the government since 1 July 1997.

At the same time, many major international print and electronic media organizations are found in Hong Kong, or make it their regional headquarters. Such entities include the *Far Eastern Economic Review,* the *Asian Wall Street Journal,* the *New York Times,* the *Washington Post,* Star TV, CNBC and many others. There can be little controversy to Hong Kong's boast, even after reversion to Chinese sovereignty, as the unchallenged communications hub and intellectual forum of East Asia, if not also the Asia Pacific.

As to other aspects of Hong Kong cultural life, distinguished former U.S. ambassador to the PRC, James Lilley, offered me a bet in November 1997 that the Hong Kong SAR Government would never allow the public screening of such inflammatory anti-China Hollywood movies as Richard Gere's *Red Corner,*[39] Brad Pitt's *Seven Years in Tibet,*[40] or Martin Scorcese's *Kundun.*[41] I took up the proffered bet on the basis that at least one, if not all three, of such films would be publicly shown within the next six to twelve months. As matters turned out, *Seven Years in Tibet* played to packed houses in Hong Kong in April and May 1998 and *Red Corner,* whilst less commercially successful in Hong Kong than in the United States, was showing publicly in the second half of 1998. *Kundun* followed shortly thereafter. Straws in the wind as these events may be, they nevertheless graphically underscore the degree of relaxation felt by the Center toward Hong Kong since the reversion to Chinese sovereignty, giving the SAR government a very comfortable margin of appreciation or high degree of autonomy in running its own affairs.

No single incident offers a better illustration of the above proposition than that concerning Radio and Television Hong Kong (RTHK) in March

1998. RTHK is a publicly funded broadcasting service modeled on the British Broadcasting Corporation (BBC). Unlike Voice of America, which observes a code eschewing criticism of the United States Government in its broadcasts, the BBC and RTHK regularly air programs critical of their respective governments. Not surprisingly, this had generated considerable controversy in the U.K. as well as in Hong Kong. In the U.K., debate has raged both within and without parliament as to whether public funds should be utilized to support programs of such nature. In Hong Kong, the debate focused on whether programs critical of the SAR government should more appropriately be aired on commercial, as opposed to publicly funded, stations. The Chinese People's Political Consultation Conference (CPPCC) delegate, Xu Simin, took that debate public in Beijing in March 1998 at the CPPCC annual conference. The reaction was remarkable, not least because no self-respecting expert would have dared make any such prediction. President Jiang Zemin and CPPCC Chairman Li Ruihuan's reaction to Xu's argument was that when Hong Kong people come to Mainland China, they should refrain from using Beijing as a platform upon which to grind their particular political axes, further that the Central Government had faith in the ability of the SAR government to run Hong Kong competently and that, should the SAR Government see fit to tolerate criticism by a publicly funded broadcasting station, that was their own business and not a matter on which the Central Government would wish to intervene.

Law and Order

From human rights, I turn to examine the other side of the coin, namely, the question of law and order. Once again, the confident expert prognosis before 1997 was that the enviable degree of social stability coupled with low crime rates, which made Hong Kong under British rule the just object of envy in East Asia, was bound to deteriorate after 1997, since Hong Kong would be engulfed by a veritable tsunami of corruption, violence, and firearms-related offences emanating from across the border. Once again, the statistics disclose an interesting story. The overall crime figures for the calendar year from 1 January to 31 December 1997 reveal Hong Kong as enjoying the lowest crime rate for 24 years, which happened also to be the lowest in the whole of East and Southeast Asia. Comparing the overall crime rate for the first four months of 1998 (1 January to 30 April) with corresponding figures covering the same period in early 1997 reveals a 3.5 percent decrease in the overall crime rate. Similarly, a comparison of the violent crime figures for the same period (1 January to 30 April 1998) with those for the first four months of 1997 reveal a 3.5 percent decrease, being a further decrease on the lowest crime rate Hong Kong has enjoyed for the last

24 years.[42] The upshot means that Hong Kong, at the time of and after reversion to Chinese sovereignty, is safer even than Singapore, long the epitome of a clean and safe, if somewhat staid, city. Present low crime rates do not mean, of course, that they will not again rise. On the contrary, it would be hardly surprising if they do, particularly given the onset of the regional economic recession and rising unemployment. However, the non-collapse of Hong Kong's civil society after reversion to Chinese sovereignty is a mark of the maturity of Hong Kong as a community subscribing to the rule of law.

The Economic Environment

The experience Hong Kong has undergone over the last 22 months underscores the point made at the outset of this chapter that the real concerns of the community after reversion to Chinese sovereignty are not so much politico-legal, but rather economic in nature. In economic terms, Hong Kong's fate is very much tied to that of the Chinese Mainland. The expert prediction before 1997 was that while Hong Kong's currency system had worked well enough and Hong Kong had accumulated an impressive amount of reserves under British rule, once Hong Kong should sever its umbilical cord from the U.K., the U.S. dollar peg was bound to break and the Hong Kong dollar would devalue. The hedge funds confidently bought this prognosis and massively shorted the Hong Kong dollar beginning on 27/28 October 1997 and continued to lay intermittent siege to Hong Kong's currency throughout the rest of 1998, all to no avail. Prescient observers have, however, considered George Soros's recant at the February 1998 Davos Conference as an early warning signal. He confessed to being wrong in his earlier evaluation of perceived weaknesses of the Hong Kong dollar and said that he realizes that China would go to the wall to protect the integrity of the Hong Kong dollar-U.S. dollar link.

Once again, the figures tell an interesting story. On 30 June 1997, Hong Kong held approximately U.S. $60 billion in hard currency reserves, making us then the second richest per capita territory anywhere in the world with, in absolute terms, the fifth largest foreign currency reserves. Today, Hong Kong's foreign currency reserves stand at U.S. $98.1 billion, making us the third biggest in absolute terms after Japan's U.S. $220 billion and Mainland China's U.S. $144 billion.[43] What explains Hong Kong's 60 percent increase in hard currency reserves? This time the paradox was generated by Sino-British mistrust. At the time of the signing of the Joint Declaration in 1984, one of China's pet fears was that Britain would spend the next 13 years surreptitiously removing Hong Kong's reserves so as to leave the piggy bank empty at the very moment of the resumption of Chinese sovereignty. Catering for this concern, the two sovereign states set up an international trust known as the Sino-British Land Fund, into which were paid the proceeds of public auctions of Hong Kong government land over the 13 transition.

Undeveloped land has long been the scarcest and one of the most precious of Hong Kong's commodities. It is also a government monopoly, previously owned by the Crown and now the SAR government. Release for public auction in 50-hectare tranches has netted a king's ransom for the public coffers over the 13-year transition, leading to an accumulated fund of close to U.S. $28 billion, which was then handed over by the two sovereign states to the SAR government in mid-July 1997. That payment boosted Hong Kong's reserve to approximately U.S. $88 billion. With post-reversion land auctions and profits made off hedge funds, Hong Kong now holds accumulated reserves of approximately U.S. $98 billion.

However, one major factor underlying the stability of the link to the U.S. dollar is to be found in a resolution reached at the fifteenth Chinese Communist Party Congress in September 1997. This was a major commitment to the securitization (or de facto privatization) of China's state-owned enterprises (SOEs), being the showpiece of the present modernization wave. At stake is not merely the success or otherwise of China's transition toward a market economy, but also the political future of Premier Zhu Rongji, who has staked his reputation on completing the program within three years. Whatever is one's evaluation of the program's likely outcome, the odds on success would be considerably shortened by efficient access to international capital markets. It is well-known that since the early 1990s, China has accessed international capital through initial public offerings (IPOs) of nonvoting shares in SOEs in various markets including New York, London, Frankfurt, Singapore, Shanghai, and Hong Kong. By the late 1990s, however, the overwhelming bulk of IPOs take place in Hong Kong. The reason is not difficult to grasp.

In New York, even when SOEs overcome the many hurdles to listing, there exists little to no liquidity after listing because there is scarcely any active trading in Chinese stocks. The New York Stock Exchange presents such a smorgasbord of available options that punters and investors alike tend to gravitate to well-known blue chips such as Microsoft or Boeing or to fashionable infotech, entertainment, or biotech stocks. There is little reason for the average American player to be attracted to relatively unknown foreign public sector shares. The same phenomenon is replicated in London and in Frankfurt. Singapore presents a wholly different set of problems for China, since Singapore is highly regulated but happens also to be a tiny market, which tends therefore to be more vulnerable to manipulation. Listing in Shanghai is often an ersatz experience because, so long as the renminbi remains not freely convertible, a Shanghai listing constitutes no real access to international capital. Hong Kong remains, therefore, at least for the time being, the market of choice. If the dollar peg breaks and the currency devalues, not only would Hong Kong and China stand to lose a lot of money, it would effectively jeopardize the SOE securitization program.

Finally, the fact of China being predominantly a continental economy, coupled with the non-convertiblity of the renminbi, underlies Chinese refusal to compete with its Southeast Asian neighbors in successive currency devaluations. Further, the commitment made by China to preserve the integrity of Hong Kong dollar-U.S. dollar link constitutes one major key to unlocking the door to entry by China into the World Trade Organization (WTO). The U.S. government's vocal appreciation for and support for Chinese (as opposed to contrary Japanese) conduct on this score, and its translation into valuable international political capital, are not lost on China. Throughout the Asian financial crisis, Hong Kong has fared better than virtually any other economy in East and Southeast Asia, with the possible exception of Mainland China and Taiwan. Asia has, by no means, even begun to emerge from the woods, but the future looks brighter today than it has for the last 22 months, particularly since Premier Zhu Rongji's visit to the United States in mid-April 1999. This is likely to pave the way for long-awaited Chinese accession to the WTO, to take place possibly before the end of the year. Such a development would yield a considerable windfall for Hong Kong, China, and the United States, not to mention world trade in general.

Conclusion

This chapter has sought to underscore the inescapable fact that the first two years of Hong Kong's reversion to Chinese sovereignty has been replete with paradoxes, both in terms of the behavior of players in the region and the reaction of those at the center, such as to turn earlier expert predictions on their heads. I have attempted to partially explain the mystery underlying such paradoxes in the hope that such explanation might help facilitate the making of more accurate prognoses in the years that lie ahead for Hong Kong as a part of China. I suspect, however, that this may be yet another short-lived triumph of optimism over experience. Indeed, Hong Kong, being the perennial graveyard for political prognosticators, is likely once again to doom that naïve hope to failure.

Notes

1. This essay is a revised and updated version of an oral presentation made by the author at a multi-disciplinary conference entitled "Hong Kong: A Year after Reversion," held at Lingnan College on 2 June 1998.

2. Views expressed in this chapter are personal ones that may not necessarily co-incide with those of the HKSAR.

3. Per contra Indonesia, Malaysia, and Thailand in Southeast Asia, and Korea and Japan in Northeast Asia.

4. For example, the Hong Kong dollar came under siege from predatory hedge funds speculating on the breaking of the currency peg to the U.S. dollar (at HK$7.8 to U.S.$1) beginning in October 1997 and continuing intermittently through to the end of 1998.

5. The elections of the First Legislature of Hong Kong SAR were held on 24 May 1998.

6. Functional constituencies mandated as a transitional measure (to last until 2007) under the Basic Law, being Hong Kong's post-reversion constitution, provides for doctors, lawyers, engineers, bankers, property developers, teachers, social workers to elect their own representatives to the legislature.

7. Source: Information Services Department of the HKSAR government.

8. *Hong Kong 1996,* p.5 (Hong Kong Government Publications, 1996).

9. *Hong Kong 1992,* p.28 (Hong Kong Government Publications, 1992).

10. Evidence of this attitude surfaced early after Hong Kong's reversion to Chinese sovereignty when the Hong Kong dollar came under pressure from predatory hedge funds in October 1997 speculating on a possible breaking of the peg to the U.S. dollar. China's central bank, the People's Bank of China, immediately offered to put the entirety of China's foreign currency reserves of U.S.$141 billion (being the largest in the world after Japan's) at Hong Kong's disposal to defend the integrity of the U.S. dollar peg.

11. The Magistrate's ruling on the defense submission of no case to answer, dated 17 May 1998 (Case No. WSS 3151 and 3152 of 1998, unreported).

12. Hong Kong is obliged to either promulgate or enact local legislation to underpin certain Chinese national laws (essentially those manifesting the exercise of Chinese sovereignty) applying to Hong Kong as specified in Annex III to the Basic Law. The Resolution on the Capital, Calendar, National Anthem, and National Flag of the People's Republic of China is one such specified national law that is underpinned by Hong Kong legislation enacted shortly after1 July 1997.

13. Paul Harris.

14. 491 U.S. 397 (1989).

15. The defendants lodged an appeal against the decision of the Magistrate on 1 June 1998 (Case No. MA563/98).

16. Decision reached 24 March 1999, as yet unreported except in the popular press: see, for example, the *South China Morning Post,* 25 March 1999.

17. 5 U.S. 137 (1803)

18. Article 8 of the Basic Law reads: "The laws previously in force in Hong Kong, that is, the common law, rules of equity, ordinances, subordinate legislation and customary law shall be maintained, except for any that contravene this Law, and subject to any amendment by the legislature of the Hong Kong Special Administrative Region."

19. Article 84 of the Basic Law reads: "The courts of the Hong Kong Special Administrative Region shall adjudicate cases in accordance with the laws applicable in the Region as prescribed in Article 18 of this Law and may refer to precedents of other common law jurisdictions."

20. The first paragraph of Article 39 of the Basic Law reads: "The provisions of the International Covenant on Civil and Political Rights, the International Covenant on Economic, Social and Cultural Rights, and international labor conventions as applied to Hong Kong shall remain in force and shall be implemented through the laws of the Hong Kong Special Administrative Region."

21. There are approximately 120 common law jurisdictions in the world today if the United States count as 50 jurisdictions.

22. Article 90 of the Basic Law of the Hong Kong Special Administrative Region (1990) 29 I.L.M. 1511; Court of Final Appeal Ordinance, Laws of Hong Kong, Cap. 484, section 6.

23. Basic Law, *id.,* Article 92; CFA Ordinance, *id.,* section 12.

24. The Privy Council is considered, technically, a "local" court when it hears appeals from any given jurisdiction. See Peter Wesley-Smith, *Constitutional and Administrative Law,* 2d. ed. (Hong Kong: Longman Asia, 1994), p.140.

25. Basic Law, *supra* note 8, Article 82; CFA Ordinance, *supra* note 8, section 16(1).

26. From 1841 to 1997.

27. *ICCPR* 6 I.L.M. 368 (1967).

28. *ICESCR* 6 I.L.M. 360 (1967).

29. *Cheung Lai Wah v Director of Immigration* [1998] 1 HKC 617 (Court of Appeal judgment of 2 April 1998) and [1998] 1 HKLRD 772 (Court of Appeal judgment of 20 May 1998) and *Chan Kam Nga v Director of Immigration* [1998] 1 HKLRD 752.

30. Basic Law, *supra* note 8, Article 24(3).

31. [1998] 1 HKC at 638–639, 652–653 and 663–664.i.

32. [1998] 1 HKC at 647–648, 656 and 666.

33. Case decided 29 January 1999, as yet unreported except in the popular press: see, for example, the *South China Morning Post,* 30 January 1999.

34. That is the subject of a separate article shortly to be published.

35. China has since acceded to these two international covenants. See below.

36. The Convention Against Torture and Other Forms of Inhumane and Degrading Punishment (CAT), the International Convention on the Elimination of Racial Discrimination (ICERD), the Convention on the Elimination of all Forms of Discrimination Against Women (CEDAW), the Convention on the Rights of the Child (CRC).

37. CEDAW 19 I.L.M. 33 (1980).

38. The most recently updated figures available to the author.

39. Adapted from a novel about the old Soviet system of justice, the screenplay purports to portray an American businessman in China facing a trumped-up murder charge.

40. Adapted from Heinrich Harrer's memoirs of his wartime exploits as a German captive who, having escaped from a British POW camp in India, made his way into Tibet and became tutor to the Dalai Lama in Lhasa.
41. Disney movie purporting to portray the childhood of the Dalai Lama.
42. These are the most recent figures available to the author at the time of revising this paper.
43. Source: Hong Kong Monetary Authority figures for June 1998.

CHAPTER 5

Strategic Development of the Hong Kong SAR: Social Policy, EIB Model, and Implications

Beatrice Leung

At the ceremony celebrating the establishment of the Hong Kong Special Administrative Region (SAR) on 1 July 1997, Mr. Tung Chee-hwa, the first Chief Executive of the SAR government, declared that Hong Kong "opens a new chapter in its history," charged with the responsibility of contributing to the modernization and the ultimate reunification of China (Tung 1997a, p.1). Under the slogan of "a Future of Excellence and Prosperity for All," Tung presented a vision of Hong Kong in his speech, viz.: "A society proud of its national identity and cultural heritage; a stable, equitable, free, democratic, compassionate society with a clear sense of direction; an affluent society with improved quality of life for all; a decent society with a level playing field and fair competition under the rule of law; a window for exchanges between China and the rest of the world; a renowned international financial, trading, transportation and communication center; a world class cultural, education and scientific research centre" (pp. 6–7).

In the real world, however, what ought to be accomplished is not necessarily what can be accomplished. While Asia's financial crisis was emerging in the summer of 1997, the overall situation was somewhat stable in Hong Kong. When the local stock market experienced a downturn in early September, the community's response was mild. Tung's release of his first policy speech, on 8 October, was still well received, although he had set for the community such controversial goals as building 850,000 flats in the coming decade and achieving a home-ownership rate of 70 percent by 2007, which might adversely affect the stability of the property market. Then, the intensification of Asia's financial turmoil in late October and again early January (1998) has resulted in the local economic downturn, which, along with the stock market crash and substantial devaluation of properties, reinforces the problem of unemployment. The negative changes in Hong Kong's economic environment have been accompanied by such crises as the outbreak of avian influenza, continual mishaps (dosage blunders) in public clinics, unnecessary fishery intoxication by "red tides," together contributing to the objectification of a bleak picture of the socio-economic development after reversion and making observers wonder if strategic development can be enhanced.

In such a context, some observers even conclude that due to, for example, Tung's appointment to the Chief Executive office, various crises have emerged and problems have escalated.[1] This, however, is a logical fallacy: the fact that Y happens after X occurs is not a basis for concluding X being the cause of Y. What really has gone wrong? How well can Tung achieve his strategic goals? This chapter represents an effort to furnish a basis for addressing those questions. Basing on literature review and document analysis, this study uses three major policy problems faced by the community as illustrative examples, namely, education, housing, and caring for the elderly, which are ranked high in Tung's priority list (Tung 1997a, p.3). The ultimate objective is to identify the most important factors that influence policy-making, both before and after the establishment of the HKSAR government, to construct a model for enhancing a better understanding of the strategic policy-making capacity of the SAR government and for drawing practical implications.

Social Policy Issues as Examples

A comprehensive analysis of various social policy issues is beyond the scope of this chapter and should be reserved for future pursuit. Further, the sole purpose of the analysis here is to review the major problems in social policy-making in order to furnish a basis for drawing practical implications, rather than fact-finding and interpretation, which have been well performed by many scholars; thus, detailed information about the policies to be examined will be kept to a minimum, only enough to substantiate the arguments to be presented.

Education

While Tung believes that the foremost task for the SAR government is "to enhance Hong Kong's economic vitality and sustain economic growth,"

"[E]ducation is the key to the future of Hong Kong" (Tung 1997a, p.3), providing a basis for further economic development. Delineating a mission for educators, Tung states that the education system "must cater for Hong Kong's needs, contribute to the country, and adopt an international outlook" (p. 5). He further states in his policy speech, among other things, "mother tongue teaching" will be pressed forward, $5 billion will be allocated to establish a Quality Education Fund, and curriculum reform at the tertiary level will be considered (Tung 1997b, pp.29–35).

These policy preferences are then materialized. Regarding "mother tongue teaching," the Education Department announced in December its instruction medium guidelines, allowing 100 of the 400 secondary schools to teach in English with the rest in Chinese. Subsequently, 14 schools' applications for appeal were endorsed. While school administrators, teachers, and students (as well as their family members) of these 114 "privileged" schools are cheerful, their counterparts are not.[2] While the Education Department is criticized for making a hasty decision with major impacts on the quality of education in Hong Kong,[3] commentary often overlooks the fact that the issue of changing of the medium of instruction has been examined and analyzed, and again examined and analyzed for more than 10 years (Lau 1997, p.106–118). More importantly, critics have failed to undermine the rationale for introducing "mother tongue teaching," which is that students learn much better if they are taught in their native language.

From a rational policy analyst's perspective, the "mother tongue teaching" is a perfect case of mediocre policy analysis, and it is going to reduce Hong Kong's capacity to maintain its strategic position in the Greater China region and in the global context. The policy is mediocre because the Education Department did not bother to conduct comparative policy analysis to learn from other countries' experiences. In, for example, the United States, student learning and academic achievement are also a national problem; yet American students are taught in English, their native language! Any responsible policy-makers, whose policy decisions would be based on comparative analysis, would then look into the matter deeply to identify the real causes of learning barriers and deficiencies, such as learning and entertainment habits, changing cultural values, peer influence, school-family cooperation, etc., rather than finding an easy way out by forcing most students to learn only through Chinese, just because they have not received proper training in mastering English in their primary school years.

The policy decision will cost Hong Kong dearly. Thus far, the effective usage of English by Hong Kong people has helped them to be more competitive than their counterparts in other Asian areas, in particular Mainland China and Taiwan. This competitiveness lies in their ability to communicate with foreigners in English and their proven capacity for absorbing cutting-edge knowledge transmitted through the English medium. The Education department's decision will deprive many Hong Kong youngsters of a precious opportunity to acquire a facility in English at an early age. While the department has reiterated that funding for programs to separately facilitate language training would follow—either out of goodwill or as political expediency for calming the critics—the effectiveness of those programs are unproven, thus compelling youngsters to take a risk of missing the prime time for language mastery, a risk not chosen at their own wish and with their consent. More importantly, as youngsters in China are fast learning English, and in view of the nation's speedy economic development, Hong Kong may soon loose the value as the bridge between China and the rest of the world, thanks to the government's instruction medium policy, which in fact contradicts Tung's principle that the education system "must cater for Hong Kong's needs" (Tung 1997a, p.5).

In view of the example above, one wonders if the $5 billion Quality Education Fund (QEF) will achieve its stated objectives of improving education quality by funding innovative projects and giving cash awards for schools' value-added performance.[4] While QEF was formally established in March (1998), and by the end of April the Fund had already received 2,300 applications for grants, requesting a total amount of $2.1 billion, it is immature to make judgment until it is revealed how and why grants for what purpose(s) are approved for what justifications; by then, researchers will be in the position to determine if the QEF is a genuine effort to enhance excellence or just another example of mediocrity.

Meanwhile, one also wonders if the reconsideration of curriculum reform at the tertiary level will not be as mediocre as the medium of instruction policy. There are signs that it might well be. Specifically, the Vice-Chancellors[5] of eight tertiary institutions have jointly presented a request for speedy curriculum change from the current three year degree program to a four-year program. The rationale of this "four instead of three" proposal is primarily pedagogical—improving higher education quality—and the proposers have tried to play down the financial implications of the curriculum change, which to many is a ground for rejection.[6] Again, rational analysis has not been a basis for consideration: the real issue is not university students spending three or four years in classrooms, but a comprehensive reform of the tertiary education system based on the British model. What has been overlooked is a genuine credit unit system that permits students to complete their undergraduate training in two to four years by taking more courses in a single semester and enrolling in summer session—which is not arranged in local universities, thus leaving all teaching facilities idle in the summer and depriving students of the opportunity to complete their training sooner. If, as part of a colonial tradition, the Vice Chancellors of Hong Kong's tertiary institutions could overlook such fundamental issues as the efficient use of resources and effective student learning, it is hard to believe that to fostering educational excellence in the post-colonial era, as Tung hopes for, is an easy undertaking.

Housing

While education has been a problematic policy area characterized by mediocrity, so too is housing. Tung's visits to cage homes and squatters, as well as his perception of the escalating prices of homes being irrational, drew his attention to the problem of housing in the community. He is thus determined to cope with the problems by setting the goals of increasing overall housing supply by no less that 85,000 flats per year to achieve a home ownership rate of 70 percent by 2007 (Tung 1997a, p.5). The increase of land supply (as a long-term solution to the problem of irrationally rapid appreciation of properties), the Home Starter Loan Scheme (a five year program with a price tag at the amount of HK$18 billion for approximately 6,000 families per year with incomes not exceeding HK$70,000 a month to enjoy low-interest loans of up to HK$600,000 for making down payments for the flat they intend to purchase), and the "rent-to-buy" scheme for tenants of public estates (a program for low-income families to purchase their own homes at the price that even pro-poor-people politicians cannot complain) are the programs for achieving Tung's objectives of stabilizing the property market and increasing the rate of home ownership.

The outbreak of Asia's financial turmoil and the subsequent downturn of the local economy have made Tung's goal achievement difficult, thus forcing the government to back down from the firm position of building an average of 85,000 flats per year (see for example *South China Morning Post,* or *SCMP,* 1 April 1998, p.6). Complaints against the Home Starter Loan Scheme are also plentiful notwithstanding the fact that more than 10,000 applications for the low-interest loans were submitted (*SCMP,* 21 March 1998, p.4). One policy analyst, Dr. Jane Lee, Chief Executive of the Hong Kong Policy Research Institute, could not help but comment: "The effect of the Asian financial turmoil is serious and its impact on the housing market is obvious. However, the Government has failed to adjust its housing policy accordingly. . . . The current concern about the Government's housing policy aptly reflects two problems which afflict the administration: weaknesses in the bureaucracy's self-improvement mechanism and policy adjustment mechanism" (*SCMP,* 21 April 1998, p.21).

Lee's comment is valid. But like other critics, she has yet to underscore the fundamental problem—non-rationality due to mediocrity. Specifically, Tung's efforts are likely to be in vain for one simple reason: his good faith policies are in fact altering the basic principle of the economic system—

individuals are responsible for their own well-being, and their decisions are made in an undistorted market context. The rate of home ownership should be decided by the market; so should tenants of public estates be compelled to work hard enough to save up the money needed for purchasing their own flats. Tung's policy has effectively changed the rules of the game, making housing a social issue rather than an economic one, thus negatively affecting individuals' motivation to excel, and positively promoting politicians and organized groups to pressure the government to redistribute more benefits, in turn upsetting the capitalist system in Hong Kong, which is constitutionally preserved by the Basic Law. His policies would inevitably result in disaster: the plunge of the property market is documented.

The market was not perfect, but governmental interference is warranted only when defects are artificially created. When demand for housing is greater than supply, the price will go up. The pitfall of Tung's policy was the neglect of the real cause of the problem. Specifically, the rapid appreciation of properties since 1996, as in 1994, was due partly to the past tight land-supply policy, but mostly to speculations together with concerted efforts by speculators, property developers, and real estate agents to objectify a reality of ever-price-rising. The result was many people's blind subscription to the purchase-now-or-never craze, with these people being the victims when the market crashed, as in early 1994 and again late 1997/early 1998. Rational and responsible policy-makers would easily identify the fundamental cause and act accordingly—levy a windfall profit tax at, say, 80 percent on profits derived from the transaction of property owned for less than two years. Such a simple and straightforward mechanism would effectively reduce speculators' motivation to artificially disturb the market system, or at least extend the time span of the up-and-down cycle of the market, thus giving prospective home-owners more time to consider their decision. Nevertheless, such a mechanism was rejected by the government in 1994 on grounds that such a windfall profit tax would complicate the current simple tax system, a justification that was hardly convincing in view of the various taxes levied on gasoline, wine and liquors, tobacco, etc., which together did not make the Hong Kong tax system simple. When there is not a will, there are always justifications. But the truth is still there: the unwillingness of the bureaucracy to introduce a windfall profit tax to make the market work is in fact an invisible hand to help distort the market.

When policy-makers lack the knowledge and abilities, or are irresponsibly unwilling to identify the fundamental causes of problems, the effectiveness of their remedial actions is inevitably limited. The welfare policy is no better.

Care for the Elderly

Based on the belief that senior citizens have contributed to Hong Kong's success, Tung argues that they "deserve respect and care from the community." Thus, under his leadership, the government will "develop a comprehensive policy to take care of the various needs of [our] senior citizens and provide them with a sense of security, a sense of belonging and a sense of worthiness" (Tung 1997a, p.5). Tung thus set up a Commission for the Elderly "to carry out a comprehensive assessment on the longer term demand for elderly housing and residential care services and draw up a strategy for both the private and public sectors to meet the needs" (Tung 1997b, p.39). Meanwhile, he instructed the Social Welfare Department to increase the monthly payment to elderly Comprehensive Social Security Assistance recipients by HK$380, a decision well received by both the elderly and politicians using care for the elderly as a prime campaign slogan. Then, the Executive Council recently endorsed a new Senior Citizens Residence Scheme submitted by the Housing Society, of which provision of subsidized housing for the elderly's middle-income group is targeted and the first 500 flats will be completed by 2001 (*SCMP*, 24 January 1998, p.41). In addition, an interdepartmental effort to achieve joint development of elderly facilities has been organized to construct three major complexes that put all under the same roof the various health, recreational, and residential facilities, which in theory will resolve the logistical problem of service delivery about which the recipients—the elderly—complained (*Ming Pao*, 7 April 1998, p.A18). All these developments, along with other programs such as the increase of tax expenditure for those living with their parents, the Mandatory Provident Fund Schemes approved by the Provisional Legislative Council, and the establishment of the Elderly Commission for developing strategies to meet the pressing needs of the aging population, confirm Tung's willingness and eagerness to turn Hong Kong into a caring and compassionate society.

Nonetheless, one cannot help but wonder if the continual increase of welfare benefits for the needy would not create an unsustainable financial burden on the SAR government and, at the same time, create a disincentive to the citizenry to take care of their problems as individual responsibility is seemingly no longer at the center of welfare policy-making. The latter is an empirical issue awaiting confirmation, a task that the Social Welfare Department should have undertaken in order to determine beforehand if the drastic increase of welfare benefits is a rational decision that would not result in unintended negative consequences. The former problem, that of the government's financial burden, is obviously one of alarming proportions: whereas HK$6.2 billion was allocated to social welfare in 1991, the figure shot up to HK$23.2 billion in 1997,[7] an increase of 274 percent over the six year period, or 46 percent a year. Given that the population is aging and welfare benefits are essentially mandated commitments on public spending that cannot be easily cut back, the continual increase of welfare spending at a soaring rate may well become a time bomb that cannot be dismantled without triggering major social disturbance.

How Tung handles the two issues is crucial as the economic development of and social stability in Hong Kong are at stake. Unfortunately, as of today, there are no signs indicating Tung's concern for the two issues, except that he is seemingly confident in the capacity of the community to sustain impacts from Asia's financial turmoil and to maintain satisfactory economic growth, thus generating sufficient resources for supporting the government's ever expanding welfare programs. If he is wrong, the consequences could be disastrous. This possibility draws our attention to a fundamental issue: why cannot government policy-making be based on rational, comparative analysis to ensure that policies made will not backfire or put the community at unnecessary risk?

All in all, the discussion above points to the existence of serious problems in the three substantive policies that are ranked high in Tung's priority list. The primary cause of those problems is non-rationality due to mediocrity, and the consequences of policy predicament are likely to be serious. How well Tung can cope with those problems remains to be seen. More educational speculations can be made after an examination of the relationship between contextual factors and policy-making.

The Policy-Making Context

Many analysts have published their research on Hong Kong since the British and Chinese governments began to negotiate on the future of Hong Kong in 1982. Their studies provide a rich mine of observations about various issues in the transitional era. Nevertheless, a careful meta-analysis reveals one fundamental problem of the field of Hong Kong studies, which is the confusion of paradigmatic approaches arising from a disagreement among scholars with regard to the question of which sets of variables are more important in shaping and reshaping the changes in Hong Kong: the progressive or regressive.

The EIB Framework[8]

Current literature on comparative politics indicates that certain variables at certain times and in certain contexts are salient in their impacts upon the policy-making system of a particular country (Bryant and White 1983; Hope 1985; Lindenberg and Crosby 1984), and that social development and policy-making are rarely influenced by a single factor (Balassa 1977; Grindle 1980; Lewis and Wallace 1984; Wilding 1997). We should not treat a specific set of variables as the primary variables to account for the dynamics of policy-making in Hong Kong, but examine all major variables and their relationships with the dependent variable being investigated. However, research employing such a comprehensive approach may be unmanageable and unfeasible.

The position of this paper is that modeling is essential and researchers should consider applying Ferrel Heady's framework for the purpose. Heady (1984) argues that there should be a basic agreement on the central concerns in comparative administrative studies: "These concerns include (1) the characteristics and behavior of public administrators . . . (2) the institutional arrangements for the conduct of large-scale administration in government . . . and (3) the environment or ecology of administration. *This combination of concerns,* proceeding from the more circumscribed to the more comprehensive, *provides a basic framework both for the analysis of particular national systems of public administration and for the comparisons among them*" (Heady 1984, pp.3–4). (Italics added.) With reference to Heady's argument, scholars would perceive that there are three levels of analysis: Environmental, Institutional-organizational, and Behavioral (EIB). We may call this the EIB framework. With this framework, variables that are prominent in their impacts upon the independent variable being investigated can be readily classified.

The focus of analysis of the EIB framework conceptualized by Heady (1984), however, is the administrative system; Hong Kong scholars have a wider concern. There is the need to refine the framework so as to maximize its utility. It is suggested here that the focus of analysis at the environmental level should be redefined, changing from "the relationship of the administration subsystem to the political system of which it is a part and to society in general" (Heady 1984, p.4) to the relationship of Hong Kong to China, of which it is a part, and to the international community in general. Moreover, the institutional arrangements for administrative decision-making and policy implementation should be redefined to include the overall constitutional and political infrastructural properties that prescribe and shape the patterns and outcomes of public policy-making. Furthermore, the foci of study at the behavioral level should include the characteristics/behaviors of administrators, political groups, and ordinary citizens.

Based on literature review and document analysis, this study, guided by the refined EIB framework, examines the impacts of various types of primary variables upon policy-making in Hong Kong. The analysis through the EIB perspective results in the establishment of a number of propositions.

The Ecological Analysis

Changes or lack of changes do not occur in a vacuum. The case of Hong Kong is illustrative. When we examine the development of Hong Kong with reference to major events occurring in China and elsewhere, we can easily note that major changes in the territory happening at specific points in time are hardly incidental. While analysis of confidential documents and interviews with major policy-makers may help furnish a basis for establishing causal relationships among changes in China and at the global level and changes in the territory, experienced researchers are prone to point out that this type of research may be unfeasible, due to access to confidential documents and the absence of major participants' interests to reveal "facts," or may generate unreliable information that cannot be cross-verified. Thus, while confirmation by interested researchers is awaiting, we may treat the associational relationships identified as causal; by doing so, we may establish two propositions for further analysis. Firstly, (Proposition 1) *since 1949, major political or economic changes in China is a sufficient condition for the occurrence of drastic changes in Hong Kong.* And secondly, (Proposition 2) *since 1949, changes in China's working relationship with Western democracies, the United States in particular, is a necessary condition for the occurrence of changes in Hong Kong, with the nature and impacts of those changes defined and created by the leadership of Hong Kong's sovereign state.*

Environmental Instability.   While it is a highly risky business to make predictions about what might happen at the global level and in China and what will be the effects on Hong Kong, research by scholars in the field of international politics has furnished a basis for us to make some educated speculations. Relevant to our pursuit here are the works by Holm and Sorensen (1995) and Miller (1995).

A group of researchers (Holm and Sorensen 1995) assessing the changes since the end of the Cold War underscored the inevitable instability of world order. One of their major findings was that globalization and the end of the Cold War have induced negative impacts upon the incentives of leaders of major powers to get directly involved in conflicts in various regions of the world, which are typically characterized by competition for regional supremacy and divisions of new nation-states. Miller (1995) notes that the Cold War era is characterized by bipolarity and that after the end of the Cold War era is the breakdown of bipolarity, which is followed by the increasing multipolarity, making it difficult for mankind to manage regional crises and raising the probability of unintended conflicts or inadvertent war at the regional level. Several years have passed by since the end of the Cold War, political turmoils, interstate wars, and intrastate military conflicts are still being observed, defying a general expectation that a new harmonious world order is to emerge. In short, (Proposition 3) *as the world order is still in flux, disorder may surface when conditions that foster stability and international and regional cooperation change.*

The implication of the proposition above is salient for Hong Kong as it is increasingly dependent on China since her economic takeoff. Major changes in China, progressive or regressive, would induce substantial impacts upon Hong Kong, for better or worse. In view of China's current faithful implementation of the "one country, two systems" policy, continual domestic policy change in the Mainland, be it political, economic, or social, will positively enrich the overall profile of Hong Kong. Yet, there is no assurance that incidents like the Tiananmen military crackdown (1989) or hyperinflation (1993–94) will not occur again.

Moreover, some China scholars have expressed concern for China's growing strength. For almost two decades since the beginning of Deng Xiaoping's economic reforms, China has had one of the most dynamic economies in the world with an average growth rate of over 10 percent a year. Yet, this continual economic growth may make China uncooperative in foreign policy, as Denny Roy argued: "China is more prone to using force . . . and will be likely to remain so after its economy has grown, because the Chinese government is authoritarian, unstable, wants to redress the status quo, and can mobilize large military forces. . . . China is also harder to deter . . . because it is less vulnerable to economic coercion, and will be even less dependent on outside suppliers as its economy continues to develop" (1994, pp.165–6). The People's Republic of China does have a record of use of force for resolving interstate disputes. Equally true is that her leadership is not abusive. In fact, as economic development has become the national priority since late 1970s, China is becoming more self-restrained in interstate conflicts and more cooperative at the international level. Such a trend is likely to continue as the post-Deng leadership, which is ideologically more pragmatic and technologically more competent, has successfully consolidated its power and is ready to concentrate its efforts on further economic development.

Nevertheless, China is still sitting on at least one time bomb: Taiwan's continual move to reinforce her international independent nation-state status may eventually provoke Beijing's leaders to launch a limited or full-scale offensive to achieve the goal of reunification, thus upsetting the region's stability or even inducing intervention by the United States, causing a major direct confrontation in the region which, if it were to get out of control, could lead to a Third World War.

The consequences for Hong Kong would be disastrous. Unfortunately, anything can happen after the breakdown of bipolarity.

With the above understanding, we may entertain some speculation that the ecological impacts upon Hong Kong in the post-transition era may change, moderately or drastically, depending on the degree and extent of changes in China and/or at the global level. Those speculations would become more educated when we take into account the characteristics of the variables at the institutional and behavioral levels as conceptualized by the EIB framework.

The Institutional Analysis

As mentioned, institutional variables include the major constitutional and political infrastructural properties, as well as the arrangements for policy-making and implementation that prescribe and shape the patterns and outcomes of policy-making in Hong Kong. Major changes, as well as their causes, of the constitutional, political, and administrative infrastructural properties in past years are well documented by Hong Kong scholars.[9] While there have been changes, there is one major institutional property, namely, unitary command, which remains relatively unchanged and which asserts salient impacts upon Hong Kong's capacity to enhance strategic development.

Unitary Command.   The status of Hong Kong as a British colony was attached with the prescription of concentration of policy-making power being reserved to Britain and her representative in Hong Kong, the Governor. When governors made decisions in the name of "Governor-in-Council"—responsible and balanced decision-making based on advice from Executive Council members, allegedly the cream of the crop in Hong Kong—they were in fact exercising their right of supremacy within the executive-led infrastructure, no matter how responsive the decisions were to public opinion or demand for actions from politicians and/or politically active citizens. Indeed, under the Letters Patent (and the Royal Instructions), the Governor, backed by the British government, was ultimately the one who could decide on the fate of any legislation by either withdrawing a government bill or refusing to give his consent to a private bill introduced by individual legislator(s) and passed by the Legislative Council. As one observer (Lau 1997, p.52) notes, "[U]p to the very last moment of British rule, the power of governing still firmly resided in the hands of the colonial administration headed by the Governor and staffed by civil servants. . . ."

The unitary command remains a key property of the SAR polity: "the first Chief Executive of the HKSAR, like his colonial counterpart . . . possesses relatively great constitutional powers" that permit Tung to play "a leading role, albeit in varying degrees, in the administrative, judiciary, and legislative arms of the government" (Wong 1997, pp.29–30). To some extent, the Chief Executive is even more powerful than his colonial counterpart. For example, Article 74 of the Basic Law stipulates that "[B]ills which do no relate to public expenditure or political structure or the operation of the government may be introduced individually or jointly by members of the Council." Yet, it also prescribes that "written consent of the Chief Executive shall be required before bills relating to government policies are introduced." Before 1997 introduction of private bills had been a powerful weapon for legislative councilors to force their will on the colonial administration; after 1997, the supremacy of the Chief Executive is further reinforced by being empowered to accept or reject private bills. Then, Articles 49, 50, and 51 also provide a mechanism for the Chief Executive to reject any bills passed by the Legislative Council or push his bills and budget through the legislative process by posing the threat of or actually dissolving the Legislative Council, although he can do so only once in each term of his office (Article 50), and he may have to resign if the bill in dispute is passed, after the dissolution of the Legislative Council, by a two-thirds majority of all the members of the newly elected Legislative Council (Provision 2, Article 52). Incorporated in these provisions are checks and balances elements, but the main burden is put on legislators; unless there is overwhelming support from the community for them to directly confront the Chief Executive and to bear the risk of Council dissolution, which is unlikely in a pluralistic community like Hong Kong, the Chief Executive is comfortably taking charge under the Basic Law protection.

To some observers, while Tung is invested with great constitutional powers, such powers are constrained at various levels. According to Wong (1997), "Tung's power as Chief Executive is subject to constraints from at least seven sources, namely those arising from the Chinese government, Hong Kong society, the method by which he was elected, the Preparatory Committee, the Legco, the civil service bureaucracy, and public sources" (p. 47). Moreover, "[o]f these seven constraints, the Chinese government is the most important . . . Tung is expected to give it the most prominent position in formulating policy" (p.47). Such an understanding leads Wong to conclude that "Tung is attempting to identify himself with the Chinese government in exchange for autonomy in governing Hong Kong" (p. 48).

In a world of interdependence, there is no such a thing as absolute power; it is all a matter of relative comparison, longitudinal or cross-sectional. Thus, it is obvious that Tung's power in policy-making may be paramount but not unlimited. As such, observers may feel free to characterize Tung as the yesman "hand-picked" by leaders in Beijing to represent their interests in Hong Kong; yet, their conception is inaccurate, misleading, and biased. It is inaccurate because the Communist leadership's interest in Hong Kong is economic and political: make use of Hong Kong's potentials and strategic values to contribute to China's economic development and set an example of effective reunification under the "one country, two systems" principle for inducing a peaceful reunion with Taiwan (Ching 1997), rather than controlling Hong Kong for a taste of power.

It is misleading because China's control over Hong Kong is fully guaranteed by the constitutional document: the Basic Law specifies in detail what Hong Kong can and cannot do, and empowers the Standing Committee of the National People's Congress—a body under the strict political control of the CCP—to interpret the Basic Law (Article 158), which is practically a complete monopoly of the power to vet laws passed by the Hong Kong legislature or administrative decisions made by the Chief Executive; whenever necessary, the Standing Committee of the National People's Congress may declare a state of emergency in Hong Kong and the Central People's Government, which is in essence the administrative instrument of CCP, may issue an order applying relevant national laws in Hong Kong (Article 18), which are legitimate acts that can be politically motivated.

It is biased because it strives to express only a partial picture. Often when scholars are themselves biased, their analysis will also be value-laden, which adversely affects the impartiality of their findings. In the case of advocating selection of the Chief Executive by direct election or criticizing the current method being a "small circle" activity without legitimacy, scholars often fail to point out the fact that during the long colonial period "Hong Kong governors had been appointed in London without any consultation with Hong Kong's populace" (Ching 1997, p.53). By the same token, when arguing that the Chief Executive must appease China, maybe at the expense of the community, those scholars typically omit the fact from their analysis that, as compared with his colonial counterpart, Tung is obviously in a more independent position under the Basic Law arrangements: while Tung may stand firm against interference from China that is not permitted or implied by the Basic Law, the colonial governor could not because the source of the Governor's power was Letters Patent, which prescribed his faithful protection of the interest of the United Kingdom, implying that the changing of the values and policy preference of the leadership in London could result in the change of policies in Hong Kong, which had to be faithfully implemented by the Governor, a power monopolizer from the local perspective but truly a yes-man from the London's viewpoint.[10]

In short, the crucial factor that affects the power of the Chief Executive is constitutional, rather than political, influence from China. Thus, a more balanced assessment of the Chief Executive's monopoly of policy-making should begin with the constitutional prescription. From such a perspective, Tung has no need to attempt to identify himself with or appease the Chinese government in exchange for autonomy in governing Hong Kong, but can merely be himself—a relatively altruistic person with national pride who requires no brain-washing or indoctrination to support the principles upheld and policies formulated by the current pragmatic CCP leadership, which aims to achieve a politically stable and economically prosperous China, and exercises his constitutional powers to run Hong Kong to achieve whatever goals he sets for the community, as long as those goals are within the parameters set by the Basic Law.

One important implication from the above analysis of China's control over Hong Kong through the Basic Law is that when the CCP leadership perceives the need to actively or even aggressively intervene, it has the means to do so. It is, however, unlikely to happen at the moment. As Ching (1997, p.64) rightly points out, "[O]n the whole, China will try to keep its commitment." After all, "China did not spend two years negotiating the Joint Declaration, five years drafting the Basic Law, and many more years negotiating other agreements with the idea that it would tear them all up on July 1."

Regardless of the probability, however, the possibility itself deserves our attention. Thus, based on the analysis of the unitary command property in Hong Kong, we would conclude that (Proposition 4) due to the constitutional prescription and unitary nature of the Hong Kong polity, both before and after 1997, the Governor or the Chief Executive monopolizes public policy-making under a condition that the values and policy preference of the political leaders of their respective sovereign state remain unchanged.

Administrative Leadership. There two implications of the presence of the unitary command property, prevailed in the colonial era and alive and well in the conditional reunification era.

Firstly, as the policy-making power has been highly centralized, leadership values and styles could and can assert substantial influence on policy reformulation. As such, (Proposition 5) under the executive-led structure, the policy preference of the Governor or the Chief Executive is a sufficient condition for drastic policy change.

A major implication with theoretical and practical significance derived from the understanding above is that the conceptual framework and cognitive styles that major policy-makers in Hong Kong bring to the decision processes have been the determining factors affecting what policy to change for what reasons to achieve what values with what impacts on the development of Hong Kong. Thus, policy analysis or advocacy must also be attentive to the psychology of decision-making and must incorporate in its pursuit the working knowledge and values of major decision-makers, rather than merely the logic or instrumental rationality of decision-making. Such an understanding is in line with the analysis in the preceding section of Tung

Chee-hwa's altruistic tendency and pragmatic approach to China's development, which furnishes a basis to more objectively account for a high degree of consistency between Tung's and the current CCP leadership's political values and policies and to reject such intended or unintended misinterpretation of Tung's political preference as appeasement or Koutou. Moreover, such an understanding draws out attention to the pressing need to systematically examine the psychological make-up and value preferences of major decision-makers in Hong Kong so as to effectively make educated speculations about the current or potential conflicts among Tung, senior civil servants, pro-democracy politicians, and business elites.

And, secondly, when the Governor in the colonial era, for whatever reasons, was willing to share his policy-making power with senior government officials, the bureaucracy was empowered, a phenomenon seemingly inevitable as the governors in the Hong Kong history were rarely interested in or committed to the making and monitoring of ordinary or procedural policy decisions, which were then inevitably made by bureaucrats, senior or even petty. And when such Governors as Sir Edward Youde (1982–86) and Chris Patten (1992–97) were busy pursuing negotiations with China and pushing through pet political reform programs, respectively, the making of even substantively important policies would be left to senior civil servants. The power-sharing phenomenon has been objectified in the conditional re-unification era as Tung Chee-hwa is behaviorally inclined to maintain a harmonious working relationship with the bureaucracy, and while he works from 7 A.M. to 11 P.M., he is still limited by the natural human cognitive capacity to make all policy decisions and inevitably has to share his power.

As such, (Proposition 6) under the executive-led structure, the willingness of the Governor or the Chief Executive to share policy-making power is a sufficient condition for the monopoly of the power by the bureaucracy.

The Behavioral Factors

As mentioned, the foci of study at the behavioral level should include the characteristics and behavior of public administrators, political groups, and ordinary citizens. Given the unitary nature of the Hong Kong polity, which makes the administrative leadership fundamentally instrumental in effecting policy change, the need to examine the bureaucracy, or in a more favorable term the civil service, becomes obvious. Hong Kong is changing, so are public administrators. Yet, there is one major property that has remained relatively unchanged, which is the culture of mediocrity in the civil service.

The Culture of Mediocrity in the Civil Service. Researchers find it difficult to establish high-performance public bureaucracies characterized by efficiency, effectiveness, and responsiveness (Chow 1991). After all, there are many forces at work adversely affecting the operation of public bureaucracies. For example, recent research has indicated that public management paradoxes are observable in Western democracies as well as developing countries (Bryant and White 1983; Perlman 1989), and that paradoxical forces adversely affect public managers' ability to act. In the case of Hong Kong, one researcher (Chow 1991) has documented that there are various paradoxical forces acting in concert in their impacts upon the Hong Kong bureaucracy in the decolonization era, that Hong Kong public executives are unable to cope with those forces, and that, unless those paradoxes are effectively managed, administrators' ability to act and their motivation to excel are very limited. In short, (Proposition 7) due to various paradoxical forces, Hong Kong public administrators' ability to act and their motivation to excel are very limited.

While civil servants' ability to act is constrained by management paradoxes, their capacity for making and executing rational and politically feasible decisions for the public is also very limited. Specifically, as the Hong Kong civil service is modeled after the British system, which places a premium on aptitude—rather than professional administrative knowledge and skills—in the recruitment of administrators, pre-service training in public administration and/or policy analysis at either the graduate or undergraduate level has not received appreciation by the Hong Kong government, as is the case in, for example, the United States, where most public managers receive formal professional training at the pre-service or in-service stage. Inevitably, formal training in public administration/policy analysis is also not a criterion for making promotion decision, and therefore civil servants have little incentive to pursue public administration training, except those piecemeal training programs offered by the Civil Service Training and Development Institute.

All these are unfortunate because in the transitional era, civil servants must be well trained enough to enhance administrative, substantive, technological, and other contextual knowledge and competences so as to systematically study, from a comparative perspective, the nature, causes, and effects of alternative public policies and to effectively and efficiently make and execute rational decisions in a rapidly changing context. They must develop a good combination of intellectual, behavioral, and entrepreneurial abilities and traits, all of which can only be developed through formal pedagogical programs and reinforced in the workplace by well trained administrators.

More importantly, this world is now characterized by constant, rapid, and even drastic changes. Managing change requires greater skills (diagnosis, judgment, interpersonal, etc.) at coping with uncertainty and the unfamiliar. The reality of frequent change requires that civil servants be skillful at working with change, dealing with the "unknown," and managing "uncertainty" in a constructive on-going manner. They need to develop a capacity for their own learning and enhance an awareness of subtle environmental changes and a sensitivity to their future implications. Their actions must be taken more in the light of the future, with relatively less regard to the past, and in functional directions that either forestall or prepare for the occurrence of critical future events or conditions, so that relevant changes can be promptly noticed, accurately interpreted, and appropriately transformed into future responsive action. In short, effective public administrators must develop adaptive capacities in living with high levels of uncertainty, ambiguity, and confusion, and enhance the skill to manage under conditions of disorganization and unstructured reality, all of which, again, can only be developed through formal pedagogical programs and reinforced in the workplace by well trained administrators.

Based on the understanding above, it is obvious that Hong Kong public administrators must receive formal professional training in public administration and policy analysis at the pre-service or in-service stage so that they have the capacity to make and execute rational decisions and become change-conscious, future-oriented, and precedent-breaking.

Notwithstanding the implication above, it is concluded here that (Proposition 8) due to the lack of appreciation for formal professional training in public administration and policy analysis, the capacity of Hong Kong public administrators to excel, particularly in the context of rapid or drastic changes, is very limited.

In sum, given the significant role civil servants play in Hong Kong policy-making, their mediocre capacity and performance would make the task of formulating rational policies difficult, if not impossible. As such, mediocrity inevitably becomes the organizational culture, and maladministration becomes the hallmark of the polity. In sum, (Proposition 9) Hong Kong public administrators' lack of motivation, limited ability, and insufficient capacity to excel are together a necessary condition for the occurrence of various crises and problems in Hong Kong.

Obviously, the Chief Executive's statement that "[W]e are privileged to have a dedicated, honest and efficient Civil Service" (Tung 1997b, p.50) should be changed to "we are unfortunate to have inherited an undedicated, sometimes dishonest, and mostly inefficient and ineffective civil service from the colonial era." The continual establishment of non-rational policies, as discussed in the first section of this chapter, is inevitable unless major, wholesale administrative reform, directly targeting the removal of the norm of mediocrity, is introduced; all other piece-meal reform measures are likely to be non-instrumental or with limited effectiveness, particularly when they are proposed for political expedients.

The Culture of Passive Participation.   Another major property at the behavioral level that remains relatively unchanged and that may be considered
the necessary condition for reinforcing the current mediocre policy-making
system is the culture of passive participation in the community.

In the decolonization era, the British government had tried to speed up
the process of democratization in Hong Kong. Nevertheless, in terms of the
extension of voting rights, the competitiveness of the political parties, and
the level of turnout in elections, the efforts to democratize had resulted in
limited progress: in terms of the right to elect representatives, some citizens
were more equal than others as they could vote in both the functional constituency election and the geographic constituency election;[11] in terms of
party competitiveness, the coalition of liberal and radical "democrats," who
had in the past successfully employed the pro-democracy and anti-communist strategies to induce support from the confused citizenry, had been the
major winner in the Legislative Council, District Board, and Municipal
Council elections since 1991; in terms of the level of turnout in elections,
the number of voters in the Legislative Council's geographic constituency
election in 1991 and 1995 was 750,000 and 920,567, yielding a turnout
rate of 39 percent and 36 percent of registered voters, and 13 percent and
15.3 percent of qualified voters, respectively (Li 1995, pp.55–6 and Louie
1996, pp.53–4), which was hardly satisfactory. Why so? Was it because the
government was not doing enough to induce democratic participation or
was the citizenry not ready for democracy?

The work on cultural preference by Kuan and Lau (1989) offers insights
and explanations. Kuan and Lau point out that research conducted by S. K.
Lau in 1977 had recorded that an overwhelming majority of respondents
preferred "social stability to economic prosperity, if asked to choose between
the two" (p.93). Moreover, the same survey "revealed that the common people's love of freedom is not based on a belief in fundamental rights" (p.93).
Furthermore, "according to the 1985 survey, there was already a substantial
proportion of people (40.8 percent) who blamed the excessive availability of
freedom in Hong Kong for the existence of social problems" (p.93). These
findings imply that freedom in Hong Kong "is valued largely from an instrumental point of view" (p.93).

Regarding democracy as a fundamental value in Hong Kong, Kuan and
Lau note that "the people of Hong Kong were more concerned about the decline in government authority as detrimental to social stability than about
the same phenomenon as conducive to government's respect of public opinion" (p.95), and that "the ordinary men in Hong Kong would probably prefer social stability to political democracy, should they be forced to make a
choice between these two values" (p. 95). Kuan and Lau also underscore an
important cultural preference that "although modern ideals such as freedom

and democracy have made inroads into the minds of the Hong Kong Chinese, it is the pragmatic value of social stability that is most cherished by them. The deep-rooted obsession with social stability largely defines the parameter upon which ideological values such as freedom and democracy may prosper" (p.95).

To Kuan and Lau, since such fundamental values, beliefs, and preferences such as "obsession with stability," "instrumental view of government," and "the frustrated selves as citizens" are more resistant to change, it is expected that "unless major organizational efforts are mounted, the public's responses to political development in the near future will remain restrained" (p.110). Likewise, it is expected that "lacking a participant tradition and in the absence of some extraordinary developments, Hong Kong is unlikely to experience an upsurge in political participation in the future" (p.111). Therefore, Kuan and Lau continue, "political participation will be confined to specific and relatively small groups of people" in post-colonial Hong Kong (p.111).

In light of the findings presented by Kuan and Lau, we can conclude that (Proposition 10) *due to traditional values, political development in China, and the lack of a participation tradition in Hong Kong, most citizens are primarily concerned with social stability and are politically apathetic.*[12]

In a more recent report released by S. K. Lau (1997), the fabric of the Hong Kong community is found loosening. According to Lau, the community has undergone "momentous socioeconomic changes since the 1980s" (p.426). "Anxieties about an uncertain political future have already produced an exodus of people, prominent among whom are the better educated and those with professional and managerial skills. . . . Sensing their collective powerlessness, Hong Kong people are using every means to safeguard the future of themselves and their families, including illegal, illicit or shady methods. . . . The rise of interpersonal and social conflicts on the one hand, and the erosion of respect for authorities of various kind . . . have together produced a social milieu suffused with greed, querulousness, disorientation, sullenness, cynicism, small-mindedness, intolerance and nastiness. These are all signs of a *fin de siècle* mentality" (pp. 429–30).

In such a milieu, one may expect two things: those who have been politically active are likely to be as active, if not less, and those who are not may remain so or become active under the influence of the former. All in all, given a culture of passive participation, demands from the Hong Kong citizenry for government action to resolve their problems are unlikely to be mild, generating relatively little impact upon policy-making.

In short, we may conclude that (Proposition 11) *the current cultural values and behavioral tendencies in Hong Kong are a necessary condition for the existence of limited citizen demand for policy actions.*[13]

As demands from the Hong Kong citizenry for government actions to resolve their problems are likely to be mild, thus even when the SAR government, like its colonial counterpart, does formulate or reformulate policies according to the preferences expressed by the citizenry, Tung still possesses the constitutional power to make decisions contrary. The key point is leadership preference, rather than requisite.

More importantly, public opinion is moldable. As such, legitimacy of the SAR government, again like its colonial counterpart, does not rest on how well the government responds to public opinion, but on how well it can manage opinion. It is particularly true that by curbing opinion advanced by selfish motives or derived from ignorance, impatience, or hostility, and by promoting the public's appreciation for what it has done or will do, and educating the public to recognize the difficulties of policy-making and underscore the importance of cooperation, patience, and understanding, which are essential to problem-solving.[14] What is needed is thus an effective instrument of communication. Unfortunately, Radio and Television of Hong Kong (RTHK) has turned into a relatively independent public organization with little interest in actively promoting the cause of the SAR government.[15] Thus, the ability of the government to effectively launch its public opinion management program is limited. As such, unless the Tung leadership can again turn RTHK into a propaganda machine, as before 1997, or decides to construct another machine, the government is likely to remain operating in an environment plagued by public outcry and complaints.[16]

In sum, (Proposition 12) given the unitary nature of the Hong Kong polity, and the culture of passive participation, public opinion is at best a contributory condition for policy formulation or reformulation.

Related to public opinion as a major factor influencing policy-making is the political party. In Western democracies, parties do play a key role in governance and their policy preferences are often the sufficient condition for inducing major policy changes (Pomper and Pomper 1985). In the case of Hong Kong, however, "party politics is still in its embryonic form" although a number of political groups have been burgeoning since 1982 (Leung 1997, p.109). Moreover, while political parties have been active in the past elections and their pressures placed on the government have occasionally affected actual policy outcomes, partisan politics has yet to become a sufficient condition for policy change or a necessary condition for changing the policy preference of Governors Patten (1992–97) and Wilson (1987–92) and senior government officials.[17] In view of the current political development, the first Legislative Council of the SAR Region is likely to be an "ungovernable" "no-majority-party legislature" (K. Leung, 1997, p.125) and that "Chief Executive Tung Chee-hwa could find no steady and reliable support in the legislature."[18] In such a context, Tung may choose to favorably respond to demands from the legislature; he may stand firm and ram his policies through, regardless of the consequences; he may make use of his constitutional powers to enhance allies and build coalitions to rally support for his policies; he may do nothing but let legislators' assessment of the technical merits of his policies be the basis for accepting or rejecting his proposals; and he may apply a combination of the strategies above in different policy contexts. The key is Tung's personal preference on how to work with the legislature, rather than political parties being the primary variable affecting policy-making. In short, the political party is at best, like public opinion, a contributory condition.

Discussion

To sum up, the preceding analysis points to the existence of serious problems in the three substantive policies—education, housing, and caring for the elderly—that are ranked high on Tung's priority list. The primary cause of those problems is non-rationality due to mediocrity, and the consequences of policy predicament are likely to be serious. The analysis also reveals that while there are various factors at work affecting policy-making and implementation in Hong Kong, the most important factors, as either necessary or sufficient conditions, are those of the unitary command, monopoly of policy-making power by the Chief Executive and senior government officials, civil servants' lack of motivation, limited ability, and insufficient capacity to excel, and a culture of passive participation in the community.

Major practical implications have been discussed in the preceding sections. As a whole, the discussion revolves around the pressing need to pay serious attention to the ecological changes, real or potential, and their impacts upon Hong Kong, and to introduce major administrative reforms targeted at improving civil servants' knowledge, skills, and abilities, and at changing the culture of mediocrity. Whether the pressing need for wholesale reform will receive attention from the Chief Executive and senior government officials remains to be seen as leadership attention is itself a scarce political resource[19] and as the incapacity of the civil service to make rational policies is found to be a necessary condition for policy predicament, it may as well lack the genuine interest in and the professional capacity to initiate comprehensive rational administrative reform.[20]

All in all, in light of the propositions established and implications drawn, we may conclude that Tung is unlikely to be able to achieve his strategic goal. If Tung is consistently concerned with maintaining a harmonious relationship with senior government officials in order to objectify a reality of harmony and cooperation, he will be unable to slave drive them to reject individual (bureaucratic) rationality, enhanced at the expense of collectivity irrationality, and to turn things around to neutralize the norm of mediocrity and, in turn, achieve excellence by taking drastic actions to upgrade the civil service's capacity, as suggested in the preceding sections. If Tung cannot even put RTHK on a leash, making it at least a responsible broadcasting system providing Hong Kong citizenry with objective and balanced assessments of policies advocated by the government, or establish an alternative propaganda machinery to replace it, there is no way for him to achieve the goal of making Hong Kong into, in his words, "a society proud of its national identity and cultural heritage" charged with the responsibility of contributing to the modernization and the ultimate reunification of China (Tung 1997a, p.6–7). RTHK, to say the least, fails to launch major resocialization programs for many Hong Kong citizens, who have been, in the colonial era, indoctrinated to have fears for socialist China or have internalized the values of liberal democracy in utopian terms; the "ifs" continue. . . .

Given the significance of the civil service in the Hong Kong unitary command polity, the capacity for the community to cope with crises and ecological impacts, as discussed, is so limited that the overall strategic development of Hong Kong is doomful, unless drastic reform actions are taken by Tung. Otherwise, the strategic values of Hong Kong to China may soon diminish due to the speedy economic development in Mainland, which inevitably will result in the further upgrading of the nation's problem-solving capacity and the substantial improvement in economic and technological competitiveness. As Hong Kong's utility depreciates, CCP leadership's appreciation will also diminish and thus its commitment to "one country, two systems" may erode, leading to, for example, unrestrained and undue interference, intended or unintended, from Mainland officials—particularly those of various Central Government bureaus or provincial organs—and, in turn, the end of Hong Kong's autonomy.[21]

Notes

1. The typical argument is that, as Tung Chee-hwa has similar constitutional powers and efficient administrative apparatus as his counterpart in the colonial era, citizens cannot help but hold him responsible for all those problems and crises emerged since 1 July. See, for example, a commentary by Lo Chi-kin (*Hong Kong Economic Journal,* 2 May 1998, p.A18).
2. For reports, see *SCMP* (14 March 1998, pp.1 and 14).
3. For details, see, for example, a report on the contents of a symposium on the issue (*Ming Pao,* 16 March 1998, p.A4).
4. Hong Kong SAR Education Commission Report No. 7, September 1997, para. 5.1–5.10.

5. Since the colonial Governor was ex offio the Chancellor of all universities in Hong Kong, the Vice-Chancellor by tradition was in effect the President of each university.

6. For example, see *Ming Pao*'s editorial comment (14 April 1998, p.B14).

7. The amount allocated was HK$21.2 billion in the 1997–98 budget prepared by the Financial Secretary (p. 19); due to the economic downturn, the Social Welfare Department had to apply in January for additional funding of HK$2 billion (*SCMP,* 15 January 1998, p.62), making a total of HK$23.2 billion.

8. EIB refers to a framework, such as used here, that examines the environmental, institutional-organizational, and behavioral dimensions of the issues.

9. See, for example, Lee (1997), Lee and Cheung (1995), Miners (1991), and Scott (1989), as well as the various editions of the *Other Hong Kong Report* published by the Chinese University of Hong Kong Press.

10. A vivid example of how the Governor perform the yes-man task is provided by Chan (1997) in his analysis of the strategic interactions among Hong Kong, China, and Britain in the decade after the WWI.

11. This is also the case in the first Legislative Council election to be held in 1998. Given this inequality of voting rights, one cannot help but wonder why the government keeps stressing in their publicity programs that the election is a "fair" election.

12. It should be noted that the validity of the proposition that Hong Kong residents are politically apathetic may seem shaky in light of the fact that in 1989 1.5 million people marched in the street to protest for democracy in China. One simple answer is that the political culture in Hong Kong is in flux and thus anything possible can happen. A more educated explanation would include a qualification which is well documented in the field of political participation: when critical events occur, resulting in the mounting of public sentiment, extraordinary behavior may follow. Indeed, Kuan and Lau did point out that "lacking a participant tradition and *in the absence of some extraordinary developments,* Hong Kong is unlikely to experience an upsurge in political participation in the future" (1989, p.111). (Emphasis added.)

13. It should be noted that a recent report by Jermain Lam (1997) presents findings contrary to the conclusion here. Specifically, Lam finds that "the political orientation of the voters has been developing toward a more participatory albeit alienated political culture" and thus Hong Kong's political system "will gradually develop toward a more open and democratic government in order to alleviate democratic challenge as well as the crises of legitimacy and confidence" (p. 97). The validity of his findings is questionable for methodological reasons. Firstly, Lam chose the 1991 and 1995 registered voters as the subjects, making his sample a biased group: the subjects were unrepresentative of the population as the registration rates were 33.78 and 42.92, respectively (Lee and Cheung 1995, p.56); those who were willing to register tended to be more politically conscious, active, and/or assertive, a bias problem well discussed by such leading political scientists as Sidney Verba (1996). Secondly, Lam employed a mail-in strategy, which tended to induce responses

from politically assertive citizens, another unavoidable methodological flaw. And, thirdly, Lam notes that "only 61.2 percent of the respondents in the 1995 survey indicated that they would stay in Hong Kong after 1997" (p. 123). If the subjects' response was reliable and the sample was representative of the population, the 38.8 percentage would then be translated into approximately 2.3 million Hong Kong people leaving the territory between 1995 and 1997, an unimaginable social event, which of course did not occur; when restricted to only the population of the registered voters, close to one million people would leave—still an unbelievable number. Such a response inevitably reveals the problems both that the reliability of the responses was questionable and that subjects tended to give the "if possible" or socially desirable answers rather than what they really wanted and would do, again a major problem in opinion surveys, which could be controlled only by well trained survey researchers (Verba, 1996). In view of the methodological problems of his study, we have to reject the validity and reliability of his findings and draw our conclusion as is.

14. The argument here may sound reactionary, but any well trained political scientist would appreciate the reality that all parties in the political arena do try to mould public opinion to serve their own interests. The key point is whether the motive behind the effort is noble or ignoble. If one accepts a fundamental function of a responsible government as positive socialization through which the personal growth and development of citizens are to be enhanced, one should find positive opinion manipulation, as suggested here, acceptable.

15. For details, see, for example, *SCMP* (14 March 1998, p.14).

16. The RTHK issue has become a highly political one. Any government action to place it under editorial control will backfire, inducing protests from local political groups and journalists and from the international community for the alleged violation of the right to the freedom of the press. But when there is a will, there is a way. What Tung can do is to leave RTHK intact and, at the same time, establish another public education broadcasting system to, on the one hand, perform the task of managing public opinion and, on the other hand, run educational programs that are essential to the personal growth and development of the citizenry and to the improvement of the community to enhance productivity and economic competitiveness, which are in fact much needed in view of the increasing competition from not just developing nation-states in Asia but also China. Nonetheless, given the prevailing culture of mediocrity in the civil service and the overloaded Chief Executive, such an arrangement is unlikely to be entertained.

17. For a discussion of partisan politics in the decolonization era, see, for example, DeGolyer (1994), Leung (1997), and Yeung (1997).

18. A statement by Professor S. K. Lau quoted in a news report (C. K. Lau, in *SCMP*, 3 March 1998, p.19).

19. Tung stated in his 1997 policy speech that he had asked the Secretary for the Civil Service to organize a leadership program for senior officials involved in the development and implementation of a target-based management process to

achieve continual improvement in public service (1997a, p.51). The Hong Kong citizenry, however, has not been briefed on the progress of the program.

20. For a discussion of underlying motives of the colonial administration's reforms introduced in the past, see Cheung (1996), and see Lee and Cheung (1995) for an analysis of various issues of administrative reforms.

21. The concern that restrained and undue interference from Chinese officials may "kill Hong Kong" was expressed by even Larry Yung Chi-kin, Chairman of China Investment Trust and Investment Corporation (ITIC) and son of Rong Yiren, then Vice President of the People's Republic. For details, see "Meddling 'Could kill Hong Kong'" (*SCMP,* 15 December 1995).

References

Almond, Gabriel A., and G. B. Powell. 1966. *Comparative Politics: A Developmental Approach.* Boston: Little, Brown and Company.

Balassa, Bela. 1977. *Policy Reform in Developing Countries.* Oxford: Pergamon Press.

Bulter, David, Andrew Adonis, and Tony Travers, 1994, *Failure in British Government: The Politics of the Poll Tax.* Oxford University Press.

Bryant, Coralie, and Louise G. White. 1983. *Managing Development in the Third World.* Boulder, Colorado: Westview Press.

Chan, Cheuk-wah. 1997. "Hong Kong's Economic Path and Its Strategic Value for China and Britain, 1946–56: A Rational-Strategic Approach." *Issues & Studies* 33, no. 6: 88–112.

Cheng, Joseph Y. S. 1997. "China's Policy toward Hong Kong: A Taste of 'One Country, Two Systems.'" *Issues & Studies* 33, no. 8: 1–25.

Cheung, Anthony B. L. 1996. "Efficiency as the Rhetoric: Public-sector Reform in Hong Kong Explained." *International Review of Administrative Sciences* 62, no. 1: 31–47.

Ching, Frank. 1997. "Misreading Hong Kong." *Foreign Affairs* (May/June): 53–66.

Chow, King W. 1991. "Hong Kong Public Administration Under Stress: The Significance and Implications of Management Paradoxes." *International Journal of Public Administration* 14, no. 4: 676–702.

DeGolyer, Michael E. 1994. "Politics, Politicians, and Political Parties," in *The Other Hong Kong Report, 1994.* Edited by Donald H. McMillen and Man Si wai. Hong Kong: the Chinese University of Hong Kong Press, 75–102.

Gould, Derek B. 1992. "Policymaking Style in the Transitional Period in Hong Kong." *Hong Kong Journal of Public Administration* 1 (September): 120–140.

Grindle, Merilee, ed. 1980. *Politics and Policy Implementation in the Third World.* Princeton, N.J.: Princeton University Press.

Heady, Ferrel. 1984. *Public Administration: A Comparative Perspective,* 3d. ed. rev. New York: Marcel Dekker, Inc.

Holm, Hans-Henrik, and George Sorensen, eds. 1995. *Whose World Order? Uneven Globalization and the End of the Cold War.* Boulder, Colorado: Westview Press.

Hope, Kempo R. 1985. "Politics, Bureaucratic Corruption, and Maladministration in the Third World." *International Review of Administrative Sciences* no. 1:1–6.

Huque, Ahmed Shafiqul, and Grace O. M. Lee. 1996. "Public Service Training in a Changing Society: Challenges and Response in Hong Kong." *Teaching Public Administration* 16: 1–16.

Kaufman, Stuart J. 1997. "The Fragmentation and Consolidation of International Systems." *International Organization* 51, no. 2: 173–208.

Kuan, Hsin-chi, and Siu-kai Lau. 1989. "The Civic Self in a Changing Polity: The Case of Hong Kong," in *Hong Kong: The Challenge of Transformation.* Edited by Kathleen Cheek-Milby and Miron Mushkat. Center of Asian Studies, University of Hong Kong.

Lam, Jermain T. M. 1993. "Chris Patten's Constitutional Reform Package: Implications for Hong Kong's Political Transition." *Issues and Studies* (July): 55–74.

Lam, Jermain T. M. 1997. "The Changing Political Culture of Hong Kong's Voters." *Issues & Studies* 33, no. 2: 97–124.

Lau, C. K. 1997. *Hong Kong's Colonial Legacy: A Hong Kong Chinese View of the British Heritage.* Hong Kong: Chinese University of Hong Kong Press.

Lee, Grace O. M. 1997. "The Succession Crisis in Hong Kong's Civil Service." *Issues & Studies* 33, no. 8: 49–62.

Lee, Jane, and Anthony B.L. Cheung, eds. 1995. *Public Sector Reform in Hong Kong: Key Concepts, Progress-to-date and Future Direction.* Hong Kong: Chinese University of Hong Kong Press.

Leung, K. K. 1997. "Fractionalization of the 'Party' System in the Hong Kong Transition," in *Political Order and Power Transition* in *Hong Kong.* Edited by Li Pang-kwong. Hong Kong: Chinese University of Hong Kong Press: 109–26.

Lewis, David and Helen Wallace, eds. 1984. *Policies Into Practice: National and International Case Studies in Implementation.* New York: St. Martin's Press.

Li, Pang-kwong. 1995. "Elections, Politicians, and Electoral Politics," in *The Other Hong Kong Report, 1995.* Edited by Stephen Y.L. Cheung and Stephen M.H. Sze. Hong Kong: the Chinese University of Hong Kong Press: 51–66.

Lindenberg, Marc and Benjamin Crosby. 1984. *Managing Development: The Political Dimension.* London: Transaction Books.

Louie, Kin-sheun. 1996. "Elections and Politics," in *The Other Hong Kong Report, 1996.* Edited by Nyaw Mee-kau and Li Si-ming. Hong Kong: Chinese University of Hong Kong Press: 51–66.

Miller, Benjamin. 1995. *When Opponents Cooperate: Great Power Conflict and Collaboration in World Politics.* Ann Arbor, MI: University of Michigan Press.

Miners, Norman. 1991. *The Government and Politics of Hong Kong,* 5th ed. Oxford: Oxford University Press.

Ming Pao (Daily) (Hong Kong), 7 April 1998, and other issues.

Perlman, Bruce.1989. "Modernizing the Public Service in Latin America: Paradoxes of Latin American Public Administration." *International Journal of Public Administration* 12, no. 4: 671–704.

Pomper, Gerald M. and Marlene M. Pomper, eds. 1985. *The Election of 1984: Reports and Interpretations.* Chatham, N.J.: Chatham House Publishers.

Poole, Marshall S. and Andrew H Van de Ven. 1989. "Using Paradox to Build Management and Organization Theories." *Academy of Management Review* 14, no. 4: 562–78.

Quattrone, George A., and Amos Tversky. 1988. "Contrasting Rational and Psychological Analyses of Political Choice." *American Political Science Review* 82, no. 3: 719–36.

Roy, Denny. 1994. "Hegemon on the Horizon? China's Threat to East Asian Security." *International Security* 19, no. 1: 149–68.

Scott, Ian. 1986. "Policy-Making in a Turbulent Environment: The Case of Hong Kong." *International Review of Administrative Sciences* 55, no. 2.

Scott, Ian. 1989. *Political Change and the Crisis of Legitimacy in Hong Kong.* Oxford: Oxford University Press.

South China Morning Post (SCMP) (Hong Kong), various issues.

Tung, Chee-hwa. 1997a. "A Future of Excellence and Prosperity for All." Speech delivered at the ceremony celebrating the establishment of the Hong Kong Special Administrative Region (SAR), 1 July 1997.

Tung, Chee-hwa. 1997b. "Building Hong Kong For a New Era." Policy Address at the Provisional Legislative Council Meeting, 8 October 1997.

Verba, Sidney. 1996. "The Citizen as Respondent: Sample Surveys and American Democracy." *American Political Science Review* 90, no. 1: 1–7.

Warren, Mark E. 1996. "Deliberative Democracy and Authority." *American Political Science Review* 90, no. 1: 46–60.

Wilding, Paul. 1997. "Social Policy and Social Development in Hong Kong." *The Asian Journal of Public Administration* 19, no. 2: 244–75.

Wong, Timothy K. Y. 1997. "Constraints on Tung Chee-hwa's Power and His Governance of Hong Kong." *Issues & Studies* 33, no. 8: 26–48.

Yeung, Chris k. H., 1997. "Elections, Politicians, and Electoral Politics," in *The Other Hong Kong Report, 1997.* Edited by Joseph Y.S. Cheung. Hong Kong: Chinese University of Hong Kong Press: 49–70.

Yahuda, Michael. 1996. *Hong Kong: China's Challenge.* London: Routledge.

CHAPTER 6

The Hong Kong Press:
A Post–1997 Assessment

Frank Ching

The annual report of the Hong Kong Journalists' Association (HKJA), published in June 1997, just before the British colony reverted to Chinese sovereignty, had this to say:

> All the signs suggest that following the handover, freedom of expression is likely to experience quite severe external and internal pressures. . . . All the indications from China's leaders, and to a good extent from the incoming Special Administrative Region Chief Executive Tung Chee-hwa, point to freedom of expression, and of the press, being restricted in some manner after the handover.[1]

Two years later, the HKJA's 1999 annual report had a rather different tale to tell: "The picture is far different from those bleak predictions [in 1997]," it said. "Freedom of speech has survived in the Hong Kong Special Administrative Region—witness the vigils and other events held in early June to mark the tenth anniversary of the suppression of the 1989 pro-democracy movement in China—and the press remains largely free, vibrant and prepared still to undertake that most difficult of all roles for the fourth estate, the watchdog of the public interest against abuse of executive power."[2]

Still, the journalists' association injected a note of caution. "The ground rules are changing," it said, "and it is these changes which may ultimately have a damaging effect on rights and freedoms in Hong Kong, including that of freedom of expression."[3]

When dire predictions of disaster failed to materialize after Hong Kong's return to China, the prophets of gloom and doom initially sought to explain this by saying that China was deferring its crackdown on Hong Kong until after President Jiang Zemin's state visit to Washington in September 1997, or until after President Bill Clinton's return visit in the summer of 1998.

The trouble with this line of reasoning is that it assumes that China's hands-off policy toward Hong Kong is a result of American pressure, or British pressure, or pressure from some other source. The fact, of course, is that the "one country, two systems" formula, with "Hong Kong people ruling Hong Kong" while enjoying "a high degree of autonomy" for 50 years was China's own idea and was made public by China before negotiations with the British began in September 1982.

Thus, it may come as a surprise to many that more than two years after Hong Kong's reversion to China, freedom of the press continues to thrive, despite prophecies of gloom and doom from both the Western and local media before 1997, most of which continued to look at China through the lens of 1989 events at Tiananmen Square. In the early months after reversion, the press was prepared to believe the worst, and this attitude was reflected in its choice of language. Despite China's hands-off approach, the press frequently attributed actions and decisions to "the China-backed Hong Kong government" and "the China-appointed legislature," leading readers to assume that it was China, not the Hong Kong government or legislature, that was running Hong Kong.

Fortunately, there are signs that things may be changing. The state visit of President Bill Clinton to China in 1998 provided an opportunity for saturation coverage by the U.S. media of all aspects of life in the country, rather than simply the plight of human rights activists. And, in some quarters at least, the U.S. media is becoming aware of its own bias. For example, Al Neuharth, founder of *U.S.A Today*, wrote a column in the paper warning readers that much of what appears on China in the U.S. media "is simply not factual or fair."[4]

And Jonathan Alter wrote a column in *Newsweek* denouncing what he called "China breakers," both in the U.S. Congress and in the press. Such people, he said, "see China as fundamentally unchanged: a gulag of forced labor, constant repression and a small corps of brave dissidents battling a closed society." According to Alter, he himself was such a China breaker. When he went to Hong Kong in mid-1997 for the reversion ceremonies, he wrote, "I went there convinced that Beijing would quickly break its promises and crush the city under its jackboot."

But, he said, "as I interviewed Chinese in Hong Kong—many of whom had once fled the Mainland—my impression began to change. They were unfazed by the handover and actually believed that the leadership would make good on its 'one China, two systems' approach. I left still skeptical, but starting to see the complexity."[5]

And now, he concluded, "it turns out that the optimists were right," although, he quickly adds, "you wouldn't know it from the U.S. press." Few others are willing to publicly repudiate their prejudices. Because Hong Kong is now part of China, the U.S. press in general is reluctant to give Hong Kong the benefit of the doubt, because it would mean giving China the benefit of the doubt.

But, more than two years after China's resumption of sovereignty over Hong Kong, it is becoming clear, even to journalists, that Beijing is honoring its pledge to allow Hong Kong people themselves to administer Hong Kong with a high degree of autonomy under the concept of "one country, two systems." While the last two years have not been free of problems, China's clear policy of noninterference in Hong Kong's domestic affairs stands out and suggests that cautious optimism for the future of press freedom may well be warranted.

The main free-press issue that has continued to arouse controversy after the handover has to do with Radio and Television Hong Kong (RTHK), the government-funded public broadcaster. Previously, it used to be a mouthpiece for the British colonial government but, in more recent years, it has evolved into an independent broadcaster. Before the handover, the British had wanted to corporatize RTHK so that it would no longer formally be an arm of the government. Currently, its head, Cheung Man Yee, is Director of Broadcasting and is appointed by the government. RTHK operates seven radio stations and produces television programs for airing on Hong Kong's two commercial terrestrial stations.

Chinese officials argued against corporatizing RTHK before 1997, saying that the post-1997 government would need its own mouthpiece. After the handover, however, China took no overt action, leaving decisions on RTHK up to the Tung administration.

In the early months of the new administration, the emphasis was on the preservation of political stability, so both Chinese and Hong Kong officials emphasized continuity rather than change. A Chinese spokesman, Wu Hongbo, soothed foreign correspondents by announcing at a reception held at the Foreign Ministry's Commission in Hong Kong that Beijing's system of accreditation of foreign journalists would not apply in Hong Kong.[6] Similarly, Hong Kong officials gave assurances that there was no change in media policy. Specific assurances were given that RTHK would remain independent.

However, in spring 1998, when the National People's Congress was holding its annual session in Beijing, a Hong Kong member of China's top advisory body, the Chinese People's Political Consultative Conference (CPPCC), Xu Simin, created a storm when he attacked RTHK. During a CPPCC session, he attached RTHK for producing and airing programs highly critical of China and of the Hong Kong government. He lamented the fact that Chief Executive Tung was evidently powerless to rectify the situation.

Mr. Xu's aim apparently was to mobilize support for his position, but the move backfired. President Jiang Zemin warned Hong Kong members of the NPC that "local deputies to the congress should not interfere" in the affairs of the Hong Kong government. Li Ruihuan, head of the CPPCC, made similar remarks, adding that CPPCC members had no authority over the Hong Kong government.

Chief Executive Tung, who was in Beijing at the time, was a little more ambivalent. On one hand, he said, freedom of the press had to be upheld but, on the other hand, the government's policies ought to be presented in a positive light.

Such ambivalence was notably absent on the part of his deputy, Chief Secretary for Administration Anson Chan. In Hong Kong, she issued a statement unambiguously upholding RTHK's editorial independence. She said that RTHK was in any event Hong Kong's domestic affair and ought not to be discussed in Beijing. A chorus of support for RTHK's editorial independence ensued, even from members of the legislature thought to be pro-China. As a result, the Legislative Council passed a motion overwhelmingly endorsing RTHK's editorial independence.

The issue of RTHK resurfaced in the summer of 1999, after President Lee Teng-hui stirred up a hornet's nest by declaring that relations between Taiwan and Mainland China were a form of state-to-state relations. Naturally, the issue provoked discussion within the Hong Kong media. RTHK, as part of its attempt to air various points of view, invited Cheng An-kuo, Taiwan's top unofficial representative in Hong Kong, whose formal title is managing director of the Chung Hwa Travel Service, to discuss the issue. Not surprisingly, he endorsed Lee's approach.

Again, pro-China figures excoriated RTHK, saying that it should not have provided a platform for someone who represented the Taiwan government. A local NPC member, Tsang Hin-chi, urged the government-owned station to exercise self-censorship and not to provide a platform for views espousing the splitting up of China.[7] Xu Simin, the CPPCC member, who is also a magazine publisher, said that RTHK, as an official radio station, should not allow remarks expressing support for Taiwan's president to be broadcast.

On its part, RTHK explained that it was merely functioning as any other independent media outlet would, providing news, opposing viewpoints, as well as commentary on an issue of great importance.

The matter was escalated on 19 August when Chinese Vice Premier Qian Qichen responded to questions from Hong Kong reporters before meeting Secretary for Justice Elsie Leung. Qian said open support for Lee Teng-hui's call for relations between Taiwan and China to be on a state-to-state basis was a violation of China's guidelines on Hong Kong-Taiwan relations.[8] Asked whether such calls could be promoted in Hong Kong, the Chinese official said: "One shouldn't do so. Under the one-China principle, one can't promote such matters in Hong Kong." However, he did not mention RTHK, or Cheng An-kuo.[9]

Although Qian did not specifically mention the media, his remarks drew a response from the head of the Hong Kong Journalists Association, Mak Yim-ting, who said the media had the responsibility to report the facts and "we do not regard it as promotion." Another media spokesman, Ronald Chiu, of the News Executives Association, feared the Qian comments were a sign of censorship. "In Hong Kong, we are used to saying what we want to say," he declared. "The media only provides a forum for views. It would be very dangerous if we are told what we should report and what we shouldn't." RTHK, too, issued a statement, asserting that everyone had the right to express views and comment on the views of others within the law. "RTHK and the Hong Kong media have adhered to this principle and fully uphold freedom of expression and the press," it said in its statement.

Stephen Lam, the Information Coordinator, responded to the Qian remarks by criticizing Cheng An-kuo while not mentioning RTHK. Alluding to Mr. Cheng's formal title as managing director of the Chung Hwa Travel Service, Lam said: "As the head of a travel agency in Hong Kong, it is more appropriate for him to focus on economics, tourism and associated matters. It is not appropriate for him to get involved in political issues."[9] Lam, of course, knew full well that the travel service was in reality Taiwan's quasi-official presence in Hong Kong, playing a role similar to that the Xinhua News Agency did for China. The RTHK issue is likely to continue to be debated intermittently, but public sentiment clearly is on the side of maintaining RTHK's editorial independence. But China's unhappiness with RTHK had again been made evident in Ian's comments.

From the beginning, China had guaranteed the preservation of Hong Kong's rights and freedoms. In the Sino-British Joint Declaration on the Question of Hong Kong, initialed in September 1984 and signed in December the same year, China agreed that Hong Kong residents' rights and freedoms would be protected by law after the changeover. It said:

"The Hong Kong Special Administrative Region Government shall maintain the rights and freedoms as provided for by the laws previously in force in Hong Kong, including freedom of the person, of speech, of the press, of assembly, of association, to form and join trade unions, of correspondence, of travel, of movement, of strike, of demonstrations, of choice of occupation, of academic research, of belief, inviolability of the home, the freedom to marry and the right to raise a family freely."

This long list of rights and freedoms was included to assure the Hong Kong people that they would be treated differently from people in the rest of China. Specifically, while families in China had to abide by the one-child policy, couples in Hong Kong could "raise a family freely" and decide for themselves when to have children, and how many to have.

In addition to this long list of rights and freedoms, the Joint Declaration also asserted that "the provisions of the International Covenant on Civil and Political Rights and the International Covenant on Economic, Social and Cultural Rights as applied to Hong Kong shall remain in force."

These two covenants are the main United Nations instruments for safeguarding the whole panoply of human rights, ranging from civil and political rights to economic, social and cultural rights. The two covenants stem from the Universal Declaration of Human Rights and represent a codification of those rights. The phrase "as applied to Hong Kong" reflects the fact that Britain, in applying those covenants to Hong Kong, chose to make certain reservations so that certain provisions would not apply, such as, for example, the right to self-determination. By including the phrase "as applied to Hong Kong," China was indicating that it would continue those British reservations.

The preservation of rights and freedoms set out in the Joint Declaration was to a large extent reiterated in the Basic Law, Hong Kong's mini-constitution, which was adopted by China's National People's Congress in April 1990.

In Article 27 of Chapter III, titled "Fundamental Rights and Duties of the Residents," the Basic Law says: "Hong Kong residents shall have freedom of speech, of the press and of publication; freedom of association, of assembly, of procession and of demonstration, and the right and freedom to form and join trade unions and to strike."

Other rights enumerated in the Joint Declaration are reflected in other articles of the same chapter. In fact, while the Joint Declaration referred to "freedom of the press," the Basic Law expands upon this concept and talks about freedom "of the press and of publication." Thus, if anything, the rights and freedoms enumerated in the Basic Law were even more comprehensive than those listed in the Joint Declaration.

If words on paper can convince, then people both in and outside Hong Kong could have set their minds at ease in the knowledge that a free press would, indeed, continue in Hong Kong after July 1, 1997.

But Hong Kong residents and their well-wishers abroad were still not satisfied. They pointed to the fact that Chapter II of the Chinese constitution, headed "The Fundamental Rights and Duties of Citizens," contained many of the same promises of rights and freedoms, and these clearly were not enjoyed by the residents of Mainland China. Thus, Article 35 of Chapter II in the constitution declares: "Citizens of the People's Republic of China enjoy freedom of speech, of the press, of assembly, of association, of procession and of demonstration."

Yet, clearly, the press was under the control of the Communist Party, and is not free. Moreover, despite paying lip service to freedom of association, citizens of the People's Republic are not allowed to form independent trade unions. The only legal unions are the ones controlled by the government and party. Moreover, demonstrations—although legal in theory—have not been allowed to be staged in practice, usually on the grounds that traffic would be disrupted or that they would result in disorder and chaos.

So it was not surprising that there were many skeptics who felt China's promises were not worth the paper they were printed on. This was particularly the case after the Tiananmen Square military crackdown of June 1989, when unarmed, peaceful, civilian demonstrators were confronted by tanks and foot-soldiers of the People's Liberation Army, who were armed to the teeth.

The Basic Law, in fact, was a product of that atmosphere, since it was finalized less than a year after the 4 June incident. And China, fearful that Hong Kong would become a base for subversion against the Central Government in Beijing, sought to tighten its control of Hong Kong by strengthening Article 23 of the Basic Law, which says:

> The Hong Kong Special Administrative Region shall enact laws on its own to prohibit any act of treason, secession, sedition, subversion against the Central People's Government, or theft of state secrets, to prohibit foreign political organizations or bodies from conducting political activities in the Region, and to prohibit political organizations or bodies of the Region from establishing ties with foreign political organizations or bodies.

Since the 1980s, the United Nations had been involved in monitoring the human rights situation in Hong Kong. The two major human rights organizations both call for the submission of periodic reports, and the holding of hearings into those reports. One major concern in the years leading up to 1997 was China's assertion that, after 1997, it would not continue the British practice of submitting human rights reports on Hong Kong to the United Nations, on the ground that China itself was not a signatory of the two covenants. This was a major issue between Britain and China and was of great concern to human rights organizations in Hong Kong.

There was great relief, therefore, when a few months after the handover, China disclosed that, contrary to its previous position, it would continue the practice of submitting reports on the human rights situation in Hong Kong to the United Nations.

Hong Kong's first post-1997 human rights report was compiled by the Special Administrative Region government and submitted to the United

Nations by China in January 1999. The Human Rights Committee scheduled hearings on the Hong Kong report for October 1999.

In the report's section on freedom of opinion and expression, the Hong Kong government declared: "The government is committed to maintaining a free press. Its policy is to maintain an environment in which a free and active press can operate under minimum regulation—regulation that does not fetter freedom of expression or editorial independence."[10]

Although critics of the Hong Kong government, such as the Hong Kong Journalists Association, acknowledge that there hasn't been a crackdown on the press, they assert that the post-1997 government is less open than its British colonial predecessor.

Much of this may well be due to the personality of Chief Executive Tung Chee-hwa. Unlike the last British governor, Chris Patten, a seasoned politician who thrived on exposure to the media, Tung was a businessman who had always opted for a low profile. He was certainly less savvy than Patten on the sensitivities of the press and on how the media can be an instrument for his political agenda.

Patten created a new position, called the Information Coordinator, an official who was in effect his personal spokesman and adviser on media issues. Instructively, the post disappeared when Tung assumed office on 1 July 1997. And while Patten's Information Coordinator used to call up the local correspondents of the *New York Times* and the *Washington Post* to offer interviews with Patten, Tung was decidedly less comfortable in dealing with the media.

When the new Chek Lap Kok airport opened in July 1998, accompanied by widespread chaos, Chief Executive Tung was advised to go to the airport and project the image that the government was on top of the situation. However, he refused, on the ground that relevant officials were already working on the problems and that he himself would only get in the way.

A change to this attitude became evident a month or two later when it was announced that the post of Information Coordinator would be revived and, in fact, would be upgraded to D8—the same level as a policy secretary. It was certainly an acknowledgement by the Chief Executive that he needed to improve his image and to improve communications with the media. The man picked to fill the job was Stephen Lam, who had helped coordinate the handover ceremonies in 1997. Since October 1998, the new Information Coordinator has held regular press briefings and has attempted to help project a more media-friendly image for the government.

Although the Chinese and Hong Kong authorities have not taken action against the media, the Hong Kong newspaper scene has changed considerably in the last few years. In fact, within weeks of the handover, an afternoon paper, the *New Evening Post,* folded. Ironically, this was a pro-Communist paper, part of the Ta Kung Pao family. It was the only afternoon paper left in Hong Kong and folded because it had been losing money for years. Its closure was due to economic reasons, not political ones.

Moreover, *The Nineties* magazine, probably the leading magazine for Chinese intellectuals in Hong Kong and overseas, announced in April 1998 that it was ceasing publication. The magazine, which began life in 1970 as *The Seventies,* also did not succumb to political pressure. What killed the magazine was simply exhaustion. After 18 years, its founder, Lee Yee, decided that the magazine had fulfilled its purpose and, with the dawning of a new era in Hong Kong, in China, and in the world, there was no longer any need for it.

Other China-watching magazines, such as *Contemporary* and *Pai Hsing,* had gone out of business in earlier years, also for economic reasons. With the popularization of television, there is less interest in the print media. Moreover, with China opening up, and with more information available in the general media, there is less that the specialized China-watching magazines can provide.

These developments reflect the fact that the greatest threat to the Hong Kong media today comes not from governments—either in Beijing or in Hong Kong—but from market forces. The competition for market share and for advertisers has resulted in ever greater emphasis on sex and sensationalism in the mainstream newspapers. Two newspaper groups, the *Oriental Daily News* and the *Apple Daily,* now account for about 70 percent of the market, while a dozen other newspapers compete for the remaining 30 percent.

The reporting on the antics of one man, Chan Kin-hong, in the fall of 1998 illustrates the behavior of the Hong Kong media. *Apple Daily* ran a picture on its front page of the man in bed with his arms around two prostitutes, days after his wife committed suicide after killing their two children. *Oriental Daily,* too, carried saturation coverage on the 41-year-old man, calling him "human scum." Both terrestrial television networks joined in, with cameramen following Chan everywhere he went. Subsequently, it was disclosed that Chan had been paid by at least one newspaper organization for his cooperation. On November 10, 1998, *Apple Daily* covered its front page with an apology by its owner, Jimmy Lai, for the way the paper had handled the story.

The following week, full-page ads—sponsored by the Hong Kong Journalists Association, the Federation of Hong Kong Journalists, the News Executives Association and the Hong Kong Press Photographers Association—appeared in seven Chinese papers, lamenting the handling of the Chan Kin-hong case and affirming their determination to uphold professional journalistic standards.

It was only the second time the four professional associations had collaborated; the first was a few months earlier when they banded together, along with the Foreign Correspondents Club, to denounce the knife attack on popular commentator Albert Cheng. The collaboration was also striking because the HKJA is critical of China while the federation is openly pro-China.

The situation's gravity was underlined when the HKJA disclosed that 77 percent of its members believed journalistic ethics were worse or much worse than a year ago. Sensational or disgusting photographs were deemed the most serious problem, while "too much sex" and "exaggerated reports" were also of concern.

As a result of the trend towards sensationalism, there has been a drop in the media's public credibility. A survey conducted by the University of Hong Kong in January 1999 showed that 41 percent of respondents thought the media was irresponsible in its reporting, compared with about 29 percent in September 1998.[11]

Looking back, it is understandable why the Hong Kong Journalists Association, and many other people in Hong Kong and abroad, expected a crackdown by the communist authorities after the handover. A number of senior Chinese officials in the run-up to the 1997 handover repeatedly made assertions that, on the surface, suggested a determination to water down press freedom in Hong Kong. Thus, in mid-1996, Lu Ping, then Director of the Hong Kong and Macau Affairs Office under the State Council, said on more than one occasion that, after 1997, the press would not be allowed to advocate "two Chinas," "one China, one Taiwan," "the independence of Taiwan," or "the independence of Hong Kong." This, he asserted, had nothing to do with freedom of the press. Journalists, the Chinese official said, would be free to report the news, but they would not be allowed to advocate the fragmentation of China.[12]

In all likelihood, Mr. Lu was thinking of a new law regarding secession, which would presumably not permit any part of China to declare independence, nor would it allow anyone to advocate such independence. Such a law is mandated by Article 23 of the Basic Law.

Aside from Lu Ping, remarks by another, more senior, Chinese official, Vice Premier Qian Qichen, were also profoundly disturbing to the Hong Kong media. In an interview with the *Asian Wall Street Journal* in October 1996, Mr. Qian warned that the Hong Kong media "can put forward criticisms, but not rumors or lies. Nor can they put forward personal attacks on the Chinese leaders."[13]

Remarks such as these led many people to believe that China would be intolerant of the Hong Kong media. While it may sound reasonable to say that the media should not publish lies or rumors, it is often difficult for journalists to verify every fact in every article. In fact, Richard Boucher, then U.S. Consul General in Hong Kong, asserted that it is important for newspapers to publish rumors. Rumors, he said, can only be denied if they are published; otherwise they simply continue to circulate.[14]

Reading Chinese statements made in 1996 and 1997, it seems clear that there was a gradual softening in policy toward Hong Kong. In fact, in June 1997, weeks before the handover, Vice Premier Qian, in another interview, this time with the *South China Morning Post*, virtually retracted everything he had said the previous year. In this interview, he explained that, previously, "I only expressed my personal view," which was not binding on the people of Hong Kong. "The laws of Hong Kong guarantee that Hong Kong will enjoy freedom of the press," he added.[15]

As of this writing, in August 1999, the Hong Kong administration has not yet moved to implement Article 23 of the Basic Law, with its call for legislation against subversion and other political crimes. It is unclear when the Hong Kong administration intends to introduce such legislation, with officials implying that there is little urgency. Such an attitude is to be welcomed, and must reflect a similar attitude in Beijing.

It should be borne in mind that China's policy toward Hong Kong is dictated by self-interest. Beijing recognizes the value of Hong Kong to its goal of economic development. In fact, China could have taken over Hong Kong at any time since 1949, when the People's Republic was proclaimed. It simply chose not to do so. And the reason was its recognition of Hong Kong's value to China.

As for the non-observance of the Chinese Constitution, one must remember that the situation in Hong Kong is quite different from that in China proper. In Hong Kong, for example, all newspapers are privately owned; the government does not own or operate any newspapers. In China, all newspapers are ultimately controlled by the Chinese Communist Party.

Besides, China clearly did not intend Hong Kong to be run like the rest of China. If it did, it would not have spent two years negotiating the Joint Declaration with Britain, five years negotiating and drafting the Basic Law, and many more years negotiating other agreements with Britain, primarily in the Sino-British Joint Liaison Group, the transitional body created by the Joint Declaration.

If China had simply wanted to absorb Hong Kong, it did not need to enunciate the concept of "one country, two systems" with "Hong Kong people running Hong Kong" while enjoying a "high degree of autonomy." It did not need to announce that the Hong Kong lifestyle would remain unchanged for 50 years and that Beijing would not send officials to run Hong Kong, or that foreigners can serve on the Hong Kong legislature and continue to serve in the Hong Kong government, the Hong Kong judiciary, and the Hong Kong police force, or that the border between Hong Kong and China would remain, or that Hong Kong would retain its own currency, or that the currency would remain convertible, or that Hong Kong would remain a member of such international organizations as the World Trade Organization, the Asia Pacific Economic Cooperation forum, and the Asian Development Bank.

Using the situation in China to predict what the situation in Hong Kong would be like in the years after 1997 might have seemed like a straightforward exercise, but it was a dangerous one. It disregarded all the commitments China made, both at the United Nations and in other international fora, commitments that China did not have to make, but that were totally voluntary.

Moreover, a look at the Hong Kong policy of the People's Republic of China since its establishment in 1949 is also instructive. In 1949, when troops of the People's Liberation Army swept southward, they could easily have overrun Hong Kong. But they stopped at the border on instructions from the party's leaders, precisely because the new Chinese Government did not want to upset the status quo.

Similarly, during the tumult of the Cultural Revolution, when Red Guards in Beijing burned down the British legation and demanded the ouster of the British colonial government from Hong Kong, the central authorities in Beijing stopped them from marching across the border into Hong Kong.

Clearly, China's policy was one of maintaining the status quo, with the British colonial government in charge. In fact, so important was it to China to keep Hong Kong under British administration that it submitted to taunts by Moscow that the Chinese communists were tolerating colonialism on Chinese territory.

China had good reasons for not wanting to disrupt Hong Kong. For one thing, Hong Kong functioned as China's window on the outside world. From 1949 until 1971, when Beijing entered the United Nations, China was isolated from the Western world—a situation that was exacerbated by China's own policy of first allying itself with the Soviet Union and, after the Sino-Soviet split, of adopting a policy of self-reliance. Hong Kong was China's main channel to the rest of the world. Even during the Korean war, when British Hong Kong joined the United Nations embargo against China, Hong Kong was still valuable to China because smugglers in Hong Kong managed to ameliorate the effects of the embargo.

What little trade China had with the non-communist world was conducted through Hong Kong. In the aftermath of the death of Chairman Mao Zedong in 1976 and the subsequent rise to power of Deng Xiaoping, China adopted a new policy of giving priority to economic development, which made Hong Kong's role even more important. Something like a third of China's foreign-exchange earnings passed through Hong Kong. And when the country opened up for foreign investment, the vast majority of investment in China came from Hong Kong. More recently, with China seeking to raise capital abroad, it has again turned to Hong Kong. What this suggests is that Hong Kong has been of great benefit to China, and that it is in China's self-interest to keep Hong Kong pretty much as it is. However, it is also clear that China has always considered Hong Kong part of its territory and, sooner or later, China would have recovered Hong Kong.

The Chinese Government's position was that the three treaties under which Britain obtained various parts of Hong Kong in the nineteenth century had been imposed on a weak Qing dynasty through gunboat diplomacy and hence were "unequal treaties" whose validity it did not recognize. From China's standpoint, the expiration of one of these treaties on 1 July 1997 had no legal significance, since it considered the treaty itself to be null and void.

But from the British standpoint, these three treaties gave them the legal right to administer Hong Kong. The expiration of the lease would mean the loss of Britain's legal right to administer 92 percent of its colony. Hence the British, in 1979, broached the issue with the Chinese. The British hoped that China would agree to sign a new treaty formally extending British rule in Hong Kong beyond 1997. While China was prepared to turn a blind eye to the fact that part of its territory was under British administration as a legacy of history, it was not prepared to formally sign a new treaty extending colonial rule in China into the twenty-first century.

China felt it had no choice but to take Hong Kong back. Still, aware of the great value of Hong Kong to China's economic development, the Chinese decided on a policy that, they thought, would allow Hong Kong to remain as it was after the departure of the British. That was why they announced that Hong Kong would be run not by Beijing officials but by Hong Kong people themselves, and that the systems in Hong Kong would be different from those in China. Specifically, China declared that while the Mainland would practice socialism, Hong Kong would continue to remain capitalistic.

This concept of setting up Special Administrative Regions under the policy of "one country, two systems" was originally developed by China with Taiwan in mind. China's policy since 1949 was to resolve the Taiwan issue first before tackling the Hong Kong issue, in the belief that Hong Kong can be helpful in resolving the Taiwan issue. But when Britain insisted on a resolution of the 1997 issue, China decided to reverse its order of priorities, and to use Hong Kong as a model for Taiwan's future.

It is clear, then, that China's own self-interest dictates that it should handle Hong Kong delicately. Any damage to Hong Kong would be reflected in a slowing down of China's economic growth. Moreover, if China should ride roughshod over Hong Kong, it would no doubt cause Taiwan to lose whatever interest it has in reunification with China by becoming, like Hong Kong, a special administrative region.

There is yet another factor: the West. The countries of the West, in particular the United States, are important to China if it is to be successful in its attempt to rapidly develop its economy. China's hope is that, by the middle of the twenty-first century, it would have succeeded in providing its people with a standard of living appropriate to that of a moderately developed country that is on its way to becoming a developed economy.

In order to achieve such a goal, China needs peace and stability in the Asia-Pacific region. It also needs the wealthy countries of the West to continue to invest their capital in China, to transfer their technology to China, and to open their markets to Chinese products. China knows that a major deterioration in relations with the West could dramatically set back its plans for economic development. And the West, China knows, is looking at Hong Kong as a litmus test of China's intentions, and of China's willingness to honor its commitments. China therefore has good reason to abide by commitments it made regarding the preservation of Hong Kong's rights and freedoms. This includes the right to a free press.

On the other hand, however, there are still valid reasons for concern. Communist governments are known for wanting to keep a monopoly on power. They are not known for willingness to share power, or for allowing regions within the country autonomy to the extent envisaged by the Joint Declaration and the Basic Law. For this reason, it has been said that China's promise not to interfere in Hong Kong's internal affairs is similar to a left-handed person promising to only use his right hand. The promise may very well be sincere but, in the absence of restraints, the left-handed person will sooner or later forget and, without even realizing it, start using his left hand.

Any discussion of freedom of the press in Hong Kong must also include the issue of self-censorship. There is a general perception in Hong Kong that the media has, in the years leading up to 1997, been engaged in self-censorship.

Self-censorship on the part of individual journalists can be accounted for by one thing: fear. There has been a general belief that the Chinese authorities keep dossiers on journalists, and from this arose a fear that what a journalist writes may lead to reprisals in future. But, now that 26 months have gone by with no hint of any attempt at reprisals by China, such fear has diminished. In fact, this form of self-censorship has probably decreased.

But the main problem doesn't lie in individual self-censorship. The main problem lies with the owners of media corporations, who see their publications as businesses that should maximize their profits, rather than as vehicles for upholding noble principles like freedom of the press. Some of these owners are also involved in other businesses that may have China connections.

Media owners are vulnerable to economic pressure even if they have no ambition to do business in China. Even within Hong Kong, Beijing is in a position to wield economic pressure. This is because companies from the People's Republic of China now have considerable economic muscle and, by placing or withholding advertisements, they are in a position to make clear China's pleasure—or displeasure—with any particular publication. Even now it is possible, simply by leafing through a newspaper, to discover if it is favored by China by counting the number of ads placed by Chinese companies.

In conclusion, it seems safe to say that the parameters of a free press may shift somewhat after laws against subversion, secession, sedition, and treason are passed. For example, newspapers today have the legal right to advocate the independence of Hong Kong or Taiwan, but none have chosen to exercise this right. It is conceivable that, in future, they will lose this right. But the loss of a right that was previously never exercised will not, of itself, spell the end to press freedom.

China's fears of subversion emanating from Hong Kong is also likely to have subsided somewhat since the handover, as the territory has remained stable despite the continuation of anti-China protests. Even the annual 4 June candlelight vigil, which is extremely provocative to China, has been allowed to be held.

To conclude, I believe that Hong Kong papers are likely to continue to function pretty much as they have in the past. It is extremely unlikely that Beijing wants every Hong Kong newspaper to read like a Communist party mouthpiece. After all, the Chinese Government wants Hong Kong to continue to function as a sophisticated, cosmopolitan city and, to maintain this status, Hong Kong needs to have a range of publications, with a range of editorial viewpoints. Chinese officials badly want Hong Kong to continue to be seen by the rest of the world as having a free press. The Chinese also wish Hong Kong to preserve an international image. The continued presence of publications such as the *International Herald Tribune, Time, Newsweek,* and the *Far Eastern Economic Review* would certainly enhance that image.

So, while one cannot rule out the possibility of political or economic pressure on a particular publication from time to time (as is known to happen in other countries as well) this would not necessarily toll the death knell for press freedom. Economic pressure, too, can be resisted as long as there continues to be a multiplicity of economic players and Chinese-owned companies do not dominate the scene. But, in the last analysis, the most important guarantee of a free press in Hong Kong is Beijing's perception of Hong Kong's importance to China in both political and economic terms.

Moreover, one should not forget developments in the Mainland itself. The fifteenth Party Congress, held in late 1997, showed the country firmly continuing on the road to economic reform, which in the future is likely to lead to greater political openness. With China itself becoming more and more open, the prospects for a free press in Hong Kong seem reasonably bright.

Notes

1. The publication was titled: "The Die Is Cast: Freedom of Expression in Hong Kong on the eve of the handover to China," *HKJA June, 1997 Annual Report.*
2. The 1999 report was titled: "The Ground Rules Change: Freedom of Expression in Hong Kong two years after the handover to China," *HKJA June 1999 Annual Report.*
3. *HKJA 1999 Annual Report,* p.3.
4. *U.S.A Today,* 26 June 1998: "You Can't Believe All You Read About China."
5. *Newsweek,* 6 July 1998, p.31: "Don't Break the China."
6. 11 September 1997 reception, attended by author.
7. "RTHK warned over 'splittist' views," *South China Morning Post,* 11 August 1999.
8. For account of guidelines, see "China tells of rules for Taiwan ties to Hong Kong," *South China Morning Post,* 23 June 1995.
9. *South China Morning Post,* 20 August 1999, provided a detailed front-page account of the incident.
10. "Report of the Hong Kong Special Administrative Region of the People's Republic of China in the light of the International Covenant on Civil and Political Rights," December 1998, p.125.
11. *HKJA 1999 Annual Report,* p.22.
12. "Freedom of the Press at Risk," *Far Eastern Economic Review,* 27 June 1996.
13. "Qian tells Hong Kong to avoid political attacks against China," *Asian Wall Street Journal,* 16 October 1996.
14. "U.S. Warns Against Curbing Hong Kong's Media," *Asian Wall Street Journal,* 11 December 1996.
15. "Looking to the Future with Optimism," *South China Morning Post,* 18 June 1997.

PART II

The External Scene

CHAPTER 7

The Paradox of Hong Kong as a Non-Sovereign International Actor: An Update[1]

James C. Hsiung

P re-handover forecasts for the future of Hong Kong in the international domain were, as a whole, quite different from the prophecies of doom regarding the territory's domestic politics upon exiting from colonial rule in mid-1997. Many pundits, including international law experts, had expected that the post-handover Hong Kong would continue to enjoy, as before, a high international profile, endowed as it was with a legal capacity and political versatility to function as an international actor, despite its lack of sovereignty.[2]

Few had doubts about the territory's international capacity to act, as this was guaranteed by the devolution agreement embodied in the 1984 Sino-U.K. Declaration on Hong Kong's return to Chinese sovereignty and, in addition, by the SAR's Basic Law. Furthermore, one could point to a body of well-established precedents (international custom), as evidence of general practice accepted as binding in general international law.[3] On a large scale, Hong Kong's international eligibility to act may encompass: (a) its legal competence to espouse causes and make claims, both in its own right and on behalf of its residents and juridical persons, on the international plane; (b) its function as a "reversible window" for China; (c) a catalyst in Taiwan-Mainland relations ; and (d) its role as a factor in U.S.-China relations (Hsiung 1997).

However, as we shall see in this chapter, things have, likewise paradoxically, not turned out quite as expected—here, in contrast to the domestic scene, the outcomes were not as well as had been expected. Among the various possible reasons for the discrepancy between the earlier higher expectations and the subsequent worse-off outcomes, in the external dimension, was a bizarre confusion apparently bedeviling foreign judicial organs. This will become obvious from the few cases that will come under our scrutiny herewithin, as the foreign judicial organs concerned wrestled with the question of the precise legal status of the Hong Kong SAR. In an extreme case, as we shall see below, the confusion even led (misled) two U.S. federal courts to rule Hong Kong as "stateless." Consequently, a Hong Kong incorporated company (Matimak Trading), was likewise tainted with the "stateless" stigma and, as such, it lacked the capacity to sue and seek relief in the U.S. federal court system. Similar discrepancies between earlier confident forecasts and unexpected below-par turnouts of events were found in the other aspects named above (b through d) in regard to the SAR's eligibility to act internationally. If the paradox on the domestic scene is an auspicious one—in the sense that the outcome exceeded the expectations—then the paradox we find in respect of Hong Kong's external relations is an inauspicious one, for the opposite reason.

This chapter focuses on the inchoate evidence of how the HKSAR has met with circumstances or treatment by foreign countries at variance with the expectations of many pundits regarding the territory's international status and capacity to act. For an easier flow of discussion, we shall first look at some judicial cases that will demonstrate how Hong Kong's international capacity to act has been given a short shrift by foreign tribunals. Following that, we shall extend our inquiry to non-legal cases and issues.

Foreign Judicial Test

Let me preface this discussion by noting that, while the HKSAR's international capacity is defined and provided for by the 1984 Sino-U.K. Joint Declaration, the SAR's Basic Law, and general international law (especially the norms on state succession), the exact nature of this capacity has to be tested in concrete cases. Although some of the cases presented here were decided shortly before the 1 July 1997 handover date—in one case only three days before—we have reasons to believe that these decisions have set the frame of reference for future judicial deliberations, *ceteris paribus*.

Matimak Trading Co. v. Albert Khalily & D.A.Y. Kids Sportswear, Inc.

The first, and most revealing, case to be examined in this connection is *Matimak Trading Co. v. Albert Khalily & D.A.Y. Kids Sportswear, Inc.* (1997). This was a case involving non-payment by the defendant totaling U.S.$80,000 for goods delivered, in which Matimak Trading, a Hong Kong corporation, sought relief by suing in U.S. federal courts for breach of contract by D.A.Y. Kids Sportswear, Inc.

Plaintiff Matimak invoked the court's diversity jurisdiction under 28 U.S.C. Section 1332 (a)(2), which provides jurisdiction over any civil action arising between "citizens of a State [in the Union] and citizens or subjects of a foreign state." In June 1996, the District Court, for the Southern District of New York, raised the issue of its own subject matter jurisdiction. And in August 1996, after allowing the parties to brief the issue, the district court dismissed the Complaint for lack of subject matter jurisdiction. The court concluded that Hong Kong was not a "foreign state" under the diversity statute and, consequently, Matimak was not a "citizen or subject" of a "foreign state." Upon dismissal by the District Court, however, the case was appealed to the U.S. Court of Appeals for the Second Circuit (118 F.3d 76; 1997 U.S. App.LEXIS 15889).[4]

At issue was whether a Hong Kong corporation was either a "citizen or subject" of a "foreign state" for purposes of alienage jurisdiction. Matimak Trading Co. was incorporated under Hong Kong law—to be exact, under the Companies Ordinance 1984—and was entitled to the protection of Hong Kong. More precisely, the question of whether Matimak, a Hong Kong corporation, had the right to sue in the U.S. federal court system, turned on the precise legal status of Hong Kong.

In the *stare decisis* of U.S. jurisprudence, a precedent was *Cedec Trading Ltd. v. United American Coal Sales, Inc.* (556 F.Supp.722–724, & n.2 [S.D.N.Y., 1983]). In that case, the Court held that corporations of the Channel Islands, a province which was part of the United Kingdom proper, governed by British Law, and whose foreign affairs were entirely controlled by the United Kingdom, were citizens or subjects of the United Kingdom. Hence, it had the right to sue in U.S. federal courts.

Before arriving at its decision in the Matimak case, the Appeals Court, in analyzing the background of the lower-court decision under review, discussed a number of basic issues in trying to navigate what it called a "shoal-strewn area of the law." The United States Judicial Code, the Appeals Court began, tracks the constitutional language used in Art. III, See. 2, Cl. 1, of the United States Constitution—which extends the federal judicial power to "all cases . . . between a State [of the Union], or citizens thereof, and foreign states, Citizens or Subjects"—by providing diversity jurisdiction over any civil action arising between "citizens of a State and citizens or subjects of a foreign State" (28 U.S.C. Sec. 1332(a)(2)). This judicial power is referred to as "alienage jurisdiction."

Neither the U.S. Constitution nor Sec. 1332(a)(2) of the United States Code defines "foreign state." In relevant cases decided by U.S. courts, it was generally held that a foreign state was one formally recognized by the executive branch of the U.S. Government (C. Wright, A. Miler, and E. Cooper, Federal Practice & Procedure Sec. 3604 (1984)).

While both plaintiff and defendant in the present case agreed that the United States had not formally recognized Hong Kong as a foreign state, Matimak, however, contended that Hong Kong had received "de facto" recognition as a foreign state by the United States. It pointed to the latter's diplomatic and commercial ties with Hong Kong as evidence of this de facto recognition.

After reiterating the habitual judicial deference to the Executive Branch on matters of recognition of foreign states (e.g., *Iran Handicraft; Abu-Zeineh v. Federal Labs, Inc., Bank of Hawaii v. Balos,* etc.), the Appeals Court then turned to the U.S. Hong Kong Act, which Congress enacted in 1992, to govern U.S. relations with Hong Kong beyond the end of British colonial rule on 1 July 1997. While expressing a strong U.S. "interest in the continued vitality, prosperity, and stability of Hong Kong" beyond its reversion to become a special administrative region of China, the Act "makes equally clear, however, that the United States did not regard Hong Kong as an independent, sovereign political entity," the Court concluded.

The parties also quarreled over the significance of the British Nationality Act of 1981, which delineates British citizenship in detail. This Act, the Court declared, "fails to support Matimak's assertion that a Hong Kong corporation is a citizen of the United Kingdom." The Act, it added, "applies only to natural persons, not corporations." Citing Matimak as "stateless," as it was not a "citizen or subject of a foreign state," the Appeals Court affirmed the District Court's dismissal of the suit for lack of subject matter jurisdiction.

The decision by the U.S. Court of Appeals for the Second Circuit, handed down on 27 June 1997, or three days before Hong Kong was to revert to Chinese sovereignty, thus definitively concluded that no Hong Kong corporation had the right to sue in a U.S. federal court. The question arises as to whether the territory's reversion makes any difference. The answer, it appears to me, depends on whether the *Cedec Trading Ltd.* case, cited earlier, can find a parallel situation in a case involving a Hong Kong incorporated company.

The key link, which the Court found missing in the Matimak case, was that Cedec Trading was a corporation of the Channel Islands, a province which was part of the United Kingdom proper, governed by British law, and whose foreign affairs were entirely controlled by the United Kingdom. As such, Cedec Trading Co. was a citizen or subject of the United Kingdom, a state recognized by the executive branch of the United States government.

By contrast, the U.S. Appeals Court did not similarly consider Hong Kong as a part of the United Kingdom proper and, although still a British colony at the time of the Court's ruling (27 June 1997), the Hong Kong colony's autonomy relative to Channel Islands deprived Hong Kong of the same juridical link tying the latter to British sovereign. By the same token, the [future] HKSAR's high degree of autonomy, such as is guaranteed by the Basic Law, would likewise, following the Court's logic, deprive Hong Kong of its link to the Chinese sovereign after its reversion to the latter on 1 July 1997. The HKSAR's autonomy became, in effect, a liability. What a paradox!

Below, I shall offer my critique of the U.S. Court's ruling from the perspective of both the intent of the 1992 U.S. Hong Kong Policy Act and, equally, earlier cases decided by U.S. federal courts relevant to the *Matimak* case. Here, however, let me reiterate that despite the pre-1997 expectations held by many well-qualified pundits and international law experts, the ruling of the U.S. Appeals Court, upholding the lower court's dismissal of the Complaint, threatens to take away whatever certainty there may have been from Hong Kong's international eligibility to act in asserting claims or espousing causes, either in behalf of or through its juridical persons (corporations).

Jerry Lui v. the U.S.
(110 F.3d 103; U.S. Court of Appeals, lst Cir., March 20, 1997)[5]

On 19 December 1995, pursuant to 18 U.S.C.A. See. 3184 (West Supp. 1996), the United States Attorney's office filed an extradition complaint in the District Court of Massachusetts against Lui Kin-Hong (a.k.a. Jerry Lui), setting forth the United Kingdom's extradition request on behalf of Hong Kong. In response, Magistrate Judge Karol issued a warrant for Lui's arrest. Since 20 December 1995, when Lui was arrested upon disembarking from a plane at Boston's Logan Airport, he was in custody pending the completion of his extradition proceedings. At the end of these proceedings, the District Court of Massachusetts granted a writ of habeas corpus to Mr. Lui, preventing his extradition to the Crown Colony of Hong Kong pursuant to two extradition treaties between the United Kingdom and the United States. Thereupon, the United States appealed before the U.S. Court of Appeals for the First Circuit. The court of appeals (per Lynch, J.) reversed the judgment of the district court and ordered that Lui continue to be held without bail, pending extradition to Hong Kong.[6]

In this straightforward case, a couple of things merit our attention as they pertain to both Hong Kong's reversion to China and the territory's international capacity. First, although the United Kingdom requested Lui's extradition on behalf of Hong Kong, pursuant to two outstanding extradition treaties, there was only a passing reference made to Hong Kong's international capacity. This came in the Appeals Court's brief mention of a new extradition treaty that the U.S. President had signed with the incoming government of the HKSAR, and which was pending in the Senate. The court, however, did not address the complex issues related to the entry of the United States into a treaty with a not yet existent entity.

Second, the appeals court, as the district court before, correctly noted that Hong Kong was soon to revert back to China, a country with which the United States had no existing extradition treaty. Both courts were clearly concerned over the prospect that, if extradited, Lui might be tried and punished by the Chinese, who would soon take over Hong Kong. The district court had argued that Lui "cannot be extradited to a sovereign that is not able to try and punish him [a reference to the United Kingdom, which was soon to relinquish its sovereignty over the colony], any more than he could be extradited to a non-signatory nation (*Lui Kin-hong*, 957 F.Supp.at 1287)." The "non-signatory nation" was, of course, a reference to China, which had no extradition treaty with the United States. However, let us not forget, Sec. 201(b) of the U.S. Hong Kong Policy Act clearly stipulates: "For all purposes, *including actions in any court in the United States,* Congress approves the continuation in force on and after July 1, 1997, of all treaties and other international agreements, entered into before such date between the United States and the United Kingdom and applied to Hong Kong, unless or until terminated in accordance with law" (italics added). Covered by this provision was the U.S.-U.K. Extradition Treaty of 1972, which per its Art. 11(a) applied to Hong Kong (28 U.S.T. 227; TIAS, No. 8468). If the U.S. appeals court were to apply the law, the political factor (i.e., Hong Kong's forthcoming reversion to Chinese sovereignty) should have no juridical consequence and would have been immaterial. But, apparently, both the appeals court and the lower court were swayed by political exigency (i.e., Hong Kong's return to China), to the neglect of the law.

Third, both the lower court and the appeals court seemed to be blowing hot and cold at the same time. While they recognized the separate existence of the Hong Kong SAR, they also seemed to suggest that the validity of the new U.S.-HKSAR extradition treaty (assuming it was ratified with Senate advice and consent) was contingent upon the good faith of the government of the People's Republic of China in respecting the independent judicial system and laws of Hong Kong.[7] The upshot, for our interest here, was that Hong Kong's capacity to act as a non-sovereign international entity was nearly totally glossed over by both the District Court of Massachusetts and the U.S. Court of Appeals, First Circuit. Sovereignty, it seems, still was the controlling desideratum to the courts, despite all the claims regarding Hong Kong's status as a non-sovereign international actor.[8] But, as we shall see below, both courts seemed to be selectively applying the sovereignty test, conveniently putting a blind eye to the fact that after reversion, the HKSAR by dint of the Basic Law, its mini-constitution, will enjoy a high degree of autonomy in all aspects except in foreign relations and defense. Like the Channel Islands in the *Cedec* case, the HKSAR will be a part of China, hence a subject region of a recognized foreign state, by virtue of the fact that its foreign relations will be controlled by its Chinese sovereign. The U.S. district and appeals courts, however, chose not to give judicial cognizance to this crucial fact. Had they done so, their decision would have been totally different.

Regina v. Secretary of State for the Home Department, ex parte Launder (1 W.L.R. 83 9; House of Lords, UK, May 21, 1997).

Unlike the two previous cases, this case took place in the United Kingdom. The U.K. Secretary of State for the Home Department appealed a decision by the Divisional Court of the Queen's Bench Division (Q.B. Div'1 Ct., 6 August 1996), quashing a warrant for the return of applicant Ewarn Quayle Launder to Hong Kong, to face trial for corruption. The House of Lords upheld the appellant. The reasoning, however, was interesting. Technically, this case arose in a somewhat different procedural posture from that found in *Lui v. U.S.*, as discussed above. In that case, because of the provisions of the U.K.-U.S. Extradition Treaty, the issue of whether the extradited individual would face trial and judgment by the same country requesting extradition was a predominant concern. Hence, the transition in Hong Kong's status from a British colony to a Special Administrative Region of China was perforce a central issue. In the present Launder case, the issue was whether the accused would face oppression or injustice upon extradition to Hong Kong.

Despite the technical difference, the Law Lords were nonetheless in reality preoccupied with the fixation of China's takeover of Hong Kong. The court, in the final analysis, weighed the significance of the political situation in Hong Kong for the sake of determining whether the Secretary of State for the Home Department had exceeded his discretion in ordering applicant's extradition. It observed:

> The question whether it is unjust or oppressive to order the applicant's return to Hong Kong must in the end depend upon *whether the P.R.C. can be trusted* in implement of its treaty obligations to respect his fundamental human rights, allow him a fair trial and leave it to the courts, if he is convicted, to determine the appropriate punishment (1 W.L.R. [1997], at 857). [Italics added.]

Like in the *Lui* case, the controlling question turned on the P.R.C., or, to be exact, the prospect of Launder facing oppression and injustice under the Chinese, who were scheduled to take back Hong Kong after 1 July 1997. There was no consideration of Hong Kong's post-reversion autonomy, including an independent judiciary, guaranteed both by treaty and by the Basic Law, much less its international capacity as a non-sovereign actor. More than in the *Lui* case, the U.K. court's fixation with the Chinese takeover and the wisdom and merit, under the circumstances, of extraditing Launder back to Hong Kong, where he was wanted for corruption, in retrospect looked almost arrogant and preposterous, if not totally absurd. Of interest to us, nonetheless, is that the U.K. court was obsessed with the question of whether Hong Kong's rule of law would not prove illusory after the end of British rule. Even more important to our interest here is that there was no deliberation by the British court of the Hong Kong SAR's capacity to handle an extradition case externally and to ensure a fair trial of an extradited individual on corruption charges internally. More than in the *Lui* case, the international status of the HKSAR was a total non-issue in the eyes of the British Law Lords in the present case.

Attorney General (CTH) v. Tse Chu-Fai and Another (3 April 1998). [1998] HCA 25.[9]

This was an appeal to the High Court of Australia from an order by the Supreme Court of New South Wales, granting Tse Chu-Fai (a.k.a. Ronald Tse) a writ of habeas corpus, in his fight against extradition to Hong Kong. Extradition proceedings had been initiated by a request from Chief Executive Tung Chee-hwa on 14 July 1997, two weeks after the territory became a Chinese Special Administrative Region (SAR).

In response to Tung's request, transmitted through the Commonwealth Attorney General, an Australian magistrate issued a warrant for Ronald Tse's arrest, which took place on 17 July 1997. Tse's application for a writ of habeas corpus, by itself, raised a crucial question, viz.: whether Hong Kong, after it ceased to be a British colony and became an SAR of China, was still an "extradition country" within the meaning of the Australian Extradition Act of 1988 (Cth). The lower court (the Supreme Court of New South Wales), in a 35-page opinion—150 ALR 566–601—came down to a negative conclusion. Cutting through the legal verbiage and reducing the arguments into the language readily relevant to our concern here, we can summarize the choice faced by the court as between two alternative interpretations of the status of the HKSAR: (a) that Hong Kong, after 1 July 1997, was a part of the People's Republic of China; and (b) that Hong Kong, after that date, was a "territory" for the international relations of which the People's Republic of China was responsible.

The New South Wales court embraced the first interpretation. Just as Hong Kong was a colony (i.e., a part) of the United Kingdom before the date of its retrocession to China, the territory became part of China after that date, it stated. Whereas Hong Kong was an "extradition country" while it remained a part of the United Kingdom, which had the same status, the retrocession changed its previous status. For, as a part of China, a country not declared "an extradition country for the purpose of [the 1988] Act," Hong Kong ceased to be an "extradition country," the court ruled.

As has been suggested in the introductory chapter of this book, the "one country, two systems" in Hong Kong after reversion could be read as conveying two alternate statuses vis-à-vis the Chinese sovereign: (i) as an inalienable part of China, and (ii) as a non-sovereign international actor. While both can find legal support in the Basic Law, the SAR's mini-constitution,[10] the key to resolving the conflict is in Art. 13, which in unequivocal terms stipulates that China is responsible for the foreign affairs of the Hong Kong SAR.[11] But, because of the novelty of the "one country, two systems" as a governance structure, which is unprecedented, foreign judicial authorities were apparently perplexed as to which of the two possible interpretations ought to be adopted. Perhaps for this reason, the High Court of Australia, in reversing the lower court's ruling, reached its decision on a ground totally extraneous to the Basic Law, skirting the question of the "one country, two systems" model altogether. It embraced a statement by the Australian Government, dated 9 September 1997, that "Australia dealt with the PRC (China) on the footing that it was responsible for the international relations of the HKSAR." Hence, Hong Kong after 1 July 1997 became a "territory" for the foreign affairs of which China was responsible. This characterization, it should be noted, cast a different light on Hong Kong's status, comparable to its previous status under British rule. While China was not an "extradition country" within the meaning of the Extradition Act of Australia (1988), that should not ipso facto and likewise prejudice Hong Kong. Technically, "international relations," used in the Australian government's certificate to the court, was a misnomer, because according to the Basic Law, the Chinese sovereign is responsible for the "foreign affairs" of the HKSAR. But the court's intention of avoiding the labyrinth of the Basic Law was obvious.

Critique of the Matimak Decision

The U.S. Court of Appeals for the Second Circuit, before reaching the *Matimak* decision, turned for guidance to the U.S. Hong Kong Policy Act of 1992 and, as noted above, it concluded that under the Act "the United States did not regard Hong Kong as an independent, sovereign political entity."

This conclusion was technically right, but was in fact violative of the spirit of the 1992 Act, as I shall attempt to demonstrate. In doing so, my purpose is to highlight that both the appeals court and the district court, in their anxiousness to disclaim subject matter jurisdiction by characterizing Matimak as "stateless," were preoccupied with the political turn of events (i.e., Hong Kong's handover to China in 1997), but not with the law, much less with the legal capacity of HKSAR under the "one country, two systems" formula. In their neglect of the law, both courts overlooked a stricture judiciously established in an earlier case (*Upright vs. Mercury Business Machines Co.* [Supreme Court of New York, Appellate Div., 1961]), to which we shall return below, that "defendant buyer cannot escape liability merely by alleging and proving that it dealt with a corporation created by an unrecognized government," or, as in the Matimak case, by a political entity not recognized by the U.S. government as a "foreign state."

The U.S. Hong Kong Policy Act of 1992 (Public Law No. 102–383). In first introducing the bill on the Senate floor, on September 21, Senator Mitch McConnell pointed out that the coming return of Hong Kong to Chinese sovereignty in 1997 would necessitate a revision of the relevant U.S. laws that had previously guided American relations with the territory. Close U.S. relations with Hong Kong's British rulers, in the past, had precluded the necessity of formulating a specific U.S. policy for Hong Kong by Congressional action. Now, however, with the forthcoming transfer of sovereignty to the People's Republic of China and growing U.S. trade and business links with Hong Kong, he noted, actions must be taken to ensure that these interests were maintained and protected after the 1997 handover.

The bill had near unanimous support in both chambers of the U.S. Congress, as well as among government officials and experts called to testify. Nevertheless, concerns were voiced that, while the United States continues to develop its multifaceted relations with Hong Kong, caution must be taken, lest its concerns should signal to China unwanted interference into Chinese exercise of sovereignty over the territory.[12] Consistent with this cautious move, the Act simultaneously speaks of U.S. concern for Hong Kong's autonomy after reversion and calls for respect for the sovereignty of China. For example, in Section 1(2), the Act expresses a Congressional "wish to see full implementation of the provisions of the Joint Declaration" of 1984, between the United Kingdom and the PRC, which guaranteed the HKSAR's autonomy. The United States, as a matter of policy, is committed to playing an "active role, before, on, and after July 1, 1997, in maintaining Hong Kong's confidence and prosperity." The Act also spells out how this policy is to be achieved. On the other hand, at a number of places in the Act, the United States is also obliged, in efforts at continuing and expanding its relations with Hong Kong, to consult the People's Republic (e.g., Sec. 103(10)). In another instance, the United States is said to "continue to recognize airplanes registered by Hong Kong *in accordance with applicable laws of the People's Republic of China*" (italics added). These linguistic genuflections in China's direction, symbolically, bespeak U.S. respect for Chinese sovereignty over the HKSAR.

Thus, by implication, the 1992 Act, while at times highlighting the imperative of Hong Kong's autonomy, minces no words in recognizing the fact that post-reversion Hong Kong is part of China. This should be read in conjunction with the HKSAR's Basic Law, under which, the HKSAR's autonomy notwithstanding, its foreign affairs and defense shall fall under the jurisdiction of the Central Government of the People's Republic of China (Arts. 12, 13, and 14). While under Articles. 150 through 154 of the Basic Law the HKSAR may participate in international organizations, conclude international agreements with foreign parties (including sovereign states), and even sit on the PRC's delegations in international organizations in which Hong Kong is not already a member, all this, however, has to be authorized by the Central Government of the PRC. Under the U.S. Hong Kong Policy Act of 1992, the United States is given the leeway of entering into international agreements with the HKSAR. All existing laws of the United States "shall continue to apply with respect to Hong Kong, on and after July 1, 1997, in the same manner as . . . before . . . (Sec. 201 (a))." Likewise, under the Basic Law, the HKSAR also has the competence to enter into the kind of agreements as envisaged in the U.S. Hong Kong Policy Act.

All told, under both the 1992 U.S. Hong Kong Policy Act and the Basic Law, the Hong Kong SAR is considered to be both an autonomous entity distinct from China and, equally, a subject region whose foreign affairs are controlled by the PRC, its post-1997 sovereign. This situation unmistakably recalls the status of the Channel Islands within the British Empire, as enunciated in the *Cedec* case. The Hong Kong SAR is an inseparable part of the People's Republic of China; in fact, the SAR was established under Art. 31 of the PRC constitution. As such, Hong Kong is not stateless" (but a part of China), nor are its corporations, so long as they were duly incorporated under Hong Kong law.

The U.S. Court of Appeals for the Second Circuit, in upholding the lower court's decision to dismiss the *Matimak* case on the ground that it was a Hong Kong corporation, hence "stateless," was both doing an injustice to the 1992 Act in letter and spirit and, no less important, deviating from the well-established, recent (1983) precedent of *Cedec,* which the appeals court cited but quixotically dismissed as inapplicable to the case on hand.

Coming back to the *Upright* case (13 A.D.2d 36, 213 N.Y.S. 2d 417), it is instructive to bear in mind that defendant Mercury Business Machines

Co., like the D.A.Y. Kids Sportswear, Inc. in the *Matimak* case, failed to honor its liability to pay (in the amount of U.S.$27,307.45) for a shipment of business typewriters delivered to it by a foreign corporation. As in *Matimak*, the district court had dismissed the complaint on the grounds that the foreign corporation was a "creature" (i.e., creation), in fact an arm and instrument, of an entity (Communist East Germany) not recognized by the U.S. government. A non-recognized entity, from the point of law, was a non-entity. The appeals court (the Supreme Court of New York, Appellate Division) struck out the defense as insufficient. Among other things, the appeals court ruled that, despite the U.S. non-recognition of the foreign entity, *"[t]he lack of jural status for such [entity] or its creature corporation is not determinative of whether transactions with it will be denied enforcement in American courts,* so long as the government is not the suitor" (italics added). Another important, perhaps the most decisive, basis for the appeals court's reversal in the *Upright* case was found in the following passage:

> Of course, non-recognition is a material fact but only a preliminary one. The proper conclusion will depend upon factors in addition to that of nonrecognition. Such is still the case even though an entity involved in the transaction be an arm or instrumentality of the unrecognized [entity]. Thus, in order to exculpate defendant from payment for the merchandise it has received, it would have to allege and prove that the sale upon which the trade acceptance was based, or that the negotiation of the trade acceptance itself, was *in violation of public or national policy.* Such a defense would constitute one in the nature of illegality and if established would, or at least might, render all that ensued from the infected transaction void and unenforceable. [italics added]

Hence, the appeals court in the *Upright* case concluded: "Defendant buyer *cannot escape liability* merely by alleging and proving that it dealt with a corporation created by and functioning as the arm and instrumentality of an unrecognized [foreign entity]." (Italics added.)

Analysis.   Admittedly, there were a number of differences between the two cases. For one, in *Upright,* the non-recognized entity was the Communist government of East Germany during the Cold War era. Secondly, the non-recognition issue in support of dismissal was raised by the defendant buyer in its attempt to escape payment. And, thirdly, the foreign corporation coming to the U.S. courts for relief was not just a creature of the non-recognized foreign entity, but an arm and instrumentality of it.

In the *Matimak* case, on the other hand, the non-recognized entity was no less than the Crown colony of Hong Kong itself, albeit a soon-to-be Chinese SAR. Regardless of the territory's jural status, the rebuttal offered by the appeals court in the *Upright* case should hold true in the present case, namely, that the controlling desideratum was that "the lack of jural status for such [non-recognized entity] or its creature corporation is not determinative of whether transactions with it will be denied enforcement in American courts, so long as the [non-recognized entity] is not the suitor."

As to a second difference, the issue of whether Hong Kong was a recognized "foreign state" in the *Matimak* case seemed not to originate from the defendant, but from the district court in its inscrutable search for a ground for denying itself subject matter jurisdiction over the case. The defendant debtor could not possibly have found a better *amigo en curia!* A third difference is obvious: whereas in the earlier case the plaintiff foreign corporation, though suing *via* an assignee, was an official arm and instrumentality of the non-recognized foreign country, plaintiff Matimak was a private corporation incorporated under Hong Kong law. The tenuous link between the plaintiff corporation and the non-recognized foreign entity was, therefore, made even more tenuous, to the point of irrelevance.

Despite these procedural differences, however, the two cases share much in common, and the main reasons supporting the appeals court's reversal in the *Upright* case should prevail in *Matimak* as well. In the first place, to re-iterate, the lack of jural status for Hong Kong (not being a "foreign state") should not be determinative of whether commercial transactions with its private corporations would be denied enforcement in U.S. courts. At issue was whether a contract properly executed on American soil was enforceable in U.S. courts. Secondly, Hong Kong, the non-recognized foreign entity, was not a suitor in court, whereas Matimak Trading Co., a Hong Kong corporation, was. Thirdly, and most important, defendant buyer, D.A.Y. Kids Sportswear, Inc., should not be allowed to escape its liability for the payment of U.S.$80,000 for goods ordered and delivered, merely by raising a question concerning a tenuous link between Hong Kong, where Matimak was incorporated, and its non-recognition as a "foreign state." And, finally, the trade on the basis of which defendant debtor's nonpayment created an outstanding liability in *Matimak,* as in the earlier case, was not found to be "in violation of public or national policy."

In *Upright v. Mercury Business Machines Co.,* the appeals court even went further to question the tenuous link between recognition of a foreign entity and the "jural capacity" of its corporations to "trade, transfer title, or collect the price for the merchandise they sell to outsiders, even in the courts of non-recognizing nations." By contrast, however, in the *Matimak* case neither the district court nor the appeals court took heed of this crucial point about the tenuous link, possibly for one reason. And, the reason might well have been that both courts were preoccupied with the imminent transfer of sovereignty over Hong Kong in its scheduled reversion to China. If true, this would be a sad instance in which U.S. federal courts were diverted from a coolheaded deliberation on the merits according to law, for the administration of which the courts were created in the first place, by a fortuitous political turn of events looming on the horizon.

The same sad commentary may, *mutatis mutandis,* be made on two of the other cases discussed above, *Lui v. the U.S.* and the *Launder* case. As has been shown above, in both cases the respective courts were fixated with the forthcoming departure of British rule in the Hong Kong colony, and raised erroneously, I think, what at best was an oblique, if not totally irrelevant, question, viz.: what would happen if the individual offender should be extradited to Hong Kong after the Chinese takeover? Although the right decision was made in the end, whereby both Lui and Launder[13] would be extradited, the fact is that the controlling desideratum, it seems, was the anticipated political turn of events surrounding Hong Kong's return to China, which was blown out of proportion, relative to what was required by law. This judicial *faux pas* led to the courts' glossing over the promised continuance of an independent judicial capacity of the HKSAR. It also led to the neglect by the appeals court in one case *(Lui v. U.S.),* of an important provision in the 1992 U.S. Hong Kong Policy Act, to the effect that the coming handover should in no way affect the continuing in force of all existing treaty obligations applicable between Hong Kong and the United States as before 1 July 1997. Notwithstanding this provision, the issue of the uninterrupted effect of the U.K.-U.S. Extradition Treaty, as it applied to Hong Kong as before, was totally glossed over by the court.

In retrospect, the intuitive assumption that the new Chinese sovereign would ipso facto be expected to trample upon the rights of the extradited offenders, including one wanted on corruption charges in Hong Kong, and that China would deprive the HKSAR of the independence of its judiciary in violation of treaty obligations, does not look very bright in the glow of hindsight, to put it in the most charitable terms.

In the frame of reference for this book, the wide discrepancies between the foreign courts' obvious expectations (about what was going to befall Hong Kong) in the cases just examined and the actual outcomes, as seen from the afterglow of the handover, present a giant paradox.

Commentary

For fear that the uninitiated may draw the wrong conclusions from the above discussion, a clarification, it seems to me, is in order on the question of Hong Kong's international capacity. First, despite the *Matimak* setback, neither Hong Kong nor its corporations are without recourse in defense of their interests internationally. In the United States, for instance, resort to state courts is nevertheless open to Hong Kong and its corporations, in the event they have to seek legal relief for the protection of their rightful interests. The *Matimak* decision merely means that a Hong Kong incorporated entity, as Hong Kong itself, cannot sue before a U.S. *federal* court, unless the U.S. Supreme Court, in a likely event, reverses the two lower courts' decisions. Despite this limit to its access, the fact that Hong Kong has the capacity to sue before other forums attests to what we, in these pages, have called the territory's legal capacity to act as a non-sovereign international actor. The import of the cases discussed here, especially *Matimak,* simply means that Hong Kong's legal capacity to act as such is not uninhibited or above challenge, as previously assumed before its reversion. The implication here is that despite the earlier expectations to the contrary, we do seem to need a necessary adjustment in our understanding of Hong Kong's international eligibility to act, in comparison with its previous colonial incarnation.

Other evidence of the SAR's legal capacity, as such, is found in the *U.S. v. Peter Yeung* case, in which a China Mainland-born U.S. citizen was wanted for alleged conspiracy to defraud the U.S. government, smuggling clothes into the United States and importing clothes labeled with false country-of-origin tags on 91 occasions. The U.S. government's hunt for the fugitive began in 1992, after the New York District Court issued a warrant for Yeung's arrest over 183 alleged offenses said to have taken place between February 1990 and April 1991 and to involve U.S.$11 million. Yeung escaped from the United States in mysterious circumstances, surfaced in Macau, and later went to Hong Kong, where he was arrested on 30 March 1998. The U.S. extradition request was a first following Hong Kong's handover to China.[14] The Justice Department of the SAR granted the request and surrendered Yeung in June 1998. While this was the first such request, I have learned on good authority that by late March 1999 there had been 25 such requests from various countries, half of them from the United States.[15] The fact that these requests, like the one from Washington on Peter Yeung, were filed directly with the HKSAR, but not through its Chinese sovereign, is a good testimony to Hong Kong's international legal status as an autonomous entity with a capacity to act despite its lack of sovereignty.

Hong Kong's Intermediary Function between China and the Outside World

There were high hopes among many quarters that after reversion Hong Kong's role as an entrepot for China would increase and that the SAR would also double as China's conduit to the world. Hence, its intermediary function, serving as a window for China on the world and, in return, as the world's window on China, would provide a "reversible window" effect.[16]

Those hopes were based on both (a) a solid record over the past one and a half decades, in respect of Hong Kong's entrepot role and its economic ties with the Mainland, and (b) a theory of synergism, which sees an ever-deepening interdependence, generated by mutual complementarity (see below). Part of the past record showed, for example, that Hong Kong had been a very important source of China's foreign exchange earnings, accounting for up to one-third of the total prior to the 1980s (Kueh and Voon 1997, p.61). In other statistics, the Hong Kong entrepot trade rose to the highest level ever in 1996, at U.S.$120 billion, or 41 percent of China's total foreign trade. Hong Kong's foreign direct investment (FDI) in Mainland China amounted to U.S.$78.6 billion, or 59 percent of cumulative FDI in China. In return, Hong Kong was also recipient of 80 percent of China's outward FDI (Sung 1997, p.709).

The synergist view comprised a number of propositions: (i) that Beijing would not kill the goose that lays golden eggs by trying to siphon off Hong Kong's vast foreign exchange reserves (Kueh and Voon 1997);[17] (ii) that China needed a prosperous Hong Kong to be prosperous itself, (iii) Hong Kong needed an independent monetary policy to maintain its attractiveness to investors; and (iv) that China needed an independent monetary policy to maintain the macroeconomic stability necessary for its own prosperity (Ho 1997, p.230).

While the jury is still out at the time of writing, in the sense that more up-to-date statistics are yet to roll in, we are nevertheless able to offer some tentative comments on whether the pre-handover expectations have materialized on the question of Hong Kong's possible intermediary role for China vis-à-vis the world. Although this chapter does not purport to discuss Hong Kong's domestic economy, nor Hong Kong-China economic ties per se, some of the points developed below, nevertheless, do peripherally touch on these questions. We cannot, in my opinion, really get to the heart of Hong Kong's intermediary role without first examining the political parameters within which the SAR has to operate as an economic actor vis-à-vis both China and the outside world. Nor can we do without touching on, at least *en passant,* the obvious question of whether some sort of an economic integration would not emerge from the conjoining of the two dynamic economies, China and Hong Kong.[18]

Before going any further, I have to pause to observe that in the months since 1 July 1997, Hong Kong does offer evidence that it can be a useful beachhead for foreign interests seeking entry into China proper. For example, Britain's biggest bus company, First Group, announced in early 1998 that it would use its successful bid to run all Hong Kong's major public bus services as a stepping stone to the Mainland's giant public transport market.[19] Another similar instance concerns Stagecoach Holdings, billed as the world's largest bus company, which announced its plans to invest HK$1.39 billion for a potential 28 percent stake in Hong Kong-listed Mainland toll-road investor Road King Infrastructure, the first British company to gain access to the Mainland's lucrative toll-road sector. Stagecoach, which had failed in its bid for a bus franchise in Hong Kong in March 1998, sought this alternative route to gain entry into the Mainland.[20]

Furthermore, an unrelated bizarre incident, quite indicative of Hong Kong's value as a conduit to China, was the discovery of a HK$5.7 million smuggled armored troop carrier, intercepted and seized by Hong Kong customs officials from aboard a container ship bound for a Chinese destination.[21] While the incident may not be an orthodox use of the Hong Kong conduit, it nevertheless demonstrates, perhaps dramatically, the territory's prized usefulness as a channel through which to reach China.

There is no question that, contrary to the rampant pre-1997 prophecies of doom, Hong Kong's continuing economic strength (even in the midst of the Asian financial crisis), as well as its ties with China, have been facilitated by the economic freedoms guaranteed by the Basic Law, under which the SAR is, for all practical purposes, a separate economy. And, as such, the trade and investment flows between Hong Kong and China proper are treated, albeit anachronistically, as international flows. On the basis of the insufficient data on hand, I would venture to suggest that at least two lessons relevant to this discussion can be drawn from the experience of the HKSAR in its first year.

The first lesson is that, although 1997 was a year of economic prosperity for the SAR (even after controlling for the adverse effects of the financial turbulence hitting the Asia Pacific region from the fall of 1997 onward), the watershed year for Hong Kong's economy was not 1997, but 1979. In that earlier time period, when China began its epochal economic reform and opening to the outside world, Hong Kong, which had lost its hinterland during the Cold War era, regained it as a result (Sung 1997, p.705). The second lesson is what Yun-wing Sung calls the "impossibility of a formal Hong Kong-Mainland trade bloc."[22]

The first of the two lessons simply confirms the absolute importance of the Chinese hinterland to Hong Kong. Since its reversion serves to solidify, or institutionalize, the hinterland-periphery relationship, the HKSAR's continuing economic dynamism should come as no surprise. Nowhere was this hinterland advantage driven home more pungently than when China's new premier, Zhu Rongji, announced at his first news conference upon assuming office that China would "do everything within its power" to bail out Hong Kong, should the SAR be in trouble under the crushing blows of the ongoing financial turbulence hitting the region.[23] With China's U.S.$142.8 billion foreign exchange reserves lining up behind Hong Kong with its own

U.S.$95 billion, that unequivocal assurance surely helped to instill confidence among Hong Kong's residents, as well as foreign commercial interests, in the future of the SAR's economy. Some natives were already wondering aloud if the British colonial rulers would have made the same commitment, had the financial crisis hit the region while they were still ruling Hong Kong.

The second lesson needs a word of explanation. Borrowing from the experience of other regions (e.g., European Union [or EU], North American Free Trade Association [or NAFTA], among others), integration theorists may have legitimate reasons to expect that the amalgamation of Hong Kong into what could be the nascent shell of an eventual Greater China entity may lead to the tantalizing option of forming a trade bloc, beginning perhaps with a customs union, with China proper. Before the jury is in, however, a respectable economist—Professor Yun-wing Sung, of the Chinese University of Hong Kong—has already ruled out the option as unrealistic. Given the free-port status of Hong Kong, guaranteed by both the Basic Law and international agreement, Sung said, "the only way that the Mainland and Hong Kong could form a customs union would be for the Mainland to abolish all of its tariffs." This, to Sung, would be "ludicrous and utopian." In fact, for Hong Kong to enter into an economic union, after the EU model, would violate sections of the Basic Law. Creating a common market would imply that the HKSAR had given up on regulating migration from China, a consequence far beyond the territory's physical capacity to bear, given its overpopulation and already tightly overcrowded space. An economic union, Sung continued, would also require that the SAR give up its independent currency, making it impossible for Hong Kong to function as an international financial center, because the Chinese renminbi is far from freely convertible. Despite its reversion to Chinese sovereignty, according to Sung, Hong Kong is institutionally more closely integrated with most other economies than with China. The extent of Hong Kong-Mainland integration is less than that between Greece and Ireland, Sung suggests (p. 707f).

I do not doubt this analysis to be true: the bottom line, however, is that the vast gap between the commonsensical expectations of greater Hong Kong-China integration and the hard reality, which turns out to be just the contrary, as Yun-wing Sung pointed out, confirms our paradox thesis regarding post-reversion Hong Kong as a nonsovereign actor.

Another related paradox is the discrepancy between prior speculations about the HKSAR being contaminated by Chinese ideological influence and what, on the other hand, may be called, as a shorthand label, the "roaring mouse" syndrome, which finds China actually at the receiving end of influence. While this remains to be further developed with more cumulative evidence, two different existing sources have identified the syndrome. One is Yuchu Xie's study (1997, p.3) that, both because of the current ideological vacuum pervading Mainland China and also Hong Kong's cultural pluralism, representing the best of both Western and Eastern traditions, it is the SAR, not Mainland China, that is the dispenser of influence. The other exposition for this line of inquiry is an extremely thought-provoking report by Daniel Fung, then Hong Kong's Solicitor General. Rather than Hong Kong's law being subject to the pollution by China, he noted, evidence pointed to the other direction. Starting even before Hong Kong's reversion, going as far back as the 1980's, at least three law schools in China launched a campaign to study Hong Kong's law as a specialty. The Hong Kong Attorney General's Chamber (AGC) had been supplying the Legislative Affairs Commission of the National People's Congress (NPC in Beijing copies of the territory's bankruptcy law, companies legislation, and other commercial legislation. The NPC consulted these in drafting their own laws, which in part accounts for the sudden orgy of new commercial laws coming into existence in the 1990s. And, since 1992, when Shenzhen was granted its special legislative autonomy status by Beijing, the neighboring Chinese town enacted, in the following three and a half years, approximately 250 pieces of primary and subsidiary legislation, two thirds of which were copied from Hong Kong's laws.[24] I have no evidence of a let-up in this trend beyond July 1997. The reversal, as such, of the earlier speculated directional flow of ideological influence, if continued, would also be a paradox unto itself.

This point is important, because as a result of China's methodic borrowings from Hong Kong's laws and legal tradition, the outcome—as attested to by the recent flurry of commercial laws enacted by the NPC, modeled after Hong Kong's laws and practices—may very well be that China's economic structure, as well as legal system., will be brought more in line with the outside world. To the extent that this development will make China more fit to be integrated into the international mainstream, such as through membership in the World Trade Organization (WTO), then the HKSAR's alleged intermediary function between China and the outside world will be more than fulfilled, thanks to the paradoxical "roaring mouse" syndrome.

As a Catalyst for Taiwan-Mainland Liaison

Although the relations between Taiwan and the Mainland of China are not an international issue, any move by both sides, in an unlikely turn of events, toward reunification or some similar peaceful solution would most likely come at the behest of foreign supporters or, else, would excite many foreign powers at both regional and global levels. The recent glacier movement toward direct talks between Taipei and Beijing, coming on the heels of open U.S. encouragement, including President Bill Clinton's Shanghai message during his China visit,[25] is a case in point. In the event Hong Kong could play a catalyst role in bringing together the two sides in any endeavoring toward a permanent resolution of their unfinished civil war since the 1940s, would spell a crowning success for Hong Kong in its role as a non-sovereign international actor (Hsiung 1997, p.240).

Because of its strategic location, situated in the interstice between Taiwan and the Mainland, Hong Kong has been a crucial hub of air traffic and transshipment across the Taiwan Strait for passengers and cargo, ever since Taiwan opened the floodgates to visitors to the Mainland in 1987, after 38 years of an armed stand-off. The SAR would, presumably, be an ideal location for any serious bilateral Mainland- Taiwan negotiations, should both sides want to seek a breakthrough in the current impasse of no direct air or maritime links. Indeed, a breakthrough of sorts was reached whereby a Mainland cargo vessel was reported to have docked at the southern Taiwan port city of Kaohsiung in early March 1998, after a stop in Okinawa. According to news reports, the 6,269-ton Tungshun was the first Mainland ship ever to operate cross-strait cargo services via a third port since Beijing and Taipei officials held informal shipping talks in February. In fact, it was the first time officials from.both sides ever met, and they did so on third-party soil. The bigger surprise, however, was that the talks were held, not in Hong Kong, but in the more remote Bangkok.[26]

I can think of two possible reasons why Hong Kong was not chosen as the negotiating site. First, after its reversion to China, Hong Kong is no longer "foreign soil" that Taiwan could consider as neutral ground. Second, the SAR presumably would have too high a visibility for Taiwan to feel comfortable as a site for conducting sensitive talks with Beijing.

If so, then it proves that, despite earlier speculations to the contrary, Hong Kong's return to Chinese sovereignty made it less desirable as a place for official or semi-official contacts between the Mainland and Taiwan. In this light, the offer of Malaysia by Prime Minister Mahamad Mahathir, made to the visiting Taiwan Vice President, Lian Chan, during the latter's recent unofficial visit, as a possible venue for future Beijing-Taipei talks,[27] would make more sense than meets the eye. In both instances, Hong Kong's role as a medium between the opposing sides across the Taiwan Strait, while much heralded before the departure of British rule, has not lived up to its full potential. Again, a paradox.

In fact, despite its high autonomy enjoyed under the "one country, two systems" arrangement, the HKSAR does not have unimpeded freedom in respect of its dealings with Taiwan. Guidelines for Hong Kong's relations with Taiwan were spelled out in the so-called "Qian's Seven Points," named after then PRC Foreign Minister, Qian Qichen, as he outlined Beijing's policy on 22 June 1995. The policy contains essentially two parts. First, while Taiwan's pre-existing offices, plus its 3,000 private firms, are permitted by Beijing to continue their presence and activities in the Hong Kong SAR, they are bound to respect the Basic Law. Second, and more important for our interest here, any official activities between Hong Kong and Taipei must have Beijing's prior endorsement. Although this two-part policy of Beijing has been known since 1995, what is new is that, despite wishful thinking to the contrary in some quarters, the policy has been strictly carried out since Hong Kong's return to Chinese sovereignty (Wong 1997, p.9f). For the legally minded, it bears noting that, while matters relating to Taiwan are strictly speaking not "foreign" affairs, the sort of "Big Brother" watch Beijing maintains on Hong Kong in regard to the latter's relations with Taiwan, nevertheless, reminds one of the two areas in which the SAR, according to the Basic Law, is not autonomous: foreign (read: external) affairs, and defense.

In the circumstances, the room left for Hong Kong to play a catalyst role, in prompting Taiwan and the Mainland to move toward a final peaceful resolution of their division, is very restricted. That eventuality would depend on the outcome of the SAR's experiment with the "one country, two systems" model. If, over a prolonged period of time, the model has proven to be a reliable and duplicable success in the HKSAR, then the reason for Taiwan's official rejection of the model[28] as an answer to the remaining Taiwan piece in the jigsaw puzzle of Chinese [re]unification—although the model would have to be modified to suit the island's different political ecology—would lose its plausibility, hence, validity. More especially, if the success should so transform world opinion and, *a fortiori,* Washington's perception that sufficient pressures should mount on Taiwan in its wake, the island's public opinion would hardly remain unswayed. Add to this some hard statistics from the conjoining of the two economies after Hong Kong's return to China: (a) their combined GNP (U.S.$885 billion) was four times that of Taiwan; (b) their combined foreign trade at U.S.$652.3 billion, admittedly including their mutual trade as well, was three times Taiwan's trade of U.S.$215 billion; and (c) their combined foreign reserves, totaling U.S.$237.1 billion (as of September 1998), was almost three times Taiwan's U.S.$83.6 billion.[29]

Add further the World Bank's forecast that by the year 2020 the People's Republic of China will be the world's second largest economy, next only to the United States.[30] The consequential bridging of the gap between a fast-growing Mainland China and an already prosperous Taiwan (as measured by its per capita income of U.S.$12,000) of today will most likely make Mainland China less foreboding as a partner to the majority of the Taiwan compatriots. Under the circumstances, Taipei's decision-makers, by then long after Lee Teng-hui's departure from the political scene, would probably find untenable any continued rejection of the, by then, time-tested "one country, two systems" solution. The ensuing result might well prove that Hong Kong will have played a catalyst's role in that it has made Taiwan's acceptance of the "one country, two systems" model an almost foregone conclusion. But, this is the preview of an outcome compelled by pure logic, whose surety, however, is not likely to be seen in the next few years, certainly not at the time of this writing.

Other than that remote development, the predictions among many quarters for more trade and investments between Taiwan and the Mainland by way of Hong Kong, to an extent more than before, are hard to ascertain in the short run. Some scholars (e.g., Brown 1997, p.19), nonetheless, had forecast that Hong Kong would become more, not less, important to Taiwan's economy. According to data made available to me by the Far East Trade Service,[31] however, Taiwan's trade with the SAR in from January through December 1998 showed a distinct pattern of decline from the previous year, falling by 13.4 percent and 2.2 percent in exports and imports respectively. There is no reliable way of knowing whether, and to what extent, the decline was caused by the region's financial crisis, except by comparison. The same statistics show a consistent pattern of decrease across the board, and in fact the reduction with Hong Kong was the smallest by comparison. Taiwan's exports, for example, to Japan, Singapore, Malaysia, the Philippines, and Thailand, went down during 1998 from the previous year, by 20.2 percent, 33.4 percent, 24.7 percent, 13.7 percent, and 24.8 percent respectively. It can be deduced therefore that the decline in Taiwan-Hong Kong trade was related to the region's financial crisis.

As a Factor in U.S.-China Relations

The point of departure is the high stakes of the United States in Hong Kong, where it has an aggregate investment of over U.S.$13 billion, 1,000 U.S. firms, and 30,000 U.S. citizens. Its official concerns stemming from these high stakes are attested to by the U.S. Hong Kong Policy Act of 1992. In declaring its "wish to see full implementation of the provisions of the Joint Declaration" between China and the United Kingdom on Hong Kong's return to Chinese sovereignty, the Act acknowledges Hong Kong's "important role in today's regional and world economy." Concerns for the protection of U.S. interests are shown in the commitment it made for the United States to "play an active role in maintaining Hong Kong's confidence and prosperity, Hong Kong's role as an international financial center, and the mutually beneficial ties" between the peoples of Hong Kong and the United States.

In aggregate, the various provisions of the Act amount to a pledge by the United States to commit itself to replacing British influence after 1 July 1997. Save for that implicit goal, the entire Act, more especially its provisions for periodic reporting by the Secretary of State to Congress on the conditions in Hong Kong, would make little sense, and would hardly even be worth the paper on which it was written. In this sense, the HKSAR is an important factor in U.S.-China relations.

During the Cold War era, it was U.S. policy to build a cocoon around Hong Kong, so that it would not be an entrepot for the transfer of strategic goods destined for China in violation of the wishes of the Coordinating Committee for Multilateral Export Controls (COCOM), the watchdog body for the United States and its allies in enforcing the ban on strategic goods exports to Communist bloc countries. Sec. 103(8) of the Act made this concern explicit for the future: "The United States should continue to support access by Hong Kong to sensitive technologies under the agreement of the [COCOM] *for so long as the United States is satisfied that such technologies are protected from improper use or export*" (italics added). In Senator Mitch McConnell's remarks while introducing on the Senate floor his bill, which later became the 1992 Act, he made references to close British cooperation over U.S. relations with Hong Kong thus far. That cooperation presumably included U.K. support of COCOM policy and the Hong Kong laws that prohibited transfer of sensitive technologies to and from China. The justification for a special legislation, such as the 1992 Act, was that in view of Hong Kong's forthcoming reversion to China, the United States had to make sure that U.S. interests would not be left unprotected. Whereas after the end of the Cold War COCOM was deactivated, its function has been taken over by two systems identified respectively with the newly created Nuclear Suppliers Group (NSG) and the Missile Technology Control Regime (MTCR). A principal concern of the United States is that, as before, Hong Kong would not become an entrepot for China in the import-export of sensitive technologies, such as those associated with the manufacture of weapons of mass destruction. A subsequent agreement concluded by the U.S. Department of Commerce with the HKSAR government, in September 1997, was precisely meant to ensure continuance of Hong Kong's previous trade policy on the transfer of sensitive technologies. As a reward in return, the SAR was given a Q trading status, one usually reserved for a Western industrially developed trading partner (Hu 1997, p.4).

Such being the case, the discovery and seizure by the SAR's customs officials of a smuggled armored personnel carrier bound for China from Thailand via Hong Kong, as noted earlier, was thus of grave concern to the United States. Hong Kong's forceful action in levying stiff fines on the captain of the ship caught in carrying the smuggled vehicle, in accordance with a 1955 ordinance banning such smuggled goods, assured Washington that Hong Kong's past policy banning sensitive trade transshipments did not change with the transfer of sovereignty (Hu 1997, p.3).

In this connection, any incident involving even the unwitting use of Hong Kong as a point of transfer of sensitive technologies could, conceivably, constitute an irritant in Sino-U.S. relations. If, in the U.S. perception, Hong Kong—should the latter lower its guard and lose control—were to become a point of entry and departure for illicit transfer of sensitive goods and/or technologies, Washington conceivably could be forced to resort to a number of countermeasures, such as: (a) declaring that Hong Kong is no longer able to protect sensitive technologies from leaking to the Mainland, and therefore technology control will be applied to Hong Kong itself; (b) determining that Hong Kong is no longer sufficiently distinct as an entity from Mainland China to permit the United States to continue to treat Hong Kong as a separate tariff area, as defined in the 1992 U.S. Hong Kong Policy Act; and (c) slapping restrictions on China's exports to the United States.[32] In any event, any of these measures, if enforced when warranted by the circumstances, would drive Sino-U.S. relations back at least 20 years. Hong Kong, in that eventuality, would also be a loser.

The above discussion, I hope, highlights a two-part lesson, namely, just as U.S.-Hong Kong relations will affect U.S.-China relations, the state of Sino-U.S. relations will likewise have untold effects on Hong Kong, in terms of its relations with both Washington and Beijing.

Another development that should be noted before we leave this discussion of the U.S.-China-Hong Kong triadic relationship is that, after reversion, there was a potentially fractious shift in the way in which Hong Kong's export trade is counted. Unavoidably, it becomes part of China's aggregate foreign trade to U.S. Customs. As a result, China's combined trade surplus with the United States will become more exaggerated. In 1993, for example, some U.S.$8.2 billion Hong Kong exports to the United States was counted by U.S. Customs as China's exports to the United States, thus exaggerating the Chinese surplus in Sino-U.S. trade by 133 percent (Kaulman 1996, p.164).

Thus, Hong Kong as a factor in Sino-U.S. relations should not be seen as all roses. Even if there are roses, beware of the prickly thorns that come with them! Contrary to previous expectations—especially among some Chinese quarters that tended to see Hong Kong's return as nothing but a plus even in China's relations with the outside (including the United States)—this conclusion, however tentative, seems to caution us that it, paradoxically, is not necessarily true.

Concluding Remarks

As indicated in the subtitle of this essay, I intended it to be an update of a similar study I did in October 1997, or three months following the handover. At the time, I was drawing upon the prevailing views of qualified pundits and international law scholars. Otherwise I had little to go on, as available supporting evidence was scanty.

In writing this update nearly two years later, however, I began by taking stock of subsequent developments in and around the SAR across different areas, with a view to gaining a broader perspective from which to approach the topic I was to focus on. To my great astonishment, and amusement, too, I found consistent, wide discrepancies between earlier projections and subsequent outcomes, as we came to know them through the end of 1998. One finding, of particular relevance to our interest in this chapter, stands out: despite earlier prognostications about Hong Kong's international capacity to act, such as in asserting claims and espousing causes, including its corporations' eligibility to seek relief before foreign courts, events proved otherwise. In the first place, Hong Kong's *jus standi* (or ability to appear before a foreign court), through no fault of its own, was found either frustrated, as in the *Matimak* case (involving U.S. federal courts), or glossed over in other foreign judicial organs. Almost invariably, an apparently irresistible fixation over the political turn of events culminating in China's actual takeover of Hong Kong from British rule (and conceivably viewed with a jaundiced eye) seemed to have overwhelmed the foreign courts, diverting them from a cool-headed deliberation on the merits according to law, as we have noted in the *Lui* and *Luander* cases. In addition to examining the treatment, by the foreign tribunals, of the legal status of Hong Kong in these judicial cases, I have extended the scrutiny to other aspects of Hong Kong's external role or capacity to effect change, i.e., as a conduit for China and the outside world, as a catalyst in the Taiwan-Mainland tangle, and as a factor in Sino-U.S. relations.

With little exception, the search turned up evidence showing that the SAR's external capacity, as such, had been constrained, at times severely, by unforeseen extraneous factors beyond its control. In fact, as noted, its high profile and visibility proved to be a fortuitous disadvantage for the Hong Kong SAR to serve as a literal meeting ground for sensitive negotiations between the two sides straddling the Taiwan Strait. Contrary to expectations by integration theorists, too, a Hong Kong-Mainland economic union, after the EU model, has thus far been found elusive and even well-nigh improbable. Besides, any abusive but successful use by illicit smugglers of the Hong Kong conduit, for the transfer of contrabands or sensitive technologies to and from China, would not only damn the SAR's reputation, but also arouse alarm among supporters of the NSG and MTCR regimes and, most probably, poison Sino-U.S. relations. A shift in the practice of U.S. Customs, to counting the SAR's exports as part of the aggregate Chinese exports, would unduly exaggerate China's surplus in Sino-U.S. trade, thus exacerbating a long-standing irritant in the bilateral relations. In all these instances, a distinct commonality was a discrepancy, at times inexplicable, between earlier expectations and subsequent outcomes, before and after the handover.

An especially ironic development came to public knowledge in early May, 1998, involving "interviews" conducted by the British consular officers with certain local candidates for the SAR's 24 May Legco election. When the news broke, the Chinese Foreign Ministry in Beijing literally "threw the book" at the British consulate, by reminding it of the obligations incumbent upon foreign consular agents under the 1963 Vienna Convention on Consular Relations. Art. 55 of the Convention, which codifies the customary international norms on consular immunities and privileges, explicitly prohibits consular agents from interfering in the domestic affairs of their host state. In self-defense, the British Consulate-General claimed that it was within usual consular practice to keep in touch with politicians of all colors in the land of the host state. While this argument sounds as if Beijing's consular officers had an equal right to rub shoulders with IRA leaders in Northern Island, if just for the sake of keeping "informed," the matter no doubt will remain a case in contention. But, what is relevant here is the paradox it presents. The British, who had not been in the least expected to be even minimally interested in the SAR's elections following the handover, were indeed showing more interest and, by the close contact they admitted to keeping with certain of the local candidates, seemed to be more directly involved in the SAR's first elections than was Beijing, which appeared to take a deliberate lay-back position. By contrast, prior expectations seemed to have pointed to just the opposite direction: that Beijing, as the SAR's sovereign after 1 July 1997, would do the meddling in Hong Kong's domestic politics. The contrast is, in itself, a paradox.

This list of discrepancies between prior expectations and subsequent outcome could go on. But the examples are enough to verify our thesis of a superparadox surrounding the SAR's developments, no doubt certifiable in the area examined in this chapter. The biggest paradox is found in the contrast between what has happened on the domestic scene, on the one hand, and what has happened to Hong Kong's much heralded international capacity to act, on the other.

If discrepancy between expectations *ex ante* and real developments *ex post* is a paradox, as we defined the term, our discovery of paradoxes actually pervades both the domestic and the external domains, even though this chapter is focused on the latter. As has been shown, to reiterate, the biggest paradox is found in the stark contrast of developments, going totally opposite directions, between the SAR's domestic and external domains, creating an auspicious and an inauspicious paradox, respectively. These findings can be said to be truly counter-intuitive.

Notes

1. The wording of the title acknowledges an earlier article of mine by the same title (Hsiung 1997), to which this is an update.
2. See, especially, Mushkat (1997); Tang (1993); Ting (1997); and Hsiung (1997).
3. This language, suggesting the relationship of practice and custom to general international law, was borrowed from Art. 38 (1)(b) of the Statute of the International Court of Justice.
4. I am indebted to Professor Hungdah Chiu, of the University of Maryland Law School, for obtaining for me, while on leave in Hong Kong, a full copy of the decision by the U.S. Court of Appeals of the Matimak case.
5. In this discussion, I am relying on Professor James D. Wilets's summary in *American Journal of International Law* 91 no. 3: 537–541 (1997).
6. According to Daniel Fung, the HKSAR's Solicitor General, in a personal communication dated 11 March 1998, Lui would be extradited and was scheduled to arrive in Hong Kong on 22 May 1998.
7. For this point, see Wilets's report in ibid., p.540.
8. This goes against the grain of a rising literature that argues that sovereignty is in decline and that non-sovereign actors are becoming both more numerous and more important in their effects on international relations. See, for example, Robert H. Jackson, *Quasi-States: Sovereignty, International Relations, and the Third World* (Cambridge: Cambridge University Press, 1990).
9. I am grateful to Professor Andrew Bums, of the Hong Kong University's Centre for Comparative Law, for calling my attention to this and other Australian cases.
10. Articles 1 and 2 support the first interpretation, and Articles 150–57 support the second interpretation.
11. The same can also find support in Item 3 (2) of the Sino-U.K. Joint Declaration of 1984.
12. Cf. *The Making of the U.S. Hong Kong Policy Act of 1992: A Documentary Chronology* (Taipei: Friends of Hong Kong and Macau Association, 1994), p.ix.
13. Launder was extradited to Hong Kong, where his trial on thirteen corruption charges began on 20 April 1998. See *SCMP,* 21 April 1998, p.3.
14. "Extradition after Six Years on the Run," *SCMP,* 22 April 1998.
15. This information was provided to me upon request by Ms. Nina Chi, head of the Mutual Legal Assistance Unit, HKSAR Department of Justice, which is gratefully acknowledged.
16. Cf., James C. Hsiung 1998, p.239.
17. However, in Taiwan, President Lee Teng-hui was reported to be fond of echoing the refrain that Beijing would do precisely that, to whoever cared to listen through 1997. Cf., *Ming Pao Daily* (Hong Kong), 28 June 1997, p.16.
18. One example of expectations about a Hong Kong-China integration is: Dumeng Zhang 1997.
19. "Deal a Launchpad for China Market," *SCMP,* 1 April 1998, p.3.

20. "UK Bus Giant Eyes 28pc Road King Stake to Tap Mainland Sector," *SCMP,* 1 May 1998, B-1.
21. "Arms Dealer in $5.7 Million Plea," *SCMP,* 1 February 1998, p.3.
22. Ibid., p.707.
23. See report in *SCMP,* 30 March 1998, p.1. While on a European tour, Zhu further explained that one important reason why China would hold out against devaluing the Chinese renminbi was that it would badly damage the Hong Kong SAR's economy; *SCMP,* 6 April 1998, p.1.
24. Fung's address before the Heritage Foundation, in Washington, D.C., on 6 January 1996.
25. While visiting Shanghai in June 1998, Clinton announced that the United States would not support a separatist Taiwan but encouraged both Taipei and Beijing to seek a peaceful solution to their conflict through direct talks. *SCMP,* 1 July 1998, p.10.
26. "Freighter's Arrival to Herald Warning of Ties," *SCMP,* 3 March 1998, p.1, citing Taiwan's *China Times Express* as its source.
27. "Malaysia Offers Venue for Unification Talks," *SCMP,* 6 March 1998, p.12.
28. Taipei's official reason is that the "one country, two systems" model will not work in Hong Kong.
29. See data in the *Economist,* 26 September 1998, p.120.
30. *China 2020,* Washington, D.C.: The World Bank, 1997, particularly pp.21 and 104.
31. I am indebted to Mr. Chie-ming Wu, Director of the Far East Trade Service, a Taiwan outfit in Hong Kong, for making these data available to me at my request.
32. Kenneth Lieberthal, in a position paper drafted for the Asia Society (New York) for dissemination to the mass media in preparation for the forthcoming handover of Hong Kong to China in 1997, p.23.

References

Brown, Deborah. 1997. "Beijing and Taipei in Hong Kong: Confluence or Conflict?" *American Asian Review* 15, no. 2: 1–30.

Ho, Loksang. 1997. "The Economy of Hong Kong as a Special Administrative Region of China." *Asian Affairs* 24, no. 4: 227–36.

Hsiung, James C. 1997. "Hong Kong as a Non-Sovereign International Actor." *Asian Affairs: an American Review,* 24, no. 4: 237–245. Washington, D.C.: Heldref Publications.

Hu, Wei-xing. 1997. "Hong Kong in Sino-U.S. Relations." Transcripts from a forum discussion sponsored by the One Country, Two Systems Economic Research Institute, 6 December 1997. I also participated in the discussion.

Kaulman, Clyde. 1996. "Asia Pacific Economic Links and the Future of Hong Kong," in a special issue on Hong Kong of *The Annals* (September) of the American Academy of Political and Social Science.

Kueh, Y. Y., and Thomas Voon. 1997. "The Role of Hong Kong in SinoAmerican Economic Relations." *The Political Economy of Sino-American Relations.* Edited by Y.Y. Kueh. Hong Kong: Hong Kong University Press.

Mushkat, Roda. 1997. *One Country, Two International Personalities.* Hong Kong: Hong Kong University Press.

Sung, Yun-wing. 1997. "The Hong Kong Economy through the 1997 Barrier." *Asian Survey* 37: 8–709.

Tang, James. 1993. "Hong Kong's International Status." *Pacific Review 6, no.* 3: 205–15.

Ting, Wai. 1997. "The External Relations and International Status of Hong Kong." *Occasional Papers in Contemporary Asian Studies,* no. 2. Baltimore, MD: University of Maryland Law School.

Wong, Timothy Ka-ying. 1997. "Post-1997 Hong Kong-Taiwan Relations." *Policy Bulletin,* no. 2 (November), Hong Kong Institute of Policy Research.

Xie, Yuchu. 1997. "Hong Kong's Ideological Influence on China." *China's Strategic Review* 11, no. 3: 1–8 (Washington, D.C.).

Zhang, Dumeng. 1997. "The Economic Integration of Hong Kong and Mainland China After 1997." *China Strategic Review* 11, no. 3:914 (Washington, D.C.).

CHAPTER 8

Hong Kong, China, and the United States

Danny S. L. Paau

Neglected Factor in the Doomsday Prophecies

In February 1997, five months before Hong Kong's return to China, a local newspaper published a cartoon showing a battered British soldier in Roman armor, holding a broken sword and shield on which the words "Western democracy" were inscribed. He looked beat up and frustrated standing among destroyed weapons deserted alongside a torn Union Jack on the ground. Also shown approaching him was an American army contingent, led by a general, also in Roman military gear, who declared: "Let us take over from here."[1] Whether the cartoon meant to express approval or resistance to U.S. involvement, or whether the cartoonist thought the United States was interested only in promoting democracy in Hong Kong, was indeterminate. One thing was certain: Britain could be gone and China would come in, but the United States of America would also be a major factor in Hong Kong's future, as the cartoon seemed to suggest.

This points to a serious omission in most of the literature about Hong Kong's fate that had mushroomed in the countdown to the territory's return: that outside forces or influences other than those from China should also be taken into serious consideration so as to more comprehensively appreciate the issues Hong Kong needed to face. The truth is: the once popular and seemingly omniscient doomsday prophecies on the fate of the former British colony had, without exception, focused wrongly only on a China whose aspirations and priorities they failed to understand. Indeed, the very dim pictures these prognostications had painted, ranging from a general downfall of its economy, politics, law and order, to "recolonization" or downright exploitation of all its resources by China[2] failed to materialize, just as they neglected other "outside" influences. Even today, unlike the above-mentioned cartoonist, scholars concerned with Hong Kong still have not learned to look beyond the China Mainland or its supporters in the former British colony for the probable even latent sources of real troubles for the little postcolonial enclave. While their visions unfortunately tunneled toward the Sino-British political wrangling over democracy and rights issues, most observers tended to neglect other problems as well as their sources, several of which have long loomed on the horizon.

Doomsday prophecies aside, being a rapidly developing and extremely busy city, Hong Kong does nevertheless have its fair share of troubles. Approximately two years before 1997, this author had begun trying, through lectures, conference presentations, and writings to alert the overseas academic community to some of these urgent problems as well as their probable sources beyond China. To begin with, post-colonial Hong Kong was much prone to problems arising from sources other than China; from itself, from certain British "legacies," and from the international community. One important reason for problems to spring from within the territory itself was the 13-year long Sino-British confrontation and feuds, 1984 through 1997, after the inking of the Sino-British Joint Declaration on Hong Kong's return. The long, vicious political wrangling had exhausted government energy and time and had also focused attention, both local and overseas, too narrowly on political issues, democracy, human rights, and fundamental differences between the two countries. The result of this was that other problems—some very urgent, serious and difficult—had thus been neglected, sidetracking their prompt handling. Just a few months after the handover, some of these problems such as housing, speculation, employment, and others emerged, having already grown into insurmountable proportions.[3] I had elsewhere discussed certain problems caused, even allegedly "created" or made worse by questionable eleventh-hour acts of the departing British rulers. For example, the British had claimed merits for finally having bestowed democracy upon Hong Kong though only on the eve of departure. Few outsiders are, however, aware that the British had also seriously infringed on the political rights of indigenous citizens by granting equivalent rights and even preferential treatment to expatriates who needed neither to pledge loyalty nor to show commitment to Hong Kong or China. Also, the departing colonial masters had also allegedly tried to undermine the already weak national identity among the young, or at least further complicated matters by some eleventh-hour acts in education and law. In contrast, allegations that the British had also attempted to retain lucrative and powerful senior civil service positions in expatriates' (mainly British) hands[4] paled against the alleged "curious" moves in other areas. Few have yet to query these questionable activities of the previous colonial government though some concerned educators recently began to wonder why, over two years after reunification, identification with the motherland had not increased among the young, and how that would affect them in the face of Hong Kong's expected increasing integration with the Mainland. Other possible outside influences in Hong Kong other than China have, until now, not attracted academic attention.

Yet, recent incidents demanded that Hong Kong people face the reality: that problems and pressures from the outside are much harder to deal with than those from either within the territory itself or from the north (i.e., China proper). The Asian financial crisis, as well as the series of legal-political crises in recent months brought the territory to the realization that other nations, particularly the United States, could level criticisms and mount pressures and even threats on Hong Kong if they are for any reason displeased with the tiny enclave. Also, there could be no doubt that after reversion it could well become a U.S. card or a pawn against China. This essay therefore seeks to examine post-reunion Hong Kong between the two titans, particularly during the recent deterioration in Sino-U.S. relations.

It is necessary to point out that though Hong Kong felt the brunt of U.S. pressures more forcefully in recent months, American interest and involvement in the territory has been a long, continuous process, and noticeably intensified 15 years ago when Hong Kong was destined to return to China. A brief review of the territory's place between the U.S. and China in the pre-handover days is useful.

Hong Kong and Sino-U.S. Relations before 1997

As one of a very few American scholars who had conducted in-depth studies of American involvement in Hong Kong affairs, Nancy Tucker noted that the United States has displayed an intense interest in the territory ever since it was ceded to the British. The U.S. interest can be explained as Hong Kong has long been perceived, first, as a foothold and a pathway to the vast Mainland of China, second, as an intelligence-gathering station in the Cold War, and, third, as the United States's strategic business hub in Asia. Tucker pointed to the fact that the United States was the first to open a consulate office in Hong Kong in 1843 and had its citizens on the founding committee of the Hong Kong and Shanghai Bank in 1865.[5] She need not remind us, to be sure, that Hong Kong has also long been one of America's largest information collection centers, and has housed regional headquarters of most U.S. companies with an interest in Asia. Moreover, America had made efforts since World War II to promote its culture, values, prestige, and interests through programs and inroads into local higher education, possibly in competition with the British.[6] The U.S. financed education programs based in Hong Kong "to turn overseas Chinese away from China" for higher education throughout the 1950s and 1960.[7] American interests, policies, and deeds in Hong Kong before and during the early half of the Cold War were therefore defined by U.S. visions, concerns, and aspirations toward China.

The China factor still figured prominently in U.S. positions regarding Hong Kong in the 1970s. Though relations between the two nations began to thaw in the 1970s, and formal relations were finally established in 1979, Taiwan and related issues such as arms sales continued to hamper the consolidation of a friendly relationship. By 1984, Hong Kong added a new issue to U.S. China policy. The Sino-British Joint Declaration, signed that year, made it clear that China was to recover Hong Kong. As it was made known as long as 13 years before reunification would take place, the U.S. had plenty of time to ponder on the consequences of the return—with Hong Kong emerging as a new factor in the problematic Sino-U.S. relations. Besides ideological and rhetorical reasons, the prospect that a huge amount of financial and human resources, technological and business know-how would join China, prompted the United States to watch over Hong Kong keenly and to search for safeguards. Tucker noted that although the United States had played only "largely a peripheral part" in the often white hot negotiations, that "did not signal a lack of American concern, but rather a lack of formal standing in the process."[8] In reality, besides Britain, the U.S. was indeed the most visibly concerned. It had been vocal and actively involved in Hong Kong matters before the handover. As early as 1984, for example, immediately after China and Britain reached the agreement to return the prize of the 1840 Opium War, the Republican Party platform "explicitly" called for self-determination for Hong Kong, an advocacy not unlike to those thrown about for Taiwan, Tibet, and Xinjiang, frontier regions where Mainland Chinese control is weak and vulnerable to challenge. In this sense, the Republicans' call seemed to suggest that some Americans viewed Hong Kong as another frontier where China's control could be challenged. This can perhaps help explain some of the U.S. approaches to Hong Kong's return.

The Cold War had ended in the countdown to Hong Kong's return. However, Sino-U.S. relations had become not better but worse. As pointed above, U.S. interest in Hong Kong grew strong as Britain was preparing to leave. Its interest and actions on Hong Kong further intensified as Sino-American relations bottomed out in 1989. Thereafter, resumption of better relations was a drawn-out, back and forth, uphill climb. Mutual accommodation between the two nations has, until now, been difficult, and frustratingly so. The early 1990s were the most hostile years between the two countries, and American actions in preparation for the handover intensified.

On 3 August 1990, the U.S. Consulate General in Hong Kong proclaimed that territory had become a factor in America's China policies.[9] Scholars would not fail to note that, just as dealing with Taiwan after official ties were broken off following normalization with Beijing, the U.S. Congress likewise enacted laws regarding Hong Kong after its return to China. In September 1991, Senator Mitch McConnell introduced "the most comprehensive" of several laws that Congress passed to protect American interests, civil liberties, and democratization in Hong Kong.[10] A revised version of this was later enacted as the all-important U. S. Hong Kong Policy Act of 1992.

Other legal preparations also took place in anticipation of Hong Kong's changeover. In the 1990 Immigration Act, the U.S. increased the quota for Hong Kong immigrants from 600 in 1987 to 5,000, and then 10,000 per year. They were also granted the flexibility to emigrate to the United States as late as 2002. This was widely welcomed by the locals, as it was considered necessary for stemming a possible mass exodus and stabilizing the territory. Other measures were, however, received with some reservations. For example, worrying about a possible demise of press freedom as the handover approached, the U.S. announced the setting up of a special quota of 3,000 visas for journalists just prior to July 1997. At least those wary of U.S. intentions noted with suspicion that the U.S. Consulate-General had been very actively visiting media institutes to ask if they would stop reporting on the Democrats after the handover. A sample from those who became suspicious is cited below:

> These so-called questions [were] actually pressures [on local journalists] to ensure that American strategic interests in Hong Kong could be maintained just as when [it was] under British rule. An organization named the Freedom Forum Asian Center has long invited editors of newspapers to seminars, "discussing" how to report on the Legislative Council election. The goal of this . . . was very clear. It was to ensure that certain people [candidates] could "safely" enter the Legislative Council.[11]

This worry might have some grounds as the United States responded well, at least verbally, to Chris Patten's attempt to "internationalize" the issue of Hong Kong—seeking support for his unilateral program to hasten the democratization of Hong Kong, ignoring specifications in the Basic Law.[12] At any rate, American interest in the agenda of the Democrats was well demonstrated. For instance, American electioneering expertise was reported to have contributed much to the landslide victories of the Democrats, first in the 1991 Legislative Council elections and then in 1995. Also, Martin Lee, leader of the Democratic Party, often dubbed Hong Kong's "Yeltsin" and "Dalai Lama," was often invited to deliver speeches in America and give

Congressional testimonies.[13] The number of Hong Kong pilgrims seeking support in the United States were plentiful, and we need not discuss the topic further.

On the other hand, though Tucker thought that the United States lacked a "formal standing" to be involved in the long and often white-hot disputes between Britain and China over Hong Kong, the United States was often ready to support the British. Some of the debated issues could eventually be resolved through mutual understanding and accommodation, but others could not. The United States, however, seemed to support Britain, almost indiscriminately, in all its arguments with China. The truth was: not all arguments between the United Kingdom and China ended the same way, and at times the U.S. support for the British could be too "quick." China was able to come to terms with Britain on, for example, the arrangement for the Court of Final Appeal. Regarding Patten's unilateral increase of directly elected seats in the Legislative Council, however, China believed he had departed from the detailed specifications of the Basic Law. The Legislative Council, elected in 1995, followed Patten's design, and China refused to allow it to "ride the through train" to continue beyond the handover date. Instead, it set up a "Provisional Legislative Council" to temporarily fill the void until a new legislature could be elected, closely adhering to the specifications of the Basic Law. Just as in other disputes, the United States supported the British on this particular score.[14] It therefore also opposed the setting up of the Provisional Legislative Council.[15] It is necessary to point out that from 1989 through 1996, the second half of Hong Kong's countdown to reunification with China coincided with the most confrontational period since the inauguration of U.S.-Chinese diplomatic ties in 1979. As late as April 1997, the United States kept on urging or warning China to stick by the British interpretation of the 1984 Joint Declaration, particularly in the preservation of all liberties.[16] Also, not until May 1997 did the White House stop criticizing the SAR Chief Executive-elect, Tung Chee-hwa, for proposed revisions on sections of the law concerning social organizations and public security.[17] These were hurriedly revised under Governor Patten to cut government powers on the eve of the handover. Fortunately, realizing the danger of conflicts going out of control during the 1996 Taiwan Strait military confrontation, the two titans have since reviewed their policies and made great efforts to amend their relations. So, by July 1997, Sino-U.S. relations were good enough to ward off American open criticisms, at least for the moment. After reversion day, China carefully stood by its hands-off policy, and before the full impact of the Asian financial crisis arrived, it was a period of rosy anticipations. And, Hong Kong was momentarily left alone from harsh criticisms and pressures from the international community.

In sum, U.S. involvement in Hong Kong prior to reversion had not been welcomed by China. Washington had demonstrated strong suspicions about China's intentions toward the former British colony and, regardless of what was specified in black and white in the Sino-U.K. Joint Declaration and in the Basic Law, it took the side of the British, and brought much pressure to bear on the Mainland as well as on Hong Kong. In fact, regardless of its stated acceptance of Hong Kong's return in the 1992 U.S. Hong Kong Policy Act, enacted by Congress, the United States at times seemed to barely stop short of opposing China's resumption of sovereignty over Hong Kong. Therefore, there seem to be ample reasons for China to suspect that the United States intended to meddle in Hong Kong's internal affairs, just as it has meddled in the affairs of Tibet, Xinjiang, and Taiwan. The joy over Hong Kong's return was tainted by an unexpected paradox: a colonial master's departure did not bring full national control by the new sovereign. Hong Kong was to return to China, and the outside world, including the United States, was worried that it would not really exercise the high degree of autonomy it was promised. The Americans were here to fill some voids that China itself could not.

Opportune Time for Intervention?
Recent Deterioration in Sino-U.S. Relations

Since 1989, Sino-U.S. relations have been highly confrontational. Tensions and mutual distrust continued to build up, climaxing in the 1996 Taiwan Strait crisis, which brought the two giants to the brink of war. The year 1997 saw efforts to repair their mutual relations. Jiang Zemin's visit to the U.S. in late 1997 was reciprocated by Bill Clinton, who visited China in June 1998. The apparent goodwill of the two leaders was, unfortunately, not enough. The claim to have formed a "strategic partnership" brought no lasting friendship. Immediately after Jiang's Washington D.C. visit, for example, the U.S. State Department appointed a coordinator for Tibetan affairs.[18] Although the Chinese played cool and held their own in the Asian financial crisis, keeping the renminbi from devaluation and thus averting further deterioration of economic stability worldwide, a series of unpleasant events erupted to threaten the existing tenuous ties between the two, particularly in 1999. These included the ominous scandal about theft of U.S. nuclear secrets and the baffling U.S. bombing of the Chinese Embassy in Belgrade. It was against this background of the rapid deterioration in Sino-American relations that the Hong Kong SAR government's unprecedented intervention in the market in 1998 caught fire in Washington. It is therefore necessary to understand the background in which several controversies sprang up almost simultaneously, to spiral to appalling proportions in a short period of time.

What caught the eye of most scholars in the recent series of rapidly worsening Sino-U.S. disputes was, to be sure, the alleged theft of sensitive American military technology by "Chinese spies." This is, to be sure, nothing new in an atmosphere of mutual suspicion. China is well-known to have been seeking high technological inputs, both military and non-military. On the other hand, it is also true that China is known to be extremely sensitive towards possible leakage of its own military secrets to the United States, as was shown in the Hua Di case of January, 1998.[19] Before the recent crisis, in November 1997, the U.S. Congress adopted a resolution requiring the CIA and FBI to report annually on Chinese economic, political, and military espionage in the United States.[20] In February 1999, the Republican-controlled Congress decided to block the sale of satellites worth U.S.$450 million to Asia Pacific Mobile Telecommunications, alleging that the sophisticated antennae could be useful for the Chinese military, in spite of the fact that the supplier had already reportedly spent over U.S.$100 million in preparation for the sale.[21] It should also be noted that at about the same time, a number of senior level managers of Chinese enterprises in America, including some who had already been in the United States, were denied visas or the renewal of them. This created serious problems of continued supervision at many Chinese enterprises, each of which employs hundreds of American workers.[22] But, fears for espionage ran high, despite the fact that not enough evidence was available to indict Lee Wen Ho, the Taiwanese scientist at the Los Alamos Nuclear Lab whose dismissal was at the center of the whole scandal.

Another serious controversy rose over China's alleged militant postures against Taiwan, and over reports that Taiwan might join the controversial Theater Missile Defense (TMD) scheme, which would make the island a virtual military ally of the United States. In late February 1999, the U.S. Defense Department released a report claiming that the Taiwan army was ill-prepared for an attack from the Mainland. Besides, Taiwan reportedly suffered low morale owing to the recently discovered series of military scandals, and it was unable to retain the technological personnel needed for its defense program. In contrast, the China Mainland had reportedly built guided and cruise missiles. The Pentagon report predicted that the Mainland would have "overwhelming superiority" over Taiwan by 2005.[23] It was also claimed that the Mainland could launch an attack against Taiwan as early as 2,000, the year for Taiwan's presidential election.[24] All this contradicted a finding by the Pentagon, about five months earlier, that unless Taiwan declared itself a separatist independent polity, cutting off its umbilical chord to the Mainland, China would be unlikely to initiate a major war within 15 years, as its immediate goal, according to the same report, was to get rich and to acquire big power status.[25] A series of U.S. dignitaries have since then visited Taiwan, including William Perry, former Secretary of Defense. But this did not help to ease tensions between China and the United States.[26] China's response could easily be predicted: it promptly refuted the allegations of spying or theft, citing leading U.S. journals including the *Los Angeles Times, Christian Science Monitor, Wall Street Journal,* and *U.S. News and World Report.*[27] What China dreaded most, on the other hand, was the reported possibility that Taiwan might be included in the TMD program, making the island a de facto military partner of the U.S. and Japan. China considered this "the last straw" and issued stern warnings to the United States.[28] Yet, when U.S. Secretary of State Madeleine Albright visited China, she made a series of criticisms, including the allegation that China had pointed hundreds of guided missiles toward Taiwan. She did not respond to China's request for a clarification on whether Taiwan would be included in the TMD plan.[29] It was under such an atmosphere that reports like "Manila Looks to U.S. in Spratly's Dispute [with China]" appeared to further increase mutual suspicions and hostilities.[30]

Other more "recurrent" and "customary" conflicts continued. In late February 1999, the U.S. State Department released a report on China's human rights record, which had allegedly "deteriorated sharply" recently, mentioning particularly the crackdown on "organized political dissent" and alleged torture, arbitrary arrests, and detention without public trial of Tibetan nationalists.[31] China's response was furious, as can be anticipated.[32]

Sino-U.S. conflict has never really ceased since the brief honeymoon months shortly after the two nations established formal ties in 1979. American comments, criticisms, advice, or threats also began to arouse renewed attention in the 1980s, as the British were known to be preparing for their exit from Hong Kong. Though, in general, most foreign powers are satisfied with how Beijing has treated Hong Kong after its reversion, the U.S. has continued its policy of close observation on how things are going in Hong Kong. Immediately after the completion of the summit diplomacy in 1997–98, criticisms of China (and counter-criticisms) revived and intensified.[33] Also, Americans have been most vocal toward Hong Kong since the summer of 1998, when the SAR government took deliberate measures to curb externally fueled speculation in the markets. The first quarter of 1999 saw Sino-U.S. relations at their worst since the 1996 Taiwan Strait military confrontation. Similarly, when the most serious legal crises broke out in Hong Kong during the same months, the Americans delivered their seemingly most concerted and forthright criticisms. Whether or not such concurrence suggests a correlation between U.S. policies toward China and Hong Kong, the simple fact that such stern reactions came at the worst possible time demands prompt attention.[34]

Prior to examining the legal crises and U.S. reactions, let us first study American criticisms of the SAR government, last summer and later on, regarding the latter's handling of market speculation by international hedge funds. This was significant as it was the first major criticism of Hong Kong since its reversion to China.

Asian Financial Crisis and
U.S. Displeasure with Hong Kong

A scholar who had examined closely the series of U.S. Congressional debates leading to the enactment of the U.S. Hong Kong Policy Act of 1992 noted three major concerns of the United States regarding post-handover Hong Kong. In descending order, they were economic factors, democratization, and human rights. The contents of the 1992 Act, upon close scrutiny, outlined mainly "the policy of the United States which expressed exclusively economic considerations."[35] American financial interests and visions for the post-reversion era have long been well-appreciated, and thereby explored, by economists as well as historians who have studied the text of the U.S. Hong Kong Policy Act of 1992 or various documents of the hearings. Suffice it to note that, in the eyes of many observers, the trade value of Hong Kong is unquestioned. It is not only a place where the average resident purchased "an incredible $1,300 worth of U.S. products in 1991,"[36] but also where, in 1996, the Hong Kong Shanghai Bank boasted a *daily net profit* of HK$100 million (U.S.$12.8 million)![37] It is therefore not surprising that one of the key elements in the U.S. legislation regarding Hong Kong's return was to ensure that, at least in trade relations, preservation of the status quo would receive top priority. Hong Kong was to continue to be accorded the status of an economic entity independent of China. Though it had long been noted that the territory's economy had been more and more integrated with that of the Mainland, it was not the intention of either the United States or the Mainland to see Hong Kong caught in the trade conflicts of the two titans.[38] In spite of worries of local businessmen to the contrary, Hong Kong has so far been left out in the numerous Sino-U.S. trade arguments. Also, as Hong Kong considers the U.S. its most important and beneficial trade partner, there have been few disputes regarding business matters since reversion. U.S. differences with China in business matters, besides, have rarely affected Hong Kong, as the latter is accorded different and usually more privileged trade conditions than is the Mainland. Occasionally, the U.S. might urge Hong Kong to try harder to curb products piracy or might even threaten to place the territory on the observation list under the 301 Clause. On the whole, however, the U.S. has no major business conflicts with Hong Kong. After reversion, American businesses have moved fast to fill up whatever void is created by the departure of the British, who no longer enjoy the previous colonial privileges or concessions. The Asian financial crisis, however, saw the U.S. making its sternest criticisms and threats to the SAR government, which rocked the tiny territory.

There is no need, to be sure, to give a detailed account of the "landing" of the Asian financial crisis in Hong Kong (see Chapter 9). It is well-known that, although not immune to the crisis, Hong Kong has so far fared much better than most of its Asian neighbors. It is important to note, however, two important, but often overlooked, points. First, general well-being has since the affluent 1980s obscured much of the weakness and vulnerability of the Hong Kong economy, which has perhaps not adapted adequately to competition from its neighbors. Rampant speculation in the name of economic liberalism, coupled with the alleged scarcity of land, has for decades kept Hong Kong property prices, for both business and residential purposes, the highest in the world. Consequently, this has pushed up wages, cost of living, and business operating costs, resulting in what may be called a "bubble economy."[39] Even worse, as almost no trade could bring as rapid returns as speculation, ranging from stocks to properties, manufacturing enterprises have withered, many of which have relocated to the north (China proper) or to other cheaper neighboring localities. Many people in Hong Kong have long since taken to speculation, to make fast money, as their "proper" employment.[40] Factors contributing to this unique Hong Kong "work culture," are plentiful and complicated, and they should be examined in depth by economic experts. I have neither the intention nor the need to discuss them thoroughly here.[41] It is important to note that, as speculation ran wild, Hong Kong has since become one of the favorite haunts of international hedge funds, including major players from the United States. The second overlooked point is that though the doomsday prophecies before the handover convinced few locals except those who read only English materials, in the last few months before the handover, speculators actually painted an overly rosy picture about the economy. In December 1996, for example, the Hang Seng Index was around 10,000 points. After eight months, however, it had shot up to an incredible 16,000! During the same period, property prices also rose by one-third. There was no comparable break in either trade or business to justify such phenomenal market growth. While the majority of the people in Hong Kong were, by and large, resigned to their destiny, their confidence in the Mainland increased, as Hong Kong's economy was performing miraculously well. Such unfounded optimism, when examined in hindsight, seemed to suggest what in reality had been a false imagery of economic prosperity, both created and shielded by such heated speculation in the markets. Further, driven by a false sense of confidence and unrealistic optimism, the amateur speculators could become even bolder trying to make a "fast buck." The combination of all this made Hong Kong a particularly vulnerable victim of powerful international hedge funds when the Asian financial crisis loomed on the horizon. It is important to note that, with their tremendous resources, sophistication, and superior financial skills, American companies could cruise in at will and make a killing on the local markets.

By January 1998 the Hang Seng Index had dropped sharply to 7,900 points, inducing some panic. Mammoth international hedge funds, spearheaded by some American wizards,[42] swooped down on the Hong Kong dollar by October 1997, after having brought down most Asian currencies along with their national economies. The SAR suffered huge losses defending its peg to the U.S. dollar; and some usually lucrative trades such as tourism dwindled to a trickle, as most Asian markets had collapsed. In early 1998, a number of "investment" or "finance" companies, where small players speculated, were heading toward collapse. These companies had "embezzled" their clients' money in speculating, and, in the general economic downturn in Asia, they suffered huge losses. As they normally served hundreds of small players, the impact of their liquidation was devastating and widespread.[43] Both the local stock market and the property market were seemingly on their way to total ruin, and the Hong Kong dollar's collapse seemed imminent in the long hot summer of 1998. The SAR government became the convenient scapegoat, and allegations ran wild, ranging from charges of incompetence and inadequate supervision over the collapsed companies, to inaction. Politicians roasted the SAR government despite the fact that, in reality, the same high officials held over from the previous Patten administration were running the same public monitoring agencies.[44] It was under such circumstances that the SAR government unexpectedly "struck back" at the attackers on the local currency on 14 August 1998. With the avowed backing of China's central bank, the SAR government launched a surprising HK$120 billion (U.S.$15.39 billion) sally into the local stock markets, which were widely reported and need not be repeated here.[45] As soon as the government intervention became public, it was greeted by a barrage of severe criticisms from economists, lawyers, democrats, and liberals as well as the international community, particularly the United States.[46] The latter was not any less caustic in its reactions than local critics.[47] Academics and economists criticized the government for having trodden into a "forbidden" area. Others were alarmed at the huge amount of money involved, and legislators labeled as "democrats" were angry that the government had acted on its own.[48] The SAR government faced, in August 1998, its most serious crisis since its inauguration in July 1997.

It is not the aim of this essay to debate the comparative merits of economic liberalism or laissez-faire versus government interventionism.[49] Nor is it necessary to detail the strong warnings from American government and institutions to the SAR government for its alleged violation of the sacred doctrine of economic liberalism. It is important to note, however, that this time American criticisms and pressures seemed to be concerted and were backed by actual threats by some giant investors. In mid September 1998, it was reported that, showing displeasure at the SAR government's intervention and follow-up suggestions to monitor more closely the "big players," four large American investment companies including Morgan Stanley and Merrill Lynch threatened to withdraw from Hong Kong.[50] As the Hong Kong economy was struggling, these external criticisms and threats added to the tremendous pressures already weighing down on the SAR government. Fortunately, the general public welcomed its resolute, though probably very risky, action. Also, the measures seemed to have worked well.[51] At least the ambush on the Hong Kong dollar stopped immediately and Hong Kong was given a short respite. The bold action backed by two of the world's largest foreign reserves (those of Hong Kong and China) seemed to have scared off, if not fought off, the speculators, at least temporarily. Having overcome their initial worries, many journals turned around to criticize the opposing academics and politicians, the former for allegedly not basing their objection on "rational and serious" grounds, and the latter for "lacking even the basics" of knowledge about finance and economics.[52] The Hong Kong government went on to overhaul the exchange and securities supervisory mechanisms, buoyed by the approval of the general public.[53] To the relief of all, including the United States, the SAR government did not undertake any additional intervention after the 14 August 1998 episode. Also, immediately afterwards, it undertook a series of public relations efforts to explain and dispel doubts from Western nations, particularly the United States. For example, in October 1998, the Financial Secretary made a presentation at a special meeting of the G-22 group of finance ministers from industrial and emerging countries at the World Bank-IMF meeting in Washington D.C., and reportedly won some understanding.[54]

Although economists remained divided until now about whether such an intervention was necessary, appropriate, or even successful, Hong Kong has largely survived this drama. First, the much feared huge loss in a big gamble did not materialize, as international speculators have at least temporarily backed off. In fact, according to latest reports, the bold action had actually brought a net gain of HK$35 billion (U.S.$4.48 billion) in six months, by the end of 1998; and it is expected to bring in more when the economy further improves.[55] Second, much more importantly, the resolute and dramatic act and its apparent success, which surfaced so quickly, meant that the debates quickly subsided. Embarrassed and bewildered laissez-faire economists retreated from the limelight to their libraries, and politicians who had barked too quickly turned away from the unfamiliar field of money game. The silent domestic scene perhaps discouraged further barrages from the outside. The above-mentioned timely public relations efforts of the SAR officials seemed to have many a sympathetic ear, if not forgiveness. It is apparent, however, that whether or not the SAR government was right in intervening in defense of the Hong Kong dollar, it did not really please the United States. As late as January 1999, U.S. Consul General, Richard A. Boucher, still listed the local government's "intervention in the stock market" as one of the events that "raised questions about whether or not Hong Kong may be compromising a bit" on the "basic principle" of freedom.[56] He was quick, however, to add that he was aware that sometimes Hong Kong had to make "hard decisions" under "complex situations."

Yet, aside from Hong Kong's desperate move to ward off the attacks by international hedge funds, there was little that the tiny territory could do to displease the United States regarding financial matters. After all, America is so important a business partner that the Hong Kong dollar is pegged to the greenback; and the SAR government has persisted in defending the peg, even at a high price during the Asian financial crisis. Perhaps more than usual, Hong Kong had more than this one issue on which it had to respond to American criticisms. For example, though the new airport had been accused of charging very high landing prices for airplanes even before it was open, Hong Kong did not respond right away. Recently, however, in response to complaints from United Airlines and DHL, the dispatcher, the Airport Authority was quick to announce that a review was being conducted, and it hinted at possibly lowering the landing charges.[57] Furthermore, Hong Kong seeks to attract further U.S. involvement in the rejuvenation of its local economy. As pointed out above, Hong Kong's decline in competitiveness was laid bare in the financial storm, and the SAR government has put forth a very ambitious plan of economic reform-seeking, first, to push forth value-added and high technological products and, second, to strengthen Hong Kong's tourist industry. Included in the plan are, for example, proposals to build a U.S.$13 billion "cyber-port," a new pier exclusively for luxurious cruises, and hopefully also a Disney theme park in Hong Kong. It is no wonder that Professor T'ien Chang-lin, formerly President of University of California at Berkeley and now Chairman of the SAR's Commission on Innovation and Technology, claimed that the proposed undertakings would "excite financial experts in the U.S."[58] Hong Kong was quick to send business delegations to promote the economic plans and to encourage financial and technological inputs.[59] As the protracted and tough negotiations were still going on between the SAR government and Disney, all of a sudden, Microsoft's Chairman Bill Gates showed up in Hong Kong to announce a Microsoft-Hong Kong Telecom joint project. His company had just clinched a lucrative government contract to computerize unspecified "public services," and Gates announced that his company would aggressively bid for more.[60] Furthermore, Microsoft announced it would participate in the proposed cyber-port.[61] The fact that U.S. and Hong Kong business interests complement each other is clear.

As stated above, though Hong Kong's intervention in the stock market has displeased the United States and has thrown the territory against the world's lone superpower, the many Sino-American trade differences, however, have so far not affected the territory itself. This was due first to the U.S. economic policy toward Hong Kong; second, to China's low profile and hands-off policy as well as its unwillingness to drag Hong Kong into its many disagreements with America; and, third, to Hong Kong's own business value to both titans and its dexterity in maneuvering between them. In other Sino-U.S. conflicts, however, Hong Kong could not so easily navigate away from the maelstroms.

The U.S. Legal Entanglement in Hong Kong

As mentioned earlier, although the U.S. Hong Kong Policy Act of 1992 was most concerned with economic issues, nevertheless, as a scholar observed, "political rhetoric characterized"[62] its pronouncement. Also, in view of the broad support given to former Governor Patten's last minute democratization of Hong Kong and the long standing U.S. policy to push for democracy in China, it was natural that the United States continues to voice concern and to apply pressure on China regarding Hong Kong, sometimes directly on the SAR government regarding democracy and the rights issues. As discussed earlier, it has repeatedly criticized the SAR Chief Executive's Office for the proposed revision of laws pertaining to social organization and public security that had been hurriedly passed by the Legislative Council in the final weeks of British administration.[63] It strongly opposed sternly the creation of a "provisional" legislature to tackle urgent legislative matters, as China refused to allow the Legislative Council elected under Patten's formula to continue past 1 July 1997 and vowed to reelect a legislature according to specifications in the Basic Law, and pending that outcome a Provisional Legislative Council would take over in the meantime.[64] Although the criticisms were stern, they did not smolder for long. Within days of the handover, the newly installed SAR government announced that a new Legco election would be held in May 1998, barely 10 months after the Provisional Legislative Council took office. Chief Secretary for Administration Anson Chan, speaking at the Los Angeles Orange County World Affairs Council in January 1998, emphasized that the May election would be fair and open, and that no political party would be barred from running, including those individuals who had criticized the election arrangement most severely. This dispelled earlier fears that the anti-Beijing Democrats would be thrown out.[65] Later developments proved that, besides shrinking the number of seats to be directly elected to that specified in the Basic Law, democracy did not retreat with the new government. More than 60 percent registered voters turned out for the 24 May 1998 election, in spite of stormy weather, and the Democrats again won the largest number of seats of any single party, in part thanks to the coaching by experts in the art of election-eering and campaigning from the United States. In fact, democracy under the new SAR administration might have been given a chance to advance in some modest ways. Under British rule, for example, elections for Village Representatives in the New Territories were open only to "indigenous" resi-dents—those whose forefathers had lived in the village for centuries—and not to other residents though they might have lived there all their lives. In March 1999, a court ruled that such restriction amounted to violation of human rights, was unlawful, and should be abolished.[66] Hence, at least at the village level, democracy registered a modest gain, a fact that brought praise from the local media.[67]

Also, although the Provisional Legislative Council had already "undone" the last-minute revisions in laws about social organization and public secu-rity made by the departing British, there are no indications that the politi-cal rights of Hong Kong citizens have been curtailed. For example, though in theory demonstrations should be approved by the Chief of Police or the Chief Executive, no attempt, to my knowledge, has ever been made to stop a demonstration, even an unapproved one. Besides, I know of no allegation that the SAR government has tried to limit the freedom of the press. On the contrary, recently there have been public criticisms against the alleged abuse of such freedom, yellow journalism, and unnecessary invasion of privacy by local journalists.[68]

Apparently, there was no real bone to pick with Hong Kong regarding rights and democracy. It is reassuring to note that a U.S. government report made known to the public in February 1999 thus remarked that the SAR enjoyed "a high degree of autonomy," and remained "a free society with legally protected rights" thus far. It claimed that the SAR government gen-erally respected the human rights of its citizens though it alleged that there were "some degree of media self-censorship, limitations on citizens' ability to change their government and limitations on the power of the legislature to affect government policies."[69] In spite of the well-known differences in democracy and rights between China and the United States, Hong Kong after reversion has thus far managed to stay out of Sino-U.S. disputes. It was the U.S. entanglement in the recent legal crisis in Hong Kong that has placed the latter under tremendous pressure, to which we will now turn.

Before examining the U.S. entanglement in the first legal crisis in the Hong Kong SAR, which broke out at the end of January 1999, it is necessary to note that ever since its return to China, the SAR government finds local lawyers the most vocal, powerful, and persistent allies of the local politicians hostile to China. Right after return, endless lawsuits were brought to the courts. Some of the first lawsuits challenged the legitimacy of the Provisional Legislative Council and thus the legality of the laws they had passed or amended. Other lawsuits sought to bring "national institutions," such as the New China News Agency, to trial. For instance, charges were filed against the New China News Agency by radical legislator Emily Lau for failing to respond to her demand to supply her a "secret file" on herself allegedly kept by the agency. The legal profession and politicians were most sensitive in matters regarding, for example, the jurisdiction of the local courts versus that of the courts in China proper. They tried unsuccessfully, for example, to secure the extradition of Cheung Chi Keung, nicknamed the "Big Spender," who was a Hong Kong resident known to have kidnapped relatives of local tycoons and had taken hundreds of millions of dollars in ransom, but sought refuge in the Mainland until he was arrested and put on trial in Guangzhou. They criticized the Secretary for Justice Elsie Leung for failing to have the Big Spender extradited to Hong Kong. The Mainland court ruled that the criminal was caught smuggling explosives while inside the Mainland and claimed full jurisdiction for itself. The accused was found guilty and sentenced to death but some among Hong Kong's legal community charged the SAR government with having succumbed to pressures from China.[70] In another case, lawyers and politicians again joined forces to try to remove the Secretary for Justice. The owner of a local English newspaper and a related business empire was allegedly involved in efforts to inflate the figures on the size of its circulation for auditing purposes (for advertisers). After reviewing the available evidence, however, Elsie Leung decided to prosecute the senior staff of the newspaper involved, but not the owner, Sally Aw. Accusations ran wild and some claimed the owner was spared indictment because of her connections in China as well as with the SAR Chief Executive. Called before the Legco, Leung explained that her decision was based on the fact that there was insufficient evidence to prosecute. This was, in fact, sufficient reason for her decision, but perhaps unwittingly she added that she had worried that, right after Hong Kong's handover, prosecuting the owner of one of the only two local English newspapers might mislead the international community to suspect that the SAR government tried to suppress press freedom. Another additional consideration she gave was that the said newspaper owner was restructuring the finance of her business empire at the moment and a prosecution might adversely affect her efforts, and, hence, the jobs of many in her employ, should her empire collapse.[71] Leung survived a vote of no confidence in the Legislative Council on 11 March 1999 as the motion was defeated by a large margin. Tension persisted, however, between her and the politicians supported by the legal profession. All the above examples of disputes were under the watchful eyes of the American Consulate General headed by Richard A. Boucher. In an address to the Hong Kong General Chamber of Commerce on 25 January 1999, he took note of all the legal disputes, and wondered aloud whether "this series of events" had not "raised questions" about whether Hong Kong might have "compromised a bit" on "the fundamentals—freedom, rule of law, local autonomy etc."[72] This background paves the way for our discussion below of the legal crisis (some called it a constitutional crisis).

As I have pointed out elsewhere, one of the major problems the departing British government chose to ignore and left unsolved was the question of Hong Kong residents' children born on the Mainland.[73] A considerable number of Hong Kong residents had married across the border, or simply had sired children there. As Hong Kong is small and expensive, however, their families could not readily come to join the bread-earning fathers and many live in neighboring cities like Shenzhen. To complicate matters, married men in Hong Kong who sired offspring out of wedlock within China proper may, at times, only want their children to come to Hong Kong, but not the children's mothers. In any event, these dependents often have to wait for many years for permits to leave the Mainland, and for the quota to become available for immigration to Hong Kong. Many would sneak into the territory as illegal immigrants. But, illegal immigrants have no chance of acquiring proper credentials and identity papers. They cannot enjoy public facilities including government hospitals and schools, and, if caught, would be sent back immediately by force.

During British rule, though, this had caused many tragedies in which wives and children could not claim the right of abode as they were not British citizens. After reversion, the problem immediately surfaced. These dependents are Chinese citizens and Hong Kong is now a part of China. Thus, many who came to the territory on travel permits (which required them to return upon expiry) and even illegals, who sneaked in without proper permits, simply refused to leave. Right after the changeover, on 3 July 1997, thousands of children without permits flooded the Immigration Department demanding the right of abode. The Provisional Legislative Council held emergency sessions and passed revisions in the immigration laws, requiring overstaying or undocumented children to return to China and wait for their permits. With lawyers and human rights activists, they took their case to the court, supported by legal aid expenses provided by the SAR government beginning October 1997. Every time they lost they would immediately appeal to a higher level court. Eventually the case went to the Court of Final Appeal (CFA), the tribunal of highest instance. On 29 January 1999, the CFA made the "landmark ruling"[74] that these children had the right of abode. Surveys conducted when the issue first erupted in July 1997 revealed that an overwhelming majority of Hong Kong residents opposed a sudden influx of Mainland children as they worried about over-strained social services and the resource allocation implications.[75] Criticisms from the Mainland quickly turned the matter into what Standard & Poor, the American credit rating agency, labeled as "the first of controversies" between local courts and the Mainland legal authorities. Many began to worry about the ultimate consequences.[76] Many among the politicians, academics, and the local legal profession strongly opposed any "interference" from the Mainland, while the SAR government was embarrassed, tense, and anxious to defuse the issue.

The matter quickly deteriorated as local lawyers and politicians were supported first by the British Consulate in Hong Kong and then by the United States. Calling the CFA ruling a "reaffirmation of Hong Kong's autonomy in judicial matters," the British warned that any move "to restrict the independent judicial power" of the CFA would be "a matter of serious concern to us."[77] The U.S. ConsulGeneral said the same. Again, the United States displayed a concerted effort to pressure the SAR government. Local American business people climbed on their government's bandwagon. In a statement issued by Frank Martin, Chairman of the U.S. Chamber of Commerce in Hong Kong, he echoed the need to defend judicial autonomy, and warned that any attempt to overturn the CFA would undermine international confidence. Saying that, though he did not expect an exodus of American companies should the controversy drag on, he claimed that any "perceived erosion of the rule of law" would undermine international investors' confidence in Hong Kong. Acknowledging that the ruling would bring serious problems to the territory, he urged Hong Kong not to make "concessions" so as to safeguard judicial autonomy.[78] Both the Chinese Foreign Ministry and the Hong Kong SAR government asked the powers to refrain from their internal matter.[79] A Hong Kong journalist reported from Washington D.C. that the Clinton administration "this week forcefully came out in support of the Court [of Final Appeal]," and warned both Beijing and Hong Kong not to undermine it. He noted that the *Washington Post* also "editorialized against Beijing."[80] Again, the U.S. Secretary of State, Madeleine Albright, also joined the chorus. When she visited China in early March 1999, she brought the matter up and asked "whether the judicial independence of Hong Kong had been respected."[81]

To the credit of the SAR government, the Secretary for Justice was quickly sent to "liaise" with Mainland legal experts and authorities. Then, in an unanticipated move, the SAR government "applied" for a clarification from the CFA regarding parts of its judgment particularly its working relationship with the NPC. The Court did not change its judgment about the right of abode of the children of Hong Kong residents, but clarified that it accepted the supreme power, in the interpretation of laws, including the Basic Law (Hong Kong's mini-constitution), of the National People's Congress and its Standing Committee.[82] Local newspapers appreciated this unusual action, calling it the "Best Way Out"[83] and expressed relief and hoped that the controversy would be over immediately.[84] The Central Government made no further criticism or demand, and most Hong Kong citizens, other than barristers and the more radical Democrats, were glad the crisis was over. When Secretary Albright asked about the case in Beijing in early March 1999, Qian Qichen, Vice Premier in charge of Hong Kong affairs, replied that the matter was "solved."[85] Yet it would be wrong to consider U.S. pressures over. Martin Lee, head of the Democratic Party, for example, while speaking to the American Chamber of Commerce in Hong Kong, claimed that the rule of law had been undermined, that the government had pressured the highest court of the territory, that Hong Kong was having "two courts of final appeal." He threatened foreign businessmen that they might be caught one day in litigation against China-affiliated bodies if they did not defend the rule of law now.[86] It was reported as late as April 1999 that in a U.S. State Department annual report to Congress on Hong Kong, it alleged that "pressure" had been applied on the CFA regarding the "clarification." It also complained that weeks after the ruling the SAR government had not yet taken steps to establish procedures for Mainland residents to apply for Certificates of Entitlement for residency, and warned the Central Government against any further remarks on the issue.[87] This continued concern and criticism demands that we look deeper into the core of the controversy.

The so-called "Right of Abode" saga has raised many questions to those who are concerned. To parents of the Mainland-born children, this would be an emotional problem of family reunion. To lawyers and human rights activists, this was a matter of judicial independence and human rights. To the government it was a matter of technical and logistical worries. To those without children in the Mainland, it was a matter of resource straining and a question of orderly migration. To Mainland authorities, the matter could be concerns about children born out of wedlock. To those now waiting for the permit and quota, it would be a matter of fairness; of whether others should be allowed to jump the queue. To most who were worried, it was the very large number (estimated from one to three million) of people[88] and the possible further straining of Hong Kong's dwindling resources.[89] All these are real and serious problems. The core of the so-called "constitutional crisis" had, however, little to do with migration. It is necessary to remember that the Mainland authorities helped control the number of people migrating into Hong Kong because the latter might not be able to handle such large influxes. There would, in fact, be nothing for them to lose, but everything to gain, to have more comrades coming to the territory, relieving the hometown of employment pressures, and possibly creating a new source of revenue from remittances back from Hong Kong.

In this total picture, U.S. criticisms thus fell on the wrong mark. It was significant that solicitor Pam Baker, who had been helping Mainlanders in their battle for the right of abode in Hong Kong, criticized the U.S. report and denied that the SAR government had "forced" the court to clarify its ruling. This, she added, should not be surprising as the "one country, two system" framework was new and time was needed to find out and reconcile the differences.[90]

What really had caused concern among Mainland legal experts and the Central Government in the judgment was that, in sum, part of the *obiter dictum,* that is, a supplementary opinion upon a matter not essential to the decision and was therefore not binding, had been interpreted by some to mean that the local court had the power to scrutinize laws passed by the National People's Congress (NPC) and its Standing Committee and this would lead to a real constitutional crisis. One observer believed that Beijing feared that the local court seemed to have declared "independence" from laws passed by China's supreme legislative body, the NPC. Further, he claimed that Beijing feared that this could be the beginning of "independence" in other aspects: "For Beijing this big picture also includes an unceasing effort by foreign and local elements to turn Hong Kong into a 'Berlin Wall' whose collapse will have a domino effect on the Mainland."[91] In the CFA's clarification, it left the ruling on abode intact but took pains to clarify that the original ruling had not questioned the authority of the NPC's Standing Committee to make interpretations of the Basic Law which would have to be followed by Hong Kong courts, and that "the court [CFA] accepts that it cannot question that authority."[92] At that, the Central Government let the matter rest.

Yet Hong Kong, since its reunion with China, has seemed unable to avoid legal controversies. As discussed above, local radical politicians and the U.S. have not yet left the CFA issue alone. Before the dust is settled on this, however, another possible constitutional crisis appeared. On 30 March 1999, the Court of Appeal, in a case of two activists convicted of defacing the national and SAR flags, ruled that the National Flag and Emblem Ordinance and the Regional Flag and Emblem Ordinance were in breach of Article 39 of the Basic Law, which guaranteed freedom of expression. This was in spite of the fact that Annex III of the Basic Law includes the national flag law, which is binding on the SAR, and makes burning of the Chinese national flag a criminal offense.[93] Unlike the right of abode question, the appellate court's ruling could have significant political implications. How the Central Government is going to react, and whether another crisis is looming, possibly leaving room for international intervention, remains to be seen. Feeling the heat of international pressures and worrying about reactions from the Chinese sovereign, the more informed Hong Kong people have reasons to be worried about being dragged into Sino-U.S. hostilities. A journalist, for example, was concerned that because of the controversy regarding the CFA's judgment, Hong Kong could not "emerge unscathed from the rough patch ahead" in Sino-U.S. relations. He declared that the "SAR *needs neither to be killed with kindness nor made a scapegoat*" (emphasis added) by the United States in its dealings with China.[94]

Legal controversies have been dogging Hong Kong since reversion. Only until recently did these develop into a constitutional crisis between the territory and the Central Government, creating an imprimatur to foreign interventions. These are the kind of problems that not even the doomsday prophets had foreseen. The people of Hong Kong need to work out the solutions themselves.

Conclusions and Observations

By now, it should be clear that the United States has closely monitored Hong Kong since its reversion to China, and has often promptly voiced disapproval and applied pressures whenever things should "go wrong" in the territory. Regardless of what "reversion" and a "high degree of autonomy" might mean to the locals, so far, paradoxically, it has been the U.S. more than China who has filled the void created by the British departure. This was particularly reflected in two major crisis situations Hong Kong has faced: the Asian financial crisis and the "constitutional crisis" arising from the Court of Final Appeal's decision on the right of abode cases. Whether that would bring more help or hardship to tiny Hong Kong needs to be further studied. One thing is certain: China so far has *not* been the source of most of Hong Kong's difficulties since reversion.

The above discussion on selected sample cases was not a complete history of post-1997 developments in Hong Kong; nor was it a full account of U.S. involvement in local matters. It should, however, offer some glimpse of U.S. approach to Hong Kong; either as a tiny independent entity or as a new Chinese frontier region. For those who believe U.S. monitoring and intervention would be beneficial for Hong Kong, the discussion might also help ascertain the proximity (or the lack of it) between claimed goals and the actual outcome. Before entering such discussions, we should first address the concept of "internationalization," a term which the British first used to justify calling for international intervention in Hong Kong. On 2 July 1998, in a reception for Bill Clinton at the former colonial Governor's mansion, the SAR Chief Executive, Tung Chee-hwa spoke of the need of patriotism as

Hong Kong was returned to China. President Clinton satirically pointed to Tung's international education and business background and to the fact that two of his sons were U.S. citizens. He hinted, therefore, that internationalization, rather than nationalization (or renationalization) of Hong Kong, should rather be the right way. Tung was embarrassed and perhaps too timid to argue. This pointed, however, to a fundamental confusion about this basic concept among many who advocated and even invited intervention—that "internationalization" is not the same as "internationally controlled."[95] Most major cities including New York, London, Tokyo, etc., are highly internationalized as they are open to foreigners and can accommodate peoples, cultures, activities, and institutions of many origins. They are, to various degrees, subject to outside influences. Yet it would be wrong to assume that the mayors or people of any of these cities would ignore national sovereignty and accept control by foreign powers. Hong Kong should remain open-armed to all peoples, but its national sovereignty could not be infringed upon or challenged. In the aforementioned Tung-Clinton exchange, one should remember that a person with an "international background" need not accept international control of his own nation. Next, we should reexamine some parts of the U.S. involvement in Hong Kong matters.

As large U.S. companies were among the biggest players in hedge fund speculation in Hong Kong, their criticisms of and threats to the SAR government on the question of the latter's brief intervention in the stock market in August 1998 could be understood. Also, as economists were clearly divided about the comparative merits of economic liberalism vs. government intervention, the U.S. government might have protested out of its convictions. Yet the fact that, for their rescue plans during the Asian financial crisis, the World Bank, and IMF invariably demanded further openness in the Asian economies, is something that deserves to be seriously examined. The experimental intervention by the Hong Kong SAR government, if proven proper and necessary after the dust has finally settled, would point up a real need for reconsidering whether established theories are universally applicable, particularly to those the U.S. wishes to "put back onto the right track."

Regarding U.S. involvement in the SAR Court of Final Appeal case, several points deserve to be considered. First, though the issue of the right of abode was the core of the judgment, both the Central and SAR governments had not challenged the ruling. As explained above, the heart of the controversy was whether or not the local court was above the NPC, the national parliament of China. As neither Hong Kong nor its sovereign wanted a constitutional crisis, the matter was soon declared "over." The concept of "one country, two systems" is, without doubt, quite an experiment. What Hong Kong and China need is mutual understanding, accommodation, and a positive attitude to make the two different legal philosophies work together. It is a fact that local lawyers, long suspicious of the Mainland legal systems, have so far been overly defensive and non-accommodating, as can be seen in the two legal cases discussed above. In the long run, they need to find peace with the "one country" part of the formula, which envelopes the "two systems." Outside pressures would only create further difficulty, rather than help the situation.

Second, the U.S. impatiently criticized the SAR government for lack of progress toward working out a system or procedure to expedite the arrival of Hong Kong offspring on the Mainland, less than a month after the constitutional crisis had subsided. The fact is, since children born out of wedlock were granted the same right of abode as legitimate children by the Court of Final Appeal, much to the surprise of both Hong Kong and Central governments, they have not been able to establish how many children are eligible, following the Court's blanket ruling. The SAR conducted a survey in an attempt to come up with a reliable estimate. Yet, owing to the sensitive nature of the questions, especially when raised with Hong Kong fathers who have left behind illegitimate children behind on the Mainland, the survey did not get very far. Besides, the ruling regarding illegitimate children ran contrary to existing laws in Mainland China. It would be naive to hope that this could be quickly settled. The final practical consequence of this development is just mind-boggling, to say the least. Democratic nations with the highest regard for human rights, including the United States, have very meticulously worked out categories, priorities, quotas, preferences, age limits, etc., for immigration. At present, the United States even requires a financial qualification on the part of the sponsor. This alone, if similarly instituted here, would probably disqualify most of the lonely fathers in Hong Kong with families on the Mainland. In sum, the technical difficulties needed to be worked out are staggering, and outside observers such as the United States should be more sympathetic.

Third, the qualifications for Mainland children of Hong Kong descent to enter Hong Kong would likely be one of the most difficult questions to resolve. The U.S. and Hong Kong human rights advocates might have no sympathy for the hardship that Mainland authorities would have to face as this could be interpreted as preferential treatment to the Hong Kong people, extended even to their illegitimate offspring. It is a fact that the supreme judiciary, the Court of Final Appeal, already made its ruling. It is, however, still necessary for lawyers critical of the government to remember that, as early as 1993 and at an official meeting of the Sino-British Joint Liaison Group, it had been determined that only if either parent of a child in question had already satisfied the criteria for permanent resident status in Hong Kong would the child be eligible. This formal decision was reported in a note to the Legislative Council's Nationality Subcommittee by the then Security Branch on 3 January 1994. A copy of this note was also included in the evidence presented to the court that made the controversial ruling on 29 January 1999.[96]

Fourth, the British made the first criticisms on queries expressed on the judgment, and the United States followed suit. It is necessary to note that as early as March 1999 the British had warned against meddling with the case. When the Sino-British Joint Liaison Group held a meeting discussing Hong Kong, the British side declared it "did not [want to] make an issue of the CFA incident." Alan Paul, head of the British side, explained that Britain "would not infringe on matters of China's sovereignty." He even lamented that there was "very tangible evidence" that Sino-British relations were "on course and proceeding well and smoothly."[97] The question to ask is: Why could the United States not refrain from sticking its nose into the matter, and allow Sino-U.S. and U.S.-Hong Kong relations to take a normal turn for a change? The Central Government has, much to the relief of Hong Kong people, already rested the case after the clarification by the CFA.

Fifth, the general public, including very well-educated non-legal specialists, were led astray rather helplessly in the series of lawsuits challenging everything about the SAR government. Many assumed that our legal professionals were just doing what their professionalism and dedication to principles required them to do, in protecting the rule of law and judicial independence in Hong Kong. The fact that legal professionals were split among themselves threw Hong Kong into further uncertainty. Worse, they did not know how to judge. Yet more and more queries were raised toward respectable lawyers and politicians as to why, instead of seeking mutual understanding and hopefully accommodation between two different legal philosophies and systems, "some people" seemed to use the cases "to test the limit of Beijing's tolerance? This pointed to the possibility that there could have been a hidden political agenda behind the challenges. For example, a commentator asked, "If the judiciary is taken as a political tool, what is the meaning of [being judicially] independent ? Once the rule of law has become a sacrifice in political games, where would [its] dignity be?"[98] Since in the CFA case the constitutional crisis attracted U.S. intervention and pressure, our legal professionals and politicians, as well as the general public, should think hard whether or not they wish this to recur. This is important as another possible constitutional crisis regarding respect for the national and the SAR flags has already created another scandal.

Finally, U.S. involvement in the above cases caused the more alert people in Hong Kong to express fear of being caught in conflicts between the two titans; their own Chinese sovereign and the United States, who happened to differ in so many ways. Even before Hong Kong's return to China, observers were already attracted to President Clinton's open call for a hands-off policy

226 • Danny S. L. Paau

by China. Though both his exhortation and China's response were mild, some of the local media were already imploring that Hong Kong after reversion should not be victimized in the event of any conflict between China and the United States. As reported earlier in this essay, Hong Kong *"need neither be killed with kindness nor made a scapegoat"* by the United States in its dealings with China.[99]

Notes

1. *Express Daily,* 11 February 1997, A9.
2. Jamie Allen, *Seeing Red: China's Uncompromising Takeover of Hong Kong* (Singapore: Butterworth-Heinemann Asia, 1997) predicted that "a process of colonization rather than decolonization" would happen and that in about 10 years, *"at least* 650,000" immigrants would come from the Mainland (p. xx) and that the economy will perish (p. 282); See also, Mark Roberti, *The Fall of Hong Kong: China's Triumph and Britain's Betrayal* (New York: John Wiley & Sons, 1994), which predicted that Democrats like Martin Lee would be barred from running for Legislative Council election and the latter "will do whatever . . . Beijing wishes" (p. 314). The book nevertheless admitted that government tenders for work under British administration had been less than open and fair and predicted that it would worsen (p. 318). See also, Louis Kraar, "The Death of Hong Kong," *Fortune,* 26 June 1995: 40–52; "The Sinking of An Island," *Express News,* 6 May 1997, p.A2. Another book accused Britain, "the world's oldest parliamentary democracy" for "abandon[ing] 6 million free people to a regime born of revolution, which respects the use of force, not the rule of law," Bruce Bueno de Mesquita, David Newman, and Alvin Rabushka, *Red Flag over Hong Kong* (Chatham, N.J.: Chatham House Publishers, Inc.,1996), p.ix.; Mike Chinoy, director of CNN Hong Kong branch even thought at the time that Martin Lee would be arrested by the PLA after the hand-over. This, he admitted, had not happened. "'Prophecies of Foreign Media Fell Short," *Ming Pao Daily,* 16 June 1998, p.A6.
3. I had delivered lectures, for example, in several American, Australian, and local universities arguing against the overly pessimistic predictions. For the "real" problems Hong Kong needs to alert itself to, see Danny S. L. Paau, "Observations: More Urgent Problems Hong Kong Faces," in Danny Paau, ed., *Reunification with China: Hong Kong Academics Speak* (Hong Kong: Asian Research Service, 1998), pp.133–8.
4. For British "legacies" allegedly with less than benevolent intentions or "land mines" set up on the eve of the change-over, see Danny S. L. Paau, "Curious Maneuver? Certain Moves of the British Government in Hong Kong before Departure," in Danny Paau, ed., *Reunification with China: Hong Kong Academics Speak,* pp.63–80.
5. Nancy Tucker, *Taiwan, Hong Kong, and the United States, 1945–1992* (New York: Macmillan, 1994), p.198.

6. Ibid., p.227. Tucker noted the works of Anthony E. Sweeting, "the Reconstruction of Education in Post-war Hong Kong, 1945–1954: Variations in the Process of Policy Making" (Hong Kong: University of Hong Kong Ph.D. Diss., 1989) and Gerald A. Postiglione, "The United States and Higher Education: Preserving the High Road to China," paper presented at the International Conference on America and the Asia-Pacific Region in the Twentieth Century, Beijing, May, 1991. For concerns about the future trend of local universities, see Terill E. Lautz, "Higher Education in China and Hong Kong," in *United States-China Relations: Notes from the National Committee,* 25, no. 1 (summer, 1996) pp.4, 12–13.

7. Tucker, *Taiwan, Hong Kong, and the U.S.,* p.226.

8. Ibid., p.218.

9. "Hong Kong More Important to the U.S.," *World Daily* (North America), 14 August 1990, p.5.

10. This is the *United States Hong Kong Policy Act of 1992.* For a discussion, see Cindy Y.Y. Chu, "The Business Concerns, Hong Kong and Sino-American Relations," paper presented at the Hong Kong and Sino-U.S. Relations Seminar, Hong Kong Baptist University, October 1998.

11. Wong Kai Yuen, "Considering Hong Kong an American Possession," *Oriental Daily,* 10 May 1998, p.B7.

12. *Ming Pao Daily,* 11 May 1993, p.59, reported Patten's trip to the U.S., supposedly to lobby for the unconditional renewal of China's MFN status but it suspected that he just wished to "deal the internationalization card" to solicit support from the U.S. Patten had tried hard to induce American support to keep his reform intact after the change-over. See also "Hong Kong's Last Governor Delivers Remarks on the Future of Hong Kong," in *United States-China Relations, Notes from the National Committee,* 25, no.1 (summer, 1996): 1–2, 14–15.

13. Ibid., p.220.

14. In the *U.S.-Hong Kong Policy Act Report* the U.S. State Department released to the Congress in early April 1995, for example, it criticized China for delaying agreement with the British over the Court of Final Appeal arrangement and claimed that this risked damaging the interests of the 1,000-strong American businesses in Hong Kong. See Jasper Becker, "U.S. Comments Provoke China," *South China Morning Post,* 7 April 1995, p.5.

15. Besides voicing objection, the U.S. Secretary of State, Madeleine Albright, declared that she would not attend the swear-in ceremony if invited. With the exception of British Prime Minister Tony Blair, however, no other political dignitaries joined the "boycott." See *South China Morning Post,* 11 June 1997, p.1.

16. Simon Beck, "Clinton Warns Beijing on HK," *South China Morning Post,* 20 April 1997, p.11.

17. *Oriental Daily,* 11 May 1997, p.A15; for earlier criticism see the same journal, 12 April 1997, p.A15.

18. "Noises Roared After Clinton-Jiang Meeting," *Oriental Daily,* 9 November 1997, p.A2.

19. "Missile Scientist 'to be Put on Trial' [in China]," *Hong Kong Standard,* 13 March 1999, p.7. Hua Di, research associate at Stanford's Center for International Security and Cooperation, was formerly a Chinese military scientist. He was detained by the Chinese police in January 1998 when he visited home, probably because he had allegedly leaked secrets about China's ballistic missile program to Americans. He was still waiting to be put on trial more than a year after his arrest.

20. *Hong Kong Standard,* 9 November 1997, p.4.

21. "Security Risks Cited, U.S. Blocks Sale of Satellite," *Hong Kong Standard,* 24 February 1999, p.6.

22. "Chinese Enterprise Managers Refused U.S. Visa," *Yazhou zhoukan,* 22 March 1999, pp.18–19.

23. "Mainland Can 'Liberate' Taiwan after Five Years," *Sing Tao Daily,* 27 February 1999, p.B12.

24. "U.S. Predicted Taiwan Straits Crisis to Erupt Next Year," *Oriental Daily,* 1 March 1999, p.A6. Taiwan was also reported to seek purchase of four of the most advanced U.S. battleships with state-of-the-art radar systems.

25. *Ming Pao Daily,* 11 November 1998, p.A14.

26. "Ex-Defense Minister [William Perry] to Visit Taipei," *Hong Kong Standard,* 7 March 1993, p.2.

27. "U.S. Newspapers Refuted So-called 'Theft of Nuclear [Technology,'" *Wen Hui Pao Daily,* 1999.3.20. It reported that on 16 March 1999, major newspapers including the *Los Angeles Times, Christian Science Monitor,* and the *Wall Street Journal* had all published articles or commentaries to criticize irresponsible reports by some newspapers, such as the *New York Times* and for making accusations without substantiation or proof. It also claimed that the *U.S. News and World Report* had demanded on 23 January 1999, that Americans refrain from making such allegations without proof.

28. "Beijing Tells U.S. to Drop Defense Plan, 'Last Straw' Warning on Taiwan Arms," *Hong Kong Standard,* 7 March l999, p.1; see also "The Last Straw in Sino-U.S. Relations," *Yazhou zhoukan,* 15–21 March 1999, p.5. The article examined Huntington's article published in the March-April 1999 issue of *Foreign Affairs,* which claimed that the United States had been "globalizing" its military power by making alliances with regional powers to ward off possible rivalries. This had been pursued in Europe, Latin America, and the Arabian Central Asia. For a more detailed analysis of U.S. plans and actions in the Asian region, see "Secrets Behind U.S. Forced Re-alignment of [the Balance of Power in] Asia," *Yazhou zhoukan,* 23–29 November 1998, p.7.

29. "Secretary [of State] Albright Delivered Yet Another Challenge [in her] Visit," *Oriental Daily,* 3 March 1999, p.A4.

30. Gret Torode, "Manila Looks to U.S. in Spratly's Dispute [with China]," *South China Morning Post,* 10 March 1999, p.10.

31. Simon Beck, "Rights Abuses condemned in U.S. Report," *South China Morning Post,* 27 February 1999, p.7.

32. A series of articles refuting U.S. delegations appeared shortly after the accusations. See, for example, Liu Weiguo et al., "'Human Rights' in Tibetan His-

tory," *Wen Hui Pao Daily,* 11 March 1993, p.A7; Ni Xiaoyang, et al., "Dalai Continues Separatist Movements," *Wen Hui Pao Daily,* 8 March 1999, p.A6. See also Charles Snyder, "U.S. Accused of Rights 'Double Standard,'" *Hong Kong Standard,* 7 March 1999, p.2.

33. Luo Fu, "Criticism [of China] Rose Again after the Jiang-Clinton Summit," *Oriental Daily,* 9 November 1997, p.A2

34. A regular columnist believed that the United States became more vocal and interventionist because Sino-U.S. relations had deteriorated. See Charles Snyder, "Spy Row May Blow Relations Apart," *Hong Kong Standard,* 12 March 1999, p.10.

35. Cindy Chu, "The Business Concerns, Hong Kong and Sino-American Relations," unpublished paper presented at the Hong Kong and Sino-U.S. Relations Seminar co-sponsored by the History Department and the Sino-Western Relations Research Program, David C. Lam Institute for East-West Studies, Hong Kong Baptist University, in October, 1998, p.3–4. Chu concluded that the preservation of the status quo and American interests are the main motives.

36. Tucker, Taiwan, Hong Kong, and the U.S., p.221.

37. "Earning [a Profit of] a Hundred Million [Dollars] Everyday: Hong Kong Shanghai Bank Makes Daily Miracles," *Hong Kong People Ruling Hong Kong Begins, Ming Pao Daily* Special Issue on Reunification, July 1997, pp.31–2.

38. James Tien, "[Jiang's] U.S. Visit Benefits Also Hong Kong," *Oriental Daily,* 9 November 1997, p.A2. Tien as a prominent local business leader appreciated the close economic relations between Hong Kong and the Mainland. In this short essay he invested hope in the so-called "strategic partnership" to benefit Hong Kong trade with the United States. This reflects, on the other hand, how a local leader would worry that Sino-U.S. business conflicts might also adversely affect the tiny territory.

39. "Bubble Economy and Bubble Psychology in Hong Kong," *Yazhou zhoukan,* 8–14 June 1998, p.6.

40. "Shelters for Economic Storms," *Yazhou zhoukan,* 8–14 June 1998, pp.47–8.

41. See, for example, "Eliminate Bubble Economy and Political Impatience," *Yazhou zhoukan,* 12–18 October 1998, p.8. For an analysis of the vulnerability of the Hong Kong economic structure, see "Hong Kong Estimated to be Able to Feed Only About Four Million People According to Current Economic Structure" (Hong Kong : The Great China Times Research Institute, 1998).

42. Local reports alleged that George Soros's Quantum Fund and Julian Robertson's Tiger Fund were among the most active speculators. See "Fierce Counter Attack Wars [for] Refusing to be ATMs [of Speculators]," *Yazhou zhoukan,* 7–13 September 1998, pp.38–43.

43. For samples of such cases, see "The Economic Terror Drama Series in the First Year After Reunification," *Sing Tao Daily,* 1 July 1998, p.A22.

44. It is important to note that James Tien, a prominent leader of the business dominated Liberal Party, criticized the previous administration for allegedly having focused too much on political wrangling with China and had thus

neglected the economy. His comments on a radio program were reported in the *Hong Kong Standard*, 12 July 1998, p.2.

45. See "Fierce Counter Attack" cited in note 42 above. See also, Zhang Li, "Hong Kong's Counter Attack Shocking," *Yazhou zhoukan*, 7–13 September 1998, p.46.

46. See "Fierce Counterattack," cited in n. 7 above. The author reported that the United States intervened for self-seeking reasons.

47. Those among the local critics who opposed the intervention included the famous economics professor Zhang Wuchang of the University of Hong Kong. In fact, most economic scholars severely criticized the government when the news first broke out.

48. Emily Lau, for example, the most outspoken and radical of legislators generally labeled as "democrats," angrily shouted at government officials for "daring to spend such a huge amount of public money without consulting the Legco" and getting permission from it.

49. For contemporary commentaries, see articles arguing that non-interventionism could no longer deal with present-day hedge funds such as: "Theoretical Inadequacy and Awakening," *Yazhou zhoukan*, 7–13 September 1998, p.6; "The Pioneer Significance of Hong Kong's Financial Defense War," *Yazhou zhoukan*, 7–13 September 1998, p.7; "Breaking the Curse of [Economic] Liberalization and Globalization," *Yazhou zhoukan*, 14–20 September 1998, p.9; "A Third Way to Save the Whole World [from] Economic Crisis," *Yazhou zhoukan*, 12–18 October 1999, p.7. These articles were happy to cite American professors such as Milton Friedman and Paul Krugman who had reportedly argued that government intervention might be justified.

50. "Four Large American Firms Threaten to Withdraw from Hong Kong," *Hong Kong Daily News*, 16 September 1998, p.B1.

51. Most newspaper editorials and commentaries, after having expressed initial worries, turned to laud the bold actions of the government. See, for example, "Hong Kong Government's Counter Attack on Speculators Gains the People's Heart," *Yazhou zhoukan*, 14–20 September 1998, p.10.

52. Ibid.

53. "Editorial—Completely Revamp the Exchange [Institutions and] Rebuild the Financial City," *Ming Pao Daily*, 8 September 1998, p.A12. This reported that the government made 30 proposed changes to improve stocks and futures exchange mechanism; "Hong Kong Government Charged on Speculators After [Initial] Victory," *Yazhou zhoukan*, 14–20 September 1998, pp.48–9. It was alleged that certain units did not act fast enough during the massive intervention. See the report in *OrientalDaily*, 9 September 1998, p.A19.

54. Charles Snyder, "HK Wins Plaudits in Washington," *Hong Kong Standard*, p.10 October 1998, p.10. Also, the Chief Executive met with some visiting U.S. Senators in November the same year to explain the Hong Kong SAR government intervention. He further urged G-7 nations to consult regions victimized by the speculators when the advanced countries discussed means to monitor international hedge funds. See *Ming Pao Daily*, 10 November 1998, p.A9.

55. "Editorial—Foreign Reserves' Ability to Earn Catches All Eyes," *Wen Hui Pao Daily*, 27 March 1999, p.A3. See also "Foreign Reserves Earned $94.2 billion," *Ming Pao Daily*, 27 March 1999, p.A2.

56. Richard A. Boucher, *The United States' Role in Asia's and Hong Kong's Recovery*, Remarks to the Hong Kong General Chamber of Commerce, 25 January 1999 (U.S.IS, 1999).

57. "Landing Charges Likely to be Cut," *Hong Kong Standard*, 27 March 1999, Business, p.1; "*Laissez-faire* a Myth," *Hong Kong Standard*, 23 March 1999, Business, p.1.

58. M. K. Shankar, "Sentiment 'Strongly Positive,' Cyberport Puts HK on the Map," *Hong Kong Standard*, 23 March 1999, Business, p.1.

59. "Team Jet out with Message for U.S.," *Hong Kong Standard*, 20 March 1999; *Wen Hui Pao Daily*, 27 March 1999, p.A11.

60. "Microsoft Got Hong Kong Government Contract," *Ming Pao Daily*, 10 March 1999, p.A2.

61. Karen Chan, "Gates Vows Microsoft Role in Cyberport," *Hong Kong Standard*, 10 March 1999, p.1.

62. Cindy Chu, "The Business Concerns," p.2.

63. *Oriental Daily*, 12 April 1997, p.A15; Ibid., 5/18/97, A15.

64. *South China Morning Post*, 11 June 1997.

65. *Ming Pao Daily*, 14 January 1998, p.A7.

66. Lily Dizon, "Resident Rights Upheld in Vote Case," *Hong Kong Standard*, 13 March 1999, p.1.

67. "Editorial—Unjust for Indigenous Residents to Hold Privilege," *Ming Pao Daily*, 13 March 1999, p.A2.

68. There were of late a number of complaints about the abuse of press freedom by journalists. For more recent ones, see, for example, Huang Wenfang, "Proper Practices and Improper Practices," *Ming Pao Daily*, 4 April 1999, p.E7; Qiu Zhenhai, "Hong Kong: Demise of the Uncrowned King," *Wen Wei Pao*, 5 April 1999, C2. For a sample criticism of journalists' abuse of press freedom from a local singer, see Leung Hon Man, "Leung Hon Man: Media Is Horrible," *Wen Wei Pao Daily* 10 April 1999, C12. Also, Michael Wong, "Media Neglecting Social Duty," *Hong Kong Standard*, 10 April 1999, p.2.

69. Charles Snyder, "SAR Praised in U.S. Rights Report," *Hong Kong Standard*, 27 February 1999, p.10.

70. "Editorial—Justice is in People's Hearts," *Wen Hui Pao Daily*, 13 March 1999, p.A5.

71. See the full text of Elsie Leung's explanations given at Legco on March 11, 1999 published in *Wen Hui Pao Daily*, 12 March 1999, p.C2.

72. Richard A. Boucher, as reported in *Oriental Daily*, 26 January 1999, p.A20.

73. Danny S. L. Paau, "Observations: More Urgent Problems Hong Kong Faces," in Danny Paau, ed., *Reunification with China: Hong Kong Academics Speak* (Hong Kong: Asian Research Service, 1998), p.136.

74. "Editorial—Landmark Ruling," *South China Morning Post*, 30 January 1999, p.14.

75. Danny S. L. Paau, "Observations," p.136.

76. "Rating Agency Expects More Tests, Court Ruling 'First of Controversies,'" *Hong Kong Standard*, 24 February 1999, p.2.

77. Lucia Tangi and Charles Synder, "Powers Told to Back Off Controversy," *Hong Kong Standard*, 12 February 1999, p.1.

78. "Confidence in Autonomy May be Undermined Says Chamber," *Hong Kong Standard*, 12 February 1999, p.3.

79. *Ming Pao Daily*, 12 February 1999, p.A9.

80. Charles Snyder, "U.S. Takes Note as Abode Storm Rages," *Hong Kong Standard*, 12 February 1999, p.10.

81. Fong Tak-ho, "U.S. to Closely Watch Developments," *Hong Kong Standard*, 3 March 1999, p.1.

82. Cliff Buddle and Chris Yeung, "Justices Clarify Ruling: We Were Not Challenging NPC," *Hong Kong Standard*, 29 February 1999, p.1.

83. "Editorial—Best Way Out," *South China Morning Post*, 27 February 1999, p.14; "Editorial—Secretary for Justice Paved Way for CFA Retreat," *Oriental Daily*, 25 February 1999, p.A19.

84. See, for example, "Editorial—Clarify Judgment, Eliminate Argument," *Sing Tao Daily*, 27 February 1999, p.A2. Also, "Editorial—It is Time Constitutional Row Ended," *Ming Pao Daily*, 27 February 1999; English version published on 1 March 1999, p.E8.

85. Fong Tak-ho, "U.S. to Closely Watch Developments."

86. Jimmy Cheung, "Warning of Two Final Courts, Political Heat 'Hurts Morale of Judiciary,'" *South China Morning Post*, 6 March 1999, p.2.

87. Charles Snyder, "Abode Row Raises U.S. Fears on Free Judiciary," *Hong Kong Standard*, 4 April 1999, p.1.

88. "Editorial—China and Hong Kong Work Closely to Solve Migration Crisis," *Oriental Daily*, 23 February 1999, p.A19. The editorial expressed worry for the large number estimated, demanded the application of a quota system like those of Europe and America, and urged the government to repatriate those without proper permits.

89. For a tabulated summary of the various concerns, see *Hong Kong Daily News*, 30 January 1999, p.A1–2.

90. May Tam, "Political Parties Divided over U.S. Comments," *Hong Kong Standard*, 4 April 1999, p.16.

91. Jackie Sam, "Ruling Raises Independence Fears," *Hong Kong Standard*, 12 February 1999, p.11;Chen Yuen Han, "What Had the CFA Said," *Oriental Daily*, 20 February 1999, p.B19

92. Cliff Buddle and Chris Yeung, "Justices Clarify Ruling."

93. "Ruling Seen as Challenge to NPC Authority, Jurisdiction Controversy Looms," *Hong Kong Standard*, 3 April 1999, p.4.

94. Charles Snyder, "U.S. Takes Note as Abode Storm Rages," *Hong Kong Standard*, 2 December 1999, p.1.

95. For a distinction between "internationalization" and "internationally controlled," see Yu Wing Yin, "Is Hong Kong [An] Internationalized City?" *Ming Pao Daily*, 9 July 1998, p.D10.

96. Ian Wingfield, "Letter to the Editor—Clearing Up Confusion," *South China Morning Post*, 14 March 1999, p.11. Wingfield is a Law Officer of the government.

97. Alan Castro, "Welcome Words from the JLG in Such Clamorous Times," *Hong Kong Standard*, 23 March 1999, 13.

98. Cho King Hang, "Who's Creating the Hong Kong Legal Crisis?" *Yazhou zhoukan*, 15 February 1999, p.25.

99. "Editorial—Hong Kong Should not be Sacrificed in Sino-U.S. Political Conflicts," *Oriental Daily*, 31 January 1999, p.A15.

CHAPTER 9

Weathering the Asian Financial Storm in Hong Kong*

Y. Y. Kueh

Could the British Have Done Better?

Hardly had Hong Kong wound up the fanfare of celebrating its return to Chinese sovereignty, on 1 July 1997, when the Asian financial crisis erupted. The next day, in a sudden collapse, the Thai baht fell precipitously in value against the U.S. dollar. Within months, the financial disaster claimed several major "dominos" in Southeast and East Asia including, notably, Malaysia, Indonesia, and South Korea.[1] It spared no one, not the newly installed Chinese Special Administrative Region (SAR) of Hong Kong. By October 1998, it became apparent that, after years of sustained growth at an annual rate of 5–6 percent, the SAR's Gross Domestic Product (GDP) for the year would be curtailed by a hefty 5 percent. By November 1998, the unemployment rate had already reached a post-war high of 5.3 percent. Just a year or so before, a rigorous labor-importation program had been launched to relieve the enormous labor shortage pressures impinging on Hong Kong for over a decade, following increased economic integration with the Chinese Mainland.[2]

Now, how does the ongoing economic meltdown in the HKSAR stand in relation to the main theme of this book: the "super paradox" of Hong Kong? While it is obvious that the drastic economic downturn had nothing whatsoever to do with the territory's reversion, it is worth noting, however, that prior to 1997 both domestic and external apprehensions about Hong Kong's future under Chinese sovereignty generally focused on domestic affairs, rather than on its external economic relations. Specifically, there were concerns as to whether Hong Kong would truly be granted full autonomy, as promised, under the "one country two systems" model. Other fears were whether under increased Chinese influence the highly efficient public administration system in Hong Kong would be eroded, giving way to bureaucratic inertia, laxity, and corruption. In addition, there were also worries whether the integrity of the British common law-based legal system would be compromised by the socialist version of jurisprudence and justice, and whether press freedom would be impaired, and democracy and civic liberty (including the right of staging street demonstrations) would be suppressed in favor of a Chinese-style "law and social order."

While such politically inspired fears have now largely subsided, as discussed in the various chapters of this study,[3] the attack from the outside by the Asian financial storm caught all Hong Kong people by surprise. Prior to the sovereignty transfer, even the pessimists had foreseen Hong Kong's future economic relations as a major bright spot that would be free of possible political interference from its new sovereign. Most people expected the economic prosperity to continue after the handover. The reasons for the optimism were manifold, some quite obvious, others not so, as I will lay them out below.

1. Being a regional business and financial center, Hong Kong would remain as an indispensable "window" for China to access the global market of information, technology, and finance.
2. Hong Kong, as the region's headquarters for many major international banking corporations from advanced Western countries, would continue to be an important banking center for making large syndicated loans to China.
3. The SAR would itself continue to be the single most important source of foreign direct investment (FDI) flows to China, with virtually all its manufacturing plants having relocated to the Chinese hinterland in the past two decades or so.
4. While the prospects of Taiwan's reunion with the Mainland remained remote, Hong Kong would continue to serve as a natural bridge for Taiwanese compatriots investing in and trading with the Mainland— a link so much treasured by Beijing.
5. By virtue of its geographic proximity and cultural affinity, Hong Kong, with a sizable population of 6.5 million with high income and consumption power, would remain an absolutely significant client for Chinese exports of foodstuff, clothing, and similar daily necessities for local consumption—a market that has in the past several decades consistently earned China the bulk of its foreign exchange sufficient to pay for the huge Chinese trade deficits incurred from Japan and

Western Europe combined; or a market that has from 1979 through the mid-1990s helped to generate a cumulative total trade surplus more than enough to pay for all the bills from the United States for imports of machine and equipment, aircraft, and grains, and for technology transfer as well, during the same period.[4]

6. Complete with state-of-the-art facilities in banking, finance, insurance, telecommunications, and shipping, as well as a well established pool of international marketing and sourcing expertise, Hong Kong would remain for many years to come the single largest trading port for China, irreplaceable by Shanghai, Tianjin, or any other major Chinese coastal metropolis—noting that the SAR's total re-exports, combining both those of Chinese origin destined to the various overseas markets, (notably North America and Western Europe), and those from these advanced industrialized markets to the various Mainland destinations, have increased spectacularly, by 40-fold within the past two decades, to surpass domestic exports to be the absolutely overwhelming source (86 percent in 1998, as compared with 31 percent in 1980) of Hong Kong's overall exports.

7. The status of Hong Kong as a "non-sovereign" but independent actor in the global economic arena was already firmly sealed, prior to the 1997 handover, by way of international treaty arrangements, guaranteeing the SAR to retain its privileges as a separate customs entity within the World Trade Organization (thus free of potential trade restrictions and discriminations), its separate membership in the International Monetary Fund, the World Bank, Asian Development Bank, APEC, and the like (alongside China's own membership).

8. Pending a resolution of the Taiwan issue, the Chinese government would certainly be keen to promote HKSAR as a role model for Taiwan—noting that international economic linkages are crucial for both Hong Kong and Taiwan for sustaining long-term economic growth and well-being.

9. And, finally, the statutory guarantee, as given in the "Basic Law," the SAR's mini-constitution, that it would be allowed to retain its own currency under the "one country two systems" model, is by any measure unparalleled in modern world history. The guarantee means that the Hong Kong dollar would continue to be the most fully convertible currency in the world, as it has been, entirely free of any restrictions, with respect to both current account and capital account transactions.

Taken together, there is, therefore, no doubt that Hong Kong is more closely integrated with the global economy rather than with that of its motherland.

And China itself treasures enormously these particular global economic linkages of the SAR. Yet, paradoxically, it is exactly on account of its total openness to the outside world that the SAR economy was almost brought to its knees in 1998, as a result of renewed speculative attacks by international hedge funds on the Hong Kong dollar since August 1997.

Now, the question is whether the British administration would have been better able to lead Hong Kong through the Asian financial typhoon. History of course cannot be rehearsed just to verify a possible hypothesis like this. However, two important points can be made, perhaps not as a rebuttal of the familiar "doomsday prophecies" widely circulating before the handover, so much as prima facie evidence that the British would probably have had no other alternative but to respond similarly to the Asian crisis, as the new Hong Kong SAR government has done over the past year and a half.

First, the new SAR administration under Chief Executive Tung Chee-hwa is actually inherited lock, stock, and barrel from the pre-handover British administration including, notably, the powerful Chief Secretary of Administration, Mrs. Anson Chan, the outspoken Financial Secretary, Mr. Donald Tsang, and Mr. Joseph Yam, the Executive Chief of the Hong Kong Monetary Authority (the quasi-Central Bank). Were they subject to intervention or influence by the Beijing authorities in handling the crisis? There is hardly any evidence of this. Martin Lee, Chairman of the Democratic Party (the largest opposition party in Hong Kong), and the habitual critic, had openly suspected that the Chinese government must have had a hand in the heavy-handed purchases by the SAR government of all leading corporate shares in August 1998 to forestall a total collapse of the stock market. But, two days later, after a futile search for evidence, Lee was obliged to retract his allegation (SCMP, 14 September 1998).

Second, it was argued that under the new political context, the SAR government was deprived of the benefit of consulting the British government in London on how to combat the unfolding Asian crisis, while the new sovereign in Beijing was not at all experienced in dealing with matters concerning international currency attacks. However, as a prelude to his official visit to the SAR on 9 October 1998, the British Prime Minister Tony Blair wrote that he understood the decision of "the Hong Kong Administration [to stand] by its commitment to maintain the Hong Kong dollar peg, despite the buffeting in the markets over the past few months," adding: "They need to safeguard the long-term financial stability and economic success for Hong Kong." (SCMP, 9 October 1998). He also underscored "the important role [played by] their defense of the dollar peg in underpinning stability and confidence in Asia more widely" (SCMP, 9 October 1998), and said indeed that "by using the powers guaranteed in the Joint Declaration (for the defense), Hong Kong has proved to the world that its autonomy is a reality" (SCMP, 9 October 1998).

Skeptics may regard Prime Minister Blair's statements as mere diplomatic rhetoric. But, uncharacteristically, in a rare show of support for Beijing, former governor Chris Patten also "threw his weight behind Premier Zhu Rongji for his firm opposition to devaluing the yuan" (*SCMP*, 31 October 1998). Patten gave his positive comments after emerging from a discussion with Tung Chee-hwa, the HKSAR Chief Executive, during a private visit to the SAR shortly after Blair's, about the way the government responded to the backlash of the Asian financial crunch. There seems little doubt that Patten must know well that one of the main factors frequently cited for Premier Zhu's policy stance against a devaluation of the renminbi, the Chinese currency, was to support Hong Kong's defense of the U.S. dollar peg.

As a matter of fact, Pattern himself, in an interview on CNN TV, New York, on 8 September 1998, disclaimed that the British departure was a cause of Hong Kong's economic downturn and said he was "lucky" that his Governorship coincided with the territory's "good economic times" (see note 19, in Hsiung's Introduction). He made the point again while in Hong Kong on 30 October 1998.

The economic logic underpinning the HKSAR's defense of the U.S. dollar peg will become clear during the course of our discussion. But before we turn to the substantive issues, another important point should be made in this context. That is, the stakes for the Chinese government, watching Hong Kong's sustained economic prosperity degenerate into chaos on the heels of the handover, were formidable. That would be, to say the least, a harrying matter of "losing face" before the watchful eyes of the entire world.

Thus, when the first wave of speculative attacks was unleashed on the Hong Kong dollar, in late July 1997, the Chinese government stood up to the occasion, declaring, in no equivocal terms, that it would, with its massive foreign exchange reserves (at around U.S.$120 billion, it had the world's second largest, next only to Japan's), stand by the Hong Kong Monetary Authority in its defense of the U.S. dollar peg.[5] It is doubtful whether Dai would have made the same pledge of support when Hong Kong remained under British colonial rule. Equally doubtful is whether the British government in London would have similarly stood up for Hong Kong, even assuming it had the financial resources to do so, if the British were still the rulers here.

Defending the U.S. Dollar Peg:
The Achilles Heel

The existing Hong Kong currency system of linking the Hong Kong dollar to the U.S. dollar, at the fixed exchange rate of HK$7.8 per U.S. dollar, was first established in October 1983. That was the time when, amidst the political crisis associated with the Sino-British negotiations over the future of Hong Kong, public confidence in Hong Kong was drastically deteriorating, sending the Hong Kong dollar tumbling from the previous market rate of around HK$5.6 to the record low of around HK$9.0 per U.S. dollar. The new U.S. dollar peg, specifically, commits the Hong Kong government to redeem, via the authorized note-issuing banks, any amount of Hong Kong currency with the foreign exchange reserves held by the Exchange Fund at the fixed rate specified. The other integral part of the system, dubbed Currency Board, is that the note-issuing banks are entitled to issuing Hong Kong dollar notes only upon surrendering the proportionate amount of U.S. dollars to the Exchange Fund in exchange for the necessary instrument called Certificates of Indebtedness. This ensures that all Hong Kong currency in circulation is fully backed by foreign exchange reserves, the amount of which is, of course, closely tied in with the performance of the highly export-oriented economy.[6]

There are two aspects to the U.S. dollar peg mechanism that serve to ensure stability of the exchange rate and that should be briefly explained as a background for understanding the intricacies of the speculative attacks on the Hong Kong currency. The first is the process of notes arbitrage and the second of interest rate arbitrage. Notes arbitrage may set in when, for some reason, the value of the HK dollar drops to, say, 7.9 per U.S. dollar. This will prompt the note-issuing banks to surrender their Certificate of Indebtedness to the Exchange Fund in exchange for U.S. dollars at the official rate of HK$7.8, in order to cash in the premium of HK$0.1 per U.S. dollar by reselling their U.S. dollar holdings to the market. Increased U.S. dollar supply may therefore hopefully help to revert the market rate back to the official par value.

The interest rate arbitrage should work similarly in that any substantial selling of the Hong Kong currency for U.S. dollars would inevitably lead to a contraction in money supply in Hong Kong dollar terms and, hence, a rise in local interest rates. Higher interest rates may, in turn, hopefully induce capital inflow to offset the initial outflow, and consequently help to stabilize the foreign exchange market. However, both notes arbitrage and interest rate arbitrage proved to be powerless in the wake of the heavy speculative attacks on the Hong Kong dollar in 1997–1998. For one thing, notes arbitrage has actually never proven to be effective for reasons of the potentially high transaction costs involved. Moreover, the Hong Kong Monetary Authority has made it a practice to consistently intervene at around HK$7.75, instead of HK$7.8, since the early 1990s, making it impossible for the notes arbitrage to take place, as is shown in Figure 9.1.[7]

More importantly, as interest rates were forced up to record a high as a result of heavy borrowing and dumping of the Hong Kong currency (short-

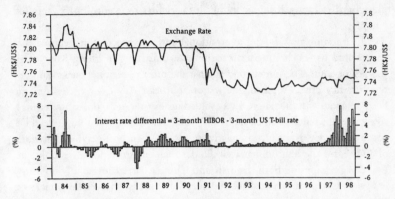

Figure 9.1   Monthly fluctuations in exchange rate (HK/U.S.$) and interest rate differential (HK/U.S.$) in Hong Kong 1984–1998

selling) by international speculators to coerce a devaluation of the HK dollar, foreign capital continued, nonetheless, to shy away from the SAR, despite substantially increased interest rate differentials between Hong Kong and the United States. During the currency attacks in January and August 1998, for example, the three-month HIBOR (Hong Kong Interbank Offered Rate) rose to an average of 12.7 percent and 11.9 percent respectively, and was 7.7 percent and 7.0 percent higher than the respective three-month U.S. T-bills rates. These interest rate differentials between the HK$ and U.S.$ were unparalleled since the mid-1984 and represented a sharp rise from the average interest premium of only 1.16 percent for the first nine months of 1997, as is also shown in Figure 9.1.

As public confidence in the U.S. dollar peg was waning, U.S. dollar deposits in Hong Kong jumped by 34 percent between September 1997 and August 1998, while HK$ deposits rose by only 2.1 percent during the same period. The highly volatile and punitive interest rates not only caused the stock and futures markets to collapse, paving the way for speculators to reap huge profits from their short positions in the HK dollar forward market and Hang Seng Index futures, but they also put enormous pressures on the local asset price bubble to burst. By late 1998, property prices in Hong Kong were generally reduced by 50 percent to 60 percent from the peak reached a year before. And the Hang Seng Index dropped from the record high of 16,673 achieved by the time of the sovereignty turnover to no more than around 6,500 in August 1998, when the controversial SAR government's rescue bid was introduced. The ensuing loss of investor and consumer confidence, coupled with a severe credit squeeze, penetrated deeply into the real sector, plunging the Hong Kong economy into its first recession in 13 years (cf. Chapter 2).

The question now is why international currency speculators believed that they would win in betting against the U.S. dollar peg in Hong Kong. This should, of course, be understood against the broader regional background of the currency crisis. With all the major Southeast Asian currencies and the South Korean won being greatly devalued against the U.S. dollar under massive speculative attacks since July 1997, the Hong Kong dollar has indeed become excessively overvalued in relative terms, and hence vulnerable. This goes, of course, for the renminbi as well. Under such circumstances, it is only natural that pressures on both the SAR and Chinese currencies rapidly built up.

Two basic questions are involved here. The first is whether the attacks on the Hong Kong dollar were just a matter of "psychic mobility" in the market place; or currency speculators were also seriously betting on a devaluation of both currencies on the grounds of real economic factors, say, the necessity for the HKSAR and Beijing governments to maintain their export competitive edge against their Asian neighbors. The second question is: why would a renminbi devaluation necessarily help to break the Hong Kong dollar peg to the U.S. dollar, as widely assumed, which was the basis for the speculators' attacks on the Hong Kong currency?

The first question posed may, by itself, seem apparent as a matter of simple conventional wisdom, but in reality the answer is by no means a straightforward one. Before we take up the issue, however, a word or two should be said about the second question. Obviously the perception that the Hong Kong dollar is intricately linked to the renminbi has arisen from the fact that the SAR economy has one way or the other become inseparable from the Chinese hinterland, as a result of increased economic integration over the past two decades, in terms of capital and trade flows, quite apart from the political reunion of July 1997.

Nonetheless, it should be noted that the "one-country two-systems" model also provides for two separate customs entities and, more important, two separate, independent currencies, viz.: renminbi and Hong Kong dollar. Could it, therefore, not be the case that a devaluation of the renminbi would only benefit Hong Kong, in the same fashion as the United States has emerged as a major beneficiary from the Asian financial turmoil? Note also that while the Chinese government makes it explicitly clear that one of the reasons for not devaluing the renminbi is to help the HKSAR government to defend its U.S. dollar peg system, the massive effort put up by the latter for the defense of the Hong Kong dollar is, however, entirely rooted, for better or worse, in considerations of its own merits in the minimization of domestic economic disturbance by forestalling, in particular, massive capital flight from the SAR, which would be destabilizing.

As a matter of fact, the HKSAR's own defense of the Hong Kong dollar against the backdrop of the Asian financial crisis has never been seen as a deliberate support of the prevailing value of the renminbi. This point should perhaps not be taken for granted, given that, dollar for dollar, (i.e., without adjusting for the purchasing power parities between HK dollar and renminbi), Hong Kong's Gross Domestic Product stood at around U.S.$173.66 billion in 1997, which is, surprisingly, some 20 percent of Mainland China's GDP of U.S.$901 billion, tiny as the former British enclave may be with a population of only 6.5 million, compared with 1.2 billion for the Mainland. More remarkably, the Hong Kong government's budget expenditure alone, totaling U.S.$22.4 billion in 1997, is equal to around 35 percent of China's national budget expenditure of U.S.$64.81 billion. In addition, by the end of 1997, the SAR's foreign exchange reserves (at U.S.$92.8 billion) were trailing quite closely behind Mainland China's U.S.$139.9 billion, which was next only to that of Japan (U.S.$220.8 billion), the world's largest foreign reserves holder.

It seems necessary to look at the two questions raised at some length, with a view to disentangling the real dynamics underlying the speculative attacks on the Hong Kong currency in 1997–98. This will provide the necessary background for examining, in the subsequent section, the drastic steps taken by the Hong Kong SAR government to fend off the massive currency attacks in August 1998.

The Chinese Connection: Asset or Liability?

Since the fate of the Hong Kong dollar has been perceived as invariably bound up with the renminbi, it is necessary to examine, first of all, whether there really are tenable economic factors underlying the international speculations for a renminbi devaluation. What is at stake, in the first place, is clearly whether China, without a devaluation, would be able to maintain its competitive edge in major Western export markets (in particular, the United States and Western Europe) against the ASEAN countries and South Korea, which have all greatly depreciated their currencies within months of the onset of the Asian currency crisis in July 1997.

In his vigorous defense of the strong position of the renminbi, Dai Xianglong, the Governor of the People's Bank of China, maintained in January 1998 that Chinese exports to the United States overlapped with those from Southeast Asia by not more than 15 percent of its total exports (*TKP*, 17 January 1998). A higher estimate, made by a senior economist at the Chinese Academy of Social Science at a CCTV (Beijing) interview in January 1999, put it between 20 and 30 percent. That is to say, the massive currency devaluations in Southeast Asia during 1997–98 may not have basically broken the established "flying geese pattern" of regional industrial specialization. Specifically, China could still take advantage of its relative abundance of labor supply and specialize in the conventional, more labor-intensive export-processing trade, without competing head-on against, say, Malaysia, which has increasingly upgraded its export structure by shifting to commodities with higher technology components, such as electronic products, including telephone sets, computers, etc. Moreover, the bulk of Southeast Asian exports comprise essentially such primary commodities as rubber, timber, palm oil, and the like, which are unavailable from China.

There is then also the question of whether a renminbi devaluation, if necessary, would really help to enhance China's export competitiveness, given that over half of the inputs required for export-processing in China consist of materials supplied from overseas, and that a renminbi devaluation would also inevitably result in higher import costs to offset any potential monetary export gains that devaluation might bring. This point seems to have been borne out by the 1998 experience of some Southeast Asian countries, which have seen their export volume declining, despite massive currency devaluation.

On the part of the Chinese authority, there was obviously also concern that a renminbi devaluation may scare away foreign direct investments (FDI), since the end effect of any devaluations of the currency may, from the perspective of foreign investors, result in a curtailment of their earlier FDI value and their profit repatriation in foreign currency terms, from their existing investments in China. The economic logic for such considerations, of course, may not be entirely straightforward. For one thing, a renminbi devaluation may, nonetheless, help to reduce investment costs for foreign investors, in the first place.[8] And, for entirely export-oriented FDIs engaged in export-processing, (as is overwhelmingly the case with investors from Hong Kong and Taiwan), it does not really matter much whether the renminbi is devalued or not, so long as the overseas export market potentials are not impaired. The reason for this is simple, though not necessarily self-apparent for people not familiar with the Greater China context of economic cooperation, as I shall explain.

"Greater China context" refers to investments that involve complementary operations in two or all three places in the Mainland-Hong Kong-Taiwan network. Specifically for this kind of investor from outside, export orders or letters of credit from overseas buyers are normally received in Hong Kong and Taiwan, instead of being channeled through the Mainland Chinese banking system. Similarly, foreign exchange expenditure on imported materials and on the necessary processing machines and equipment also normally incurs outside of China. In other words, both foreign currency receipts and expenditure on the part of Hong Kong and

Taiwan investors in China are basically shielded off from the volatility of the renminbi.

Taken together, there seem, therefore, to be no compelling economic reasons why the Chinese currency would likewise be forced to devalue amidst the Asian currency crisis. That is, of course, not to say that China is entirely spared from the adverse impact of the regional crisis, in terms of capital inflows and trade flows. FDI inflows from Hong Kong did indeed slow down in 1998. But this seemed nonetheless more a result of the severe credit squeeze in Hong Kong associated with its massive efforts at defending the U.S.-dollar-peg system (which has strongly boosted local interest rates), as well as the withdrawal of Japanese funds—upon which Hong Kong's FDI in the Chinese Mainland substantially hinges—to meet domestic difficulties. Hence, it was not really a result of the concerns about the prospects of the value of the renminbi. As a mater of fact, FDI flows from elsewhere—the United States and Europe in particular—remained quite robust throughout 1998, to compensate for the relative decline in FDI inflows from Hong Kong and other Asian economies affected by the regional financial turmoil.[9]

Chinese exports in 1998 were clearly affected by the Asian crisis as well. Enhanced price competition from the Southeast Asian countries for the "overlapping" export commodities to Western markets may partially account for this. But a more important source of export decline seems rather to be found in the impaired import capabilities of Southeast and East Asian countries, including Japan, which were all helplessly drawn, directly or indirectly, into the Asian financial whirlpool. Under such circumstances, a renminbi devaluation might not really help much in promoting Chinese exports to these countries. Besides, there seems to be considerable elbow room for the Chinese government to maneuver in (as indeed it did throughout most of 1998), such as resorting to policy alternatives like export value-added tax rebates, easy export credits, and the like, capable of arresting the potential adverse trends in Chinese exports.

On balance, China's exports for the whole of 1998 still recorded a slight increase, from U.S.$182.7 billion (1997) to U.S.$183.76 billion (1998). Its total imports, on the other hand, declined slightly from U.S.$142.36 billion (1997) to U.S.$140.17 billion (1998), in part as a result of reduced imports for export-processing by Hong Kong and Taiwan investors. Hence, the country could still score an impressive trade balance of U.S.$43.6 billion for 1998. As a result, China's total foreign exchange reserves stood at U.S.$144.9 billion by 31 December 1998, compared with U.S.$139.9 billion a year earlier, not counting gold reserves and foreign earnings held by enterprises and other non-treasury entities.[10] This balance sheet clearly provides a strong backbone for continuing confidence in the renminbi and, with it, in the Hong Kong dollar as well.

There are no clues for a possible deterioration in the internal financial balance in China that might eventually trigger a renminbi devaluation. Rather, with the inflation rate being kept in the negative for more than one and a half years (through early 1999), the Chinese government was able in 1998 to resort, time and again, to bond issuing to finance infrastructure investments, and to other fiscal and monetary reflationary policy measures for boosting domestic demand, as a means to achieving the much flouted GDP growth target of 8 percent for 1998.[11] And the public spending spree is clearly to continue well into 1999.

As a matter of fact, Zhu Rongji, the Chinese Premier, who frequently reiterated in 1997–98 that the renminbi would not be devalued for the sake of maintaining regional economic stability and, in particular, upholding Hong Kong's U.S.-dollar-peg system, made it clear for the first time—at a meeting with the visiting Laotian premier in Beijing in late January 1999—that the Chinese government would only reconsider its position on the external value of the renminbi, when the Chinese trade balance turns sour (*SCMP* 26 January 1999). With this statement, we seem to see evidence of economic considerations in Chinese policy-making. It should nonetheless be noted that in the established Chinese practice of maintaining external trade balance, imports seem to have always been used as an instrumental variable (rather than a target variable), for readjusting any imbalance. Whether this is a matter of financial prudence or political gimmick is, of course, subject to interpretation. And whether Zhu's reassurances about the robustness of the renminbi would effectively help to dispel any further doubts remains to be seen.

At any rate, with most of the Southeast Asian currencies having more or less regained their lost ground in the past several months, the pressures on the renminbi are bound to be easing gradually. Hopefully, in a year or two the dust of the financial sandstorm will be settled, and the pre-crisis currency parity among the various currencies in Southeast and East Asian countries will be basically restored.

Let us now turn to the second set of questions, if only for theoretical interest, to see whether it makes sense, in economic terms, that in the unlikely event of a renminbi devaluation, the Hong Kong dollar should also fall simultaneously. Several points may be made briefly.

First, a renminbi devaluation may actually only help to lower Hong Kong investors' overheads and expenditure on the Mainland, especially with regard to the costs of plant construction materials supplied from inside China. Managerial and wage costs may or may not be affected, depending on whether these are denominated in renminbi yuan, Hong Kong dollar, or U.S. dollar. Outlay on current input materials brought in from abroad for export-processing in China, as well as foreign exchange earnings from export sales, would clearly remain unchanged, by virtue of the particular mode of operation adopted by Hong Kong investors engaged in export-processing in China, as alluded to earlier.

Second, the bulk of import supplies from China to Hong Kong of wage foods, clothing, and other light-industrial goods retained for local consumption should, theoretically speaking, become even cheaper, in the absence of a concomitant Hong Kong dollar devaluation; although, in reality, this may not necessarily be the case. It depends really on whether Chinese suppliers are willing to pass on the price benefits to Hong Kong consumers by quoting their exports to Hong Kong in renminbi rather than Hong Kong dollar. The well established convention over the past several decades has been for the supplies to be quoted in Hong Kong dollars.[12] Under such circumstances it does not really matter whether the Hong Kong dollar would stay put in case of a renminbi devaluation. Rather, if the Hong Kong dollar were to follow the renminbi or to depreciate to an even greater extent, it would be entirely possible that Chinese suppliers would be prompted to quote higher export prices in Hong Kong dollars, in order to reach their expected level of renminbi receipts.

Third, even if a renminbi devaluation would help stimulate China Mainland's exports to the West, the Hong Kong SAR—with or without a concomitant Hong Kong dollar devaluation—would clearly stand to gain handsomely, for the obvious reason that a very substantial amount of Chinese exports are channeled through Hong Kong, taking advantage of all the necessary facilities it offers, including international marketing skills, banking, finance, insurance, and shipping.[13] Of course, any gains to be made in this respect would have to be weighed against the potential revenue losses for Hong Kong, which may result as well from reduced Chinese imports after renminbi devaluation. Around 50 percent of the SAR's exports to China comprise goods (mostly of third-country origin) for final consumption on the Mainland, rather than for export-processing. However, it should also be noted that, irrespective of what the net balance may be, a concomitant devaluation of the Hong Kong dollar would only make things worse, as it would curtail the import capability of the SAR for re-exports to the Chinese hinterland.

The fourth point concerns the potential effect on capital flow from China to Hong Kong, which has over the past 15 years or so played an increasingly important role in enhancing confidence in the territory and its economy. Would Chinese capital flow into Hong Kong be retarded if the Hong Kong dollar does not depreciate in tandem with a renminbi devaluation? This may not necessarily be true in the present context. So far as capital inflow of true Mainland origin is concerned, few cases seem to have taken the form of hard currency repatriations obtained through conversion of renminbi sources. Apart from possible direct allocation of foreign exchange from the treasury, most "expatriations" probably represent funds raised by state or quasi-state agents, (e.g., the itics)[14] from outside of the country or through direct "interception" of foreign exchange earnings in Hong Kong, by the familiar practice of "under-invoicing" exports to the SAR for re-exports to third countries at much higher prices.

Finally, the only venue that may unequivocally require a Hong Kong dollar devaluation to help mitigate curtailment in earnings following a renminbi devaluation should be the SAR's domestic exports to the Chinese Mainland. But note that such exports of Hong Kong origin make up not much more than 10 percent of Hong Kong's total exports to the Mainland (1997 figure), and the major proportion (around 75 percent) is indeed destined for further export-processing, rather than for end-users in China. More seriously, perhaps, if the Hong Kong dollar were to devalue to a greater extent than the Chinese renminbi, then it would probably only help to thwart the emerging access to the domestic Mainland market, something long sought after by investors from Hong Kong (and Taiwan as well). The devaluation of the NT dollar in Taiwan in late 1997 is a good case in point. In that particular case, as a result, many consumer goods industries of Taiwan origin operating on the Mainland—food processing in particular, which extensively relies on Japan and Taiwan for imports of spices and packaging materials—simply had to suspend their production for not being able to withstand price competition from domestic producers.[15]

Taken together, therefore, it seems quite clear that there are no hard economic reasons why a renminbi devaluation will inevitably trigger the Hong Kong dollar to follow suit immediately. This leaves us with the conclusion that the renewed speculative attacks on the Hong Kong currency have been no more than a matter of international hedge funds managers attempting to capitalize on the "herd psychology" of the small investors, who are fearful of their savings capital being wiped out overnight by a totally unanticipated depreciation of the Hong Kong dollar. These hedge fund managers have sought to do so by speculating for the devaluation of the renminbi as an inevitable domino of the Southeast Asian currency crisis.

Viewed this way, the resolve of the Chinese government against devaluing the renminbi, and the unparalleled strength of the Chinese economy (e.g., a most impressive GDP growth, negative inflation rate, negligible fiscal deficits, at least for 1998, remarkable trade surplus, and huge foreign exchange reserves), have certainly helped to greatly boost the confidence in the HKSAR government's defense of its own currency.

Paradoxically, however, the solid Chinese hinterland backbone has, in a way, proved to be a liability to the tiny SAR: while Mainland China's capital account is strictly closed to any outside manipulations, the fully convertible Hong Kong currency unfortunately became a target of international speculations, as the Southeast Asian financial turbulence was sweeping through the whole of East Asia in 1998.

Revamping the Currency Board: Has It Really Helped?

This is not the place for a detailed discussion of how the speculative attacks on the Hong Kong dollar evolved in 1997–98.[16] But, briefly, by mid-August 1998, it had become evident that a massive showdown was imminent. As Donald Tsang, the Financial Secretary, put it, the evidence was a two-pronged attack on both the Hang Seng Index futures of the local stock market, and the foreign exchange market. Specifically, international hedge funds were not only targeting the foreign exchange market by shortselling the Hong Kong dollar; but, more critically, by thus pushing up the local interest rates, they were also aiming at the much more lucrative Index futures market on which they had earlier accumulated huge short positions.

The decisive showdown came on 14 August 1998, when the HKSAR government acted decisively to rescue the local stock market. Funds from the Exchange Fund were used to buy 33 leading stocks. The Hang Seng Index was, as a result, massively pushed up by 564 points to 7,224.7, a rise of 8.5 percent. It seemed that Hong Kong had, for the first time, exerted, with such a dramatic turnaround in the Hang Seng Index, an impact on the global stock market. In London, the FTSE 100 Index closed 55.5 points higher at 5,455.0. And the Dow Jones Industrial Average was also raised by 18.03 points to 8,477.53 after mid-day in New York (*SCMP,* 15 August 1998).

It is clear that by intervening in the stock and futures markets, the SAR government opened another front in its defense of the U.S.-dollar-peg system. It no longer just relies on raising the interest rates to fend off speculative attacks on the Hong Kong currency. As Secretary Tsang saw it, the simultaneous speculative activity in the futures and currency markets was "a double play" by people who were trying to "play around" in an attempt to "wreak havoc" on the linked exchange rate system. "We will," he added, "continue this action as long as it is necessary. The linked exchange rate is paramount" and here to stay (*SCMP,* 15 August 1998).

The stock market intervention of 14 August 1998 was, of course, not entirely controversy-free, although the Financial Secretary strongly rejected suggestions that his action contravened Hong Kong's commitment to non-intervention. He explained the day after the event:

I must emphasize that our long-standing policy of non-intervention in the stock and futures markets remains unchanged. No intervention may take place unless a sufficiently clear linkage exists between the currency play and the stock and futures markets play. I must make it clear that we are not against the shorting (short selling) of Hang Seng Index futures by hedge funds. But we do not tolerate attempts by speculators to manipulate our interest rates so that they can benefit from the short positions they have built up in Hang Seng Index futures (*SCMP*, 15 August 1998).

In the subsequent months, Tsang had to make it part of his agenda for his overseas trips to explain to and convince his major foreign counterparts, including Alan Greenspan, Chairman of the Federal Reserve Bank in the United States, about the rationale and necessity of the unprecedented interventionist move. Nonetheless, the conservative U.S.-based Heritage Foundation announced in December 1998 that it would likely downgrade for 1999 the status of Hong Kong as the most free economy in the world, because of the government's unprecedented stock market intervention.

More seriously perhaps, voices were raised by local political parties and academics as well, on the question of whether the SAR government's uncompromising move, wrestling with giant international hedge funds, would not put the entire Exchange Fund at risk. As a matter of fact, the stake that the government invested in the 33 leading stocks was later officially confirmed to be a massive U.S.$15.2 billion, against the initial speculative estimate of U.S.$0.39 billion. In other words, about 16 percent of the Exchange Fund's reserves was literally put at risk in one stroke. A few more rounds of similar wrestlings might eventually see the entire Exchange Fund rapidly depleted, should the government lose the battle against the speculators.

It is probably against this background that the Hong Kong Monetary Authority was obliged to embark, three weeks later (on 7 September 1998), on a major initiative of significantly revamping the existing Currency Board system as a new defense mechanism against international currency assault. The initiative comprises a seven-point package, but essentially they all converge on two major policy measures.

The first is to rigorously reinforce the Currency Board's discipline by committing the government, under a "convertibility undertaking," to convert the Hong Kong dollars held by all licensed banks in their clearing accounts with the Exchange Fund (rather than just the legal tender in circulation) into U.S. dollars at the fixed exchange rate of HK$7.50 to the U.S. dollar, which is even higher than the official benchmark of HK$7.80 for cash arbitrage. The second is to greatly enhance the financial liquidity of the entire banking system by replacing the restrictive "Liquidity Adjustment Facility" with a "Discount Window," whereby the banks may repeatedly borrow from the Hong Kong Monetary Authority with their holdings of Exchange Fund bills and notes as collateral to cope with any drastic overnight demand for liquidity.[17]

Clearly, the new measures are all strongly backed by the government's sizable foreign exchange reserve, which, despite the massive stock market buyup by the government in August 1998, was still three and a half times larger than the SAR's entire monetary base of U.S.$23 billion. But has the new Currency Board really helped? The answer is a mixed one.

Subsequent to the launching of the seven-measure package, the prime lending rate of the Hong Kong Association of Banks did indeed decline successively from the high of 10 percent in October to 9.25 percent in early December 1998. As a result, the gap with the comparable U.S. interest rate narrowed from 3 percent to 2.5 percent, although clearly the interest rate arbitrage for the U.S. dollar peg is still very much in operation. In essence, however, the interest rate reductions in Hong Kong seem to have been prompted by a similar move by the U.S. Federal Reserve Bank, and perhaps more importantly by the withdrawal of the international hedge funds following their collapse in Russia, Brazil, and elsewhere. These, in the final analysis, seemed to be more important reasons than the augmented financial maneuverability arising from the bold Currency Board reform measures Hong Kong adopted in early September 1998, although the latter was also a contributing factor to the interest rate reductions. Similarly, both the stock market and the property market in Hong Kong have also rebounded quite remarkably, as elsewhere in Southeast Asian countries, which have likewise seen their currencies recover substantial ground against the U.S. dollar, as international speculative currency attacks subsided.

As a matter of fact, the Chief Executive of the Hong Kong Monetary Authority, Joseph Yam, himself admitted that to effectively counter the international hedge funds, a new global financial system should eventually be put in place. He said: "We face a world crisis. If Hong Kong, with its sound fundamentals and prudent financial management, can be brought to the brink of a systemic breakdown by aggressive cross-border speculation, then something must be wrong with the world financial order." (SCMP, 6 January 1999). Specifically, Yam called for the regulation of international capital flows and the eventual creation of an Asian currency to prevent a repeat of the regional economic crisis (SCMP, 6 January 1999). This echoed a point made earlier by the Harvard economist, Dani Rodrik, that it is simply inept to subject the policy of any government to "what 20 or 30 foreign-exchange dealers in London, New York or Frankfurt think" (New York Times, 4 September 1998). This also seems to be the general consensus arising from the annual joint meeting of the World Bank and IMF held in Washington D.C. in October 1998.

Impact on External Trade

This chapter is not the place for rehearsing how the real economic sphere in Hong Kong has been affected by the financial crisis. But, briefly, by early 1999 it was confirmed that the SAR's GDP was curtailed by a hefty 5.1 percent in 1998 from 1997. And earlier, by February 1999, the unemployment rate had reached a 25-year high of 6 percent. Interestingly, however, the Hang Seng Index—the barometer of the local stock market—has been hovering in the region between 10,000 and 11,000 points since early 1999. This represents roughly a 60 percent gain from the trough of October 1997, and is only about 35 percent short of a full recovery to the all-time high of 16,670 points achieved in August 1997. Obviously, in relation to the remarkable magnitude of GDP decline in 1998, and the projected modest GDP growth of only 0.5 percent for 1999, Hong Kong stocks seem to be excessively overvalued once again. There is no doubt that the renewed "overheating" has been fueled by the return of major international investors and stockbrokers.

Are the Hong Kong stock exchange and currency markets heading for another showdown? We really do not know for sure at this moment. But Joseph Yam warned explicitly that hedge funds might regroup in 1999 and renew attacks on the Hong Kong dollar's 15-year-old peg to the U.S. dollar (*SCMP*, 6 January 1999). What seemed certain, nevertheless, was that amidst the renewed financial euphoria, there were, by the first quarter of 1999, still no signs of an early recovery in the real domestic sector. And this was in spite of the respectable efforts made by the SAR government, following tenets of Keynesian economics for the first time ever in Hong Kong, to revive the economy through large-scale public spending on large infrastructure projects.

In this context, the only spot that tends to offer a ray of hope and comfort seems to be the external trade, which is always the lifeline of the Hong Kong economy. For the whole of 1998, the territory's total exports declined by a relatively modest amount of 7.4 percent, as compared with 1997.

And, more importantly, as Figure 9.2 reveals, the major sources of export decline are the weakened position of the Southeast and East Asian trading partners that were hardest hit by the Asian financial crisis. Because of massive currency devaluations, all their import capabilities were seriously impaired. By contrast, Hong Kong's most significant export markets, the United States and Western Europe, have basically remained intact. The same goes for exports to the rest of world as a whole. Total exports to the Chinese Mainland also declined by 8.7 percent from 1997 to 1998; but the reductions comprised essentially re-exports of third-country origin destined for end-users in China (rather than for export-processing). Obviously the decline was a result of a measured deliberate Chinese curtailment of imports, instituted in order to maintain a favorable trade balance to hedge against the increased pressures for a renminbi devaluation.

The more important point that should be made in this context is that virtually all the SAR's exports to the United States and Western Europe represent re-exports from China on behalf of Hong Kong manufacturers based in the Mainland engaged in export-processing. That is to say, as long as these major overseas markets can sustain their import demand, (as basically has been the case amidst the Asian financial crisis), Hong Kong will continue to enjoy the privileges of making use of the Chinese hinterland as its export-processing base for enhancing its exports and GDP growth.

A word or two should be said about the invisible trade. Hong Kong does not really have any natural resource of any scale to build its economy on, apart from the beautiful landscape and perhaps some cultural relics. Traditionally, the SAR has had to rely on imports, not only for export-processing, but also for maintaining the livelihood of its population. Visible trade has therefore been, as a rule, in the red, and it has had consistently to rely on tourism earnings to balance the trade deficits throughout the past several decades.

Unfortunately, however, the number of tourist arrivals dropped quite remarkably after the political handover to China on 1 July 1997—perhaps due to some global misperception about the nature of the historical event. Worse still, the difficulty has been greatly compounded by the Asian financial crisis, in that the massive devaluation of the Southeast Asian currencies and the persistent recession in Japan have indeed made the SAR relatively too expensive a destination for tourists from these conventional Asian sources. As a result, rigorous efforts have been made by the tourist industry to court even such unlikely tourist sources as Mainland China, to compensate for the losses from elsewhere. Nonetheless, total tourism receipts for 1998 still recorded a 14 percent decline in real terms (HKGFS, 1999, p.3). Moreover, Hong Kong's exports of financial and other business services also fell by 6.6 percent, again in real terms, for 1998, because of setback in regional demand. (p.4)

Fortunately, under the full impact of the Asian crisis, Hong Kong's import demand also weakened substantially, resulting in a large reduction in the visible trade deficit. Coupled with a still sizable invisible trade surplus, the overall trade balance could still improve to a small surplus in 1998, from the deficits recorded for 1995–97. (HKGFS, p.14) However, it is clear that for the Hong Kong government to bring the SAR economic growth back to the pre-crisis track, some extra efforts need be made, in addition to promoting conventional exports. It is against this background that an enormous initiative was made to attract the Disney Company to set up a second theme

254

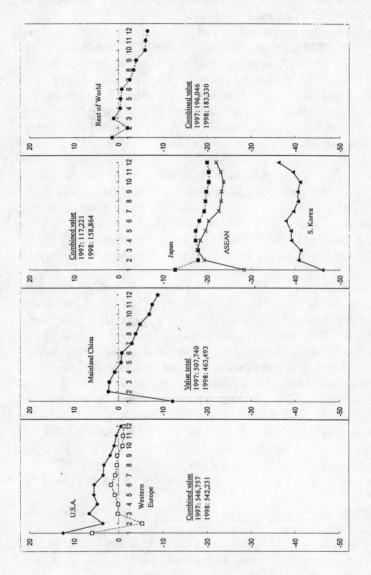

Figure 9.2 Value of Hong Kong's exports and year-on-year changes by major regions and trading partners, 1997 and January to December, 1998 (HK$ million and percent)

park in Hong Kong, next to Tokyo. Similar initiatives include the idea for the construction of a "Fisherman's Wharf" in Aberdeen (the familiar tourist attraction on the island) and, more importantly, the development of a "cyber-port" worth HK$13 billion to provide high-tech information services to business and the financial sector in Hong Kong and the region.

Concluding Remarks

The major points made in this chapter are as follows:

First, there are no fundamental economic reasons to suggest that the Chinese currency, the renminbi, would have to be devalued under the impact of the Asian financial crisis. Nor are there any strong economic factors at work that would suggest that the value of the Hong Kong dollar would fall as well against the U.S. dollar, should a renminbi devaluation take place, as is widely speculated.

Second, the speculative attacks on the Hong Kong currency in 1997–98 seem, therefore, essentially a matter of premeditated attempt made by international hedge funds to coerce Hong Kong to be the next Asian financial domino, by exploiting the volatile market sentiments following the Southeast and East Asian crisis.

Third, an important reason why the fully convertible Hong Kong dollar has become the sole focus of international currency assault within Greater China is that the capital account of the Chinese Mainland is strictly closed to outside manipulations. Nonetheless, the remarkable strength of the Mainland economy (in terms of both internal and external balances) in 1997–98 has also greatly helped to boost confidence in the Hong Kong SAR's rigorous defense of its U.S. dollar peg system.

Fourth, the unprecedented, massive government buy-up of leading stock shares in August and the significant Currency Board reform in September 1998 have proved to be effective in stabilizing both the stock and currency markets in Hong Kong. But the success was not without its external effects: it entailed the timely lowering by the U.S. Federal Reserve Bank of interest rates, not to mention the withdrawal of the international hedge funds following their defeat in Russia, Brazil, and elsewhere.

Fifth, pending a long-term solution to the global financial system, Hong Kong will likely remain vulnerable to international speculators' attack on its currency, despite its well-organized and efficient banking and financial systems, as well as prudent fiscal and monetary policies.

Sixth, while, as a result of the collapse of investors' and consumers' confidence, (coupled with high interest rate pressures and severe credit squeeze), the Hong Kong economy was nearly brought to the brink of collapse, it has been fortunate in being able to fall back on the Chinese hinterland as a base for generating exports to the resilient, single most important export markets of the United States and Western Europe.

Is a full economic recovery in Hong Kong from the Asian financial meltdown around the corner? Donald Tsang, the Financial Secretary, in his 1999–2000 budget speech delivered on 3 March 1999, envisaged a modest GDP growth of 0.5 percent in 1999, to be followed by a marked acceleration of around 3.5 percent to 4 percent in 2000, and to appreciably above that in 2001 and 2002. That is to say, it will take four to five years for the SAR economy to restore its pre-crisis pattern of GDP growth. As Tsang aptly put it, "This in turn will depend on the external economic and financial situation turning progressively more favorable" (HKGFS, p.4).

Assuming that the major United States and West European markets will be able to sustain their demand for imports from China and the SAR, the situation in the ASEAN countries, and in South Korea and Japan as well, obviously holds the key to an early recovery in Hong Kong. Nonetheless, despite encouraging signs of a rapid recovery in some of these countries, both the renminbi and the Hong Kong dollar still remain "overvalued" to Southeast Asian currencies and the Korean Won, in relation to the pre-crisis benchmarks. This undoubtedly will continue to be a source of uncertainty for the Mainland and the HKSAR.

Notes

* I am grateful to Shu-ki Tsang and Pansy Yau for their helpful comments, and to Phoebe M. W. Wong and David Ji for excellent research assistance.
1. See the Appendix to this chapter for a chronological illustration of the development of the Asian crisis, as it bears on Hong Kong.
2. Following relocation of tens of thousands of manufacturing plants from Hong Kong to the Chinese hinterland, starting especially from the mid 1980s, Hong Kong's Gross Domestic Product had continued to grow at the rate of 5 to 6 percent per year through the advent of the Asian crisis. Domestic labor supply could increase, however, by only around one percent per year during the same period. These occurred, most remarkably, against the background of a declining workforce in the manufacturing sector from some 930,000 in 1986 to not more than 250,000 by 1997, as a result of the accelerated exodus of Hong Kong manufacturers. By 1998, the manufacturing sector's contribution to GDP had fallen to a mere 8 percent from 29.6 percent in 1985 and 24.9 percent in 1991, while the share of the services sector meanwhile rapidly expanded to over 90 percent from 69.7 percent and 74.8 percent, respectively. The entire Hong Kong economy has, therefore, been radically transformed into a Manhattan-type services industry. This is, by any historical measure, a massive radical transformation.

3. In the first of its annual reports on Hong Kong released in January 1999, the European Commission also found "basic rights, freedoms and autonomy have been broadly upheld" (*SCMP,* 10 January 1999).

4. For a detailed analysis of the relative contributions of Hong Kong to China's total foreign exchange earnings and the role the SAR plays in the broader context of facilitating China's trade with the United States and the outside world in general, see Kueh and Voon, especially pp.61–83.

5. See attached chronology for this and subsequent reassurances given by Chinese leaders that China would uphold the value of the renminbi in support of the Hong Kong government's defense of the U.S. dollar-peg.

6. For an excellent full-fledged study of the Currency Board as it is adopted in Hong Kong, see Y.C. Jao, especially pp.224–39.

7. This and the following one or two paragraphs, as well as Figure 1, derive from Hang Seng Bank, 1998, p.3.

8. In a recent personal communication (March 1999), a major Hong Kong investor was resolute that he would only invest more in the Mainland, if the renminbi were to depreciate in the present context.

9. See Kueh (1999) for a detailed study in this respect.

10. These are all official statistics released at the Ninth National People's Congress held in early March 1999 (*Ta Kung Pao* [TKP], 7 and 12 March 1999).

11. The realized GDP growth rate for 1998 was confirmed, in March 1999, to be 7.8 percent.

12. The reasons for this are familiar: Chinese domestic prices are normally not comparable to the free market prices in Hong Kong or elsewhere in the world. A straightforward conversion from renminbi prices for the export goods concerned on the basis of the given exchange rate, may not therefore yield an export quotation that is comparable to those offered for similar products from other sources in the Hong Kong market. Moreover, the officially fixed exchange rates used for the conversion are normally not in accord with the purchasing power parity to complicate the matter. Hence, China, as well as all other former Soviet-type economies, being deprived of reliable scarcity price signals (due to distortions in official price-setting) are said to have been "trading in the dark," by relying on the changing world market prices.

13. This point is also strongly shared by Jeffrey Saches, the Harvard economist, at the Asian Investment Conference sponsored by Credit Suisse First Boston in Hong Kong (*SCMP,* 25 March 1999).

14. This is the familiar acronym for various incarnations of what is known as "international trust and investment corporation." The most well-known ones include CITIC (China International Trust and Investment Corp.), a national-level agency based in Beijing, and the notorious GITIC—a Guangdong provincial agency that collapsed in late 1998 prompting several major banks in Hong Kong to provide, for the first time, under such circumstances, full debt provisions in the balance statements.

15. Personal communication from a major Taiwan investor in the food-processing industry.

16. Readers interested in the process should consult the chronology given in the Appendix.
17. See Hong Kong Monetary Authority (1998) for details and Hang Seng Bank (October 1998) for a brief interpretative study.

References

Hang Seng Bank. 1998. *Hang Seng Economic Monthly.* (September/October).

Hong Kong Government Census and Statistics Department (HKGCSD), *Hong Kong External Trade* (HKET), various issues.

Hong Kong Government Financial Secretary (HKGFS), *The 1999 - 2000 Budget,* 3 March 1999.

Hong Kong Monetary Authority. 1998. *Strengthening of Currency Board Arrangements in Hong Kong.* (September). http://www.info.gov.hk/hkma/new/press/others/ 980905e.htm

Jao, Y. C. 1998. "The Working of the Currency Board: The Experience of Hong Kong 1935–1997." *Pacific Economic Review* no.3, (October): 219–41.

Kueh, Y. Y. 1999. "The Greater China growth triangle in the Asian financial crisis." Paper written for the workshop sponsored by the World Bank in Singapore, 11–13 January 1999, in preparation for the Bank's *World Development Report 1999/2000.*

Kueh, Y. Y., and Thomas Voon. 1997. "The Role of Hong Kong in Sino-American Economic Relations." *The Political Economy of Sino-American Relations: A Greater China Perspective.* Edited by Y.Y. Kueh. Hong Kong: Hong Kong University Press: 61–92.

South China Morning Post (SCMP), Hong Kong.

Ta Kung Pao (TKP) Daily, Hong Kong.

Wong, Kar-yiu, ed. 1998. *The Asian Crisis: What Has Happened and Why.* University of Washington, Seattle (October).

APPENDIX (CHAPTER 9)

Chronology of the Asian Financial Crisis in the Hong Kong SAR July 1997 to January 1999

1997
July

2   Under heavy pressure from massive selling by foreigners, the Bank of Thailand floats the baht and raises the discount rate by 2 percentage points, to 12.5 percent. By the end of the day, the currency devalues 18 percent from the rate of 25 baht per U.S. dollar.

27   First wave of speculative attack on the Hong Kong currency takes place. Overnight HIBOR (Hong Kong Interbank Offered Rate) raised from 5.13 percent to 6.38 percent p.a., and the three-month HIBOR from 6.19 percent to 6.75 percent. It was confirmed later that the Hong Kong Monetary Authority (HKMA) spent around U.S.$1billion of its reserves (of U.S.$81 billion) to defend the HK dollar

21–27   The Indonesian rupiah, Thai baht, Malaysian ringgit, and Philippine peso all slump as confidence in the region rapidly deteriorates.

28   U.S. Secretary of State, Madeleine Albright, criticizes Mohamad Mahathir, the Prime Minister of Malaysia, for blaming the West for a decline of Southeast Asian currencies.

August

19   The second wave of international speculative attack against the Hong Kong dollar takes place, sending overnight HIBOR up to 10 percent from the previous close of 6.81 percent, and the three-month HIBOR to 9 percent from 8.25 percent.

September

4   The baht falls to the all-time low of 38.4 to the U.S. dollar.

8   On the eve of his trip to the United States, the Chief Executive of the Hong Kong Special Administrative Region (HKSAR), C.H. Tung, reassures that the HK dollar will remain pegged to the U.S. dollar.

22   At the joint annual meeting of the World Bank and International Monetary Fund (IMF), Mahathir lashes out against U.S. financier George Soros and blames him for Malaysia's currency crisis.

October
18   The Central Bank of China in Taipei allows the New Taiwan Dollar (NTD) to fall, giving up defending the currency at NTD28.6/U.S. dollar.
23   The third wave of international speculative assault on the Hong Kong dollar forces the HKMA to push HIBOR overnight up to 100 percent (closing high) after reaching, at one point of the day, 280 percent, from the previous low of 6.25 percent. The three-month HIBOR also surges to 25 percent from 9.25 percent.

      The Hang Seng Index drops to 10,426.30—a 10.4 percent decline from the previous close of 11,637.77, and a total reduction of 23 percent from previous Friday's close of 13,601.01.

      Both China and the HKSAR governments subsequently promised to protect the HK dollar peg to the U.S. dollar with their massive foreign reserves (U.S.$88 billion in HK and U.S.$120 billion in China).

November
19   The Central Bank of China in Taipei indicates that it will allow the market to determine the value of the NTD in the wake of sharp depreciation of the Korean won.

December
16   President Jiang Zemin announces, in his meeting with heads of the ASEAN countries, that the renminbi will not be devalued.

1998
January
1   The fourth wave of speculative currency assault on the Hong Kong dollar forces HIBOR to climb to 12 percent from the previous close of 5.5 percent, and the three-month HIBOR to 13.5 percent from 12.5 percent
16   People's Bank of China Governor Dai Xianglong reiterates that the renminbi will remain stable in light of China's favorable balance of payments.
23   The Indonesian rupiah falls to a low of 15,000 against the U.S. dollar, leaving its banking sector on the verge of collapse.

February
2   Hong Kong stocks jump 14.33 percent, the second largest one-day points gain. Regional markets also advance.
10   The rupiah rebounds more than 30 percent at one stage, amid rumors that Indonesia is to adopt capital controls and a Hong Kong-style currency board exchange rate system
15   IMF Managing Director, Michel Camdessus, threatens to cut off Indonesia's IMF-led U.S.$43 billion bailout if the country pushes ahead with

plans to peg the rupiah to the U.S. dollar. Regional stocks and currencies fall sharply on the news the following day.

18   Financial Secretary Donald Tsang announces, in his first budget for the Hong Kong SAR, a $13.6 billion tax cut package to boost the sagging economy and comfort jittery financial markets. The unexpected tax cuts send the Hang Seng Index 4.29 percent higher.

20   U.S. credit rating agency, Moody's Investors Service, downgrades Hong Kong's short-term sovereign rating, the first since the agency assigned the same rating 14 years ago.

March

7   Dai Xianglong reaffirms that China's increased foreign reserves will help to strengthen the renminbi.

19   The Chinese Premier, Zhu Rongji, guarantees that the renminbi will not be devalued.

April

2   During his visit to the United Kingdom, Chinese Premier Zhu Rongji emphasizes that there are only pressures for the renminbi to appreciate, rather than depreciate.

3   The yen plummets to a six and a half years low as Moody's downgrades Japan's outlook for country ceilings and domestic currency ratings from stable to negative.

18   The Chinese Finance Minister, Xian Huaicheng, reconfirms that the value of the renminbi will be maintained.

May

7   Chinese President, Jiang Zemin, tells Reuter's Chief Executive that he has full confidence in keeping the renminbi stable without following the devaluation of neighboring countries.

13   Regional stock markets and currencies slide in the wake of days of intense anti-government protests and riots in various Indonesian cities. Indonesian stocks dive almost 7 percent as the rupiah crashes.

27   Government economists confirm Hong Kong's economy shrank for the first time in 13 years during the first quarter. Investor fears of a recession send the Hang Seng Index down 5.55 percent the following day.

29   HKSAR government announces a seven-point emergency economic rescue package as the economy is confirmed to have shrunk 2 percent in real terms during the first quarter of the year. The measures aim to stimulate the property, banking, and tourism sectors and enhance the financial system's liquidity.

June

3   Jiang Zemin tells U.S. President Clinton that the decision not to devalue the renminbi not only helps to maintain economic stability in China, but it also supports economic stability in Asia and the world at large.

9   People's Bank of China Governor, Dai Xianglong, warns the falling yen is having a "very negative" impact on the mainland's foreign trade, capital inflows, and economic restructuring.

10   Hong Kong shares tumble 4.91 percent to their lowest level in more than three years, as the yen's fall triggers fears about a devaluation of the yuan.

11   HKMA decides to issue statements twice a day detailing aggregate balances in accounts it holds on behalf of banks for foreign exchange dealings for the coming two days. By increasing transparency, the authority hopes to smooth out volatility in interbank rates, in a bid to stave off attacks on the currency peg.

12   Japan enters into the worst recession since World War II, as it is revealed its economy shrank at an annualized rate of 5.3 percent in the first quarter, bringing real GDP down 0.7 percent in fiscal 1997.

15   Another round of assault by international hedge funds on the Hong Kong dollar pushes overnight HIBOR up to 12.5 percent from 8.25 percent, and the three-month HIBOR from 12 percent to 14 percent, with the Hang Seng Index tumbling by nearly 6 percent from 7,915.44 to 7,462.50.

17   Heavy intervention by the U.S. and Japan in the foreign exchange market sends the yen rising more than 4 percent to 137.25 against the U.S. dollar. This follows heavy selling from hedge funds on the previous day, which put an end to the yen's dramatic slide. The yen's rise triggers a 6.35 percent rebound by the Hang Seng Index to just over 8,000 points.

18   The rupiah plunges to near its January record low of 17,000, raising prospects of a possible collapse of Indonesia's banking system.

22   China's Minister of Foreign Trade and Economic Cooperation, Shi Guangsheng, announces once again on behalf of the Chinese government that the renminbi will not be devalued.

     HKSAR Chief Executive, C. H. Tung, announces a HK$44 billion rescue package for the economy, which includes a surprise plan to freeze land sales until March 31, a pay freeze for top civil servants, and a tax exemption for companies on interest earned on locally held deposits.

July

8   HKMA announces that it is to repatriate "billions of dollars" in overseas deposits to the local banking system as part of its effort to ease the SAR's liquidity crisis.

17   China reports economic growth of 7 percent in the first six months of the year, the slowest growth rate in seven years, underlining the difficulty Beijing would have in achieving its target of 8 percent for the year.

August

7   Dai Xianglong reiterates no devaluation of the renminbi.

     HKMA Chairman, Joseph Yam, calls speculation on the Hong Kong dollar peg a conspiracy. Chief Executive C.H. Tung reiterates willingness to defend the HK dollar.

Hong Kong shares plunge to their lowest in more than three years amid renewed fears of a yuan devaluation. The Hang Seng Index teeters near 7,000 points, ending the week down 10.9 percent.

11   The yen tumbles to an eight-year low of 147.63 against the U.S. dollar, as the Economic Planning Agency says Japan is in a prolonged slump.

14   The SAR government launches a heavy, unprecedented market intervention in stock and derivative markets, which sends the Hang Seng Index soaring 8.47 percent to 7,224.69 points.

27   Sources close to the Quantum Fund, run by George Soros, confirm it has made sizeable bets against the Hong Kong dollar. Stanley Druckenmiller, the Manager of the fund, says the Hong Kong government will lose its bet on the local currency.

28   In a further massive defense of Hong Kong's currency peg, the HKMA throws HK$70 billion at the markets, bringing the amount spent in the campaign to about 15 percent of the SAR's reserves.

The government confirms the SAR has fallen into recession, announcing a 5 percent second quarter GDP contraction.

30   Taiwan authorities order a crackdown on what they say is illegal trading in the funds of George Soros.

Hong Kong shares slump 7.09 percent as the government halts its intervention.

September

2   Malaysia's Central Bank pegs the local currency at 3.80 ringgits against the U.S. dollar. The rate peg sees Malaysian share prices soar 12 percent.

5   HKMA announces a seven-point package to strengthen the currency board and deter manipulation of local interest rates by speculators.

7   The Hong Kong stock market surges 7.85 percent, surpassing the 8,000-point level, on massive short covering by funds squeezed by a resurgent Japanese yen and falling interest rates.

Malaysian shares soar 22.45 percent, recording the country's largest one-day rise. Investors chase after stocks related to associates of Prime Minister Mohamad Mahathir, who appoints himself first Finance Minister.

10   Jiang Zemin says stability of the renminbi has greatly helped to stabilize Asian economies.

14   State Council spokesman, Bai Hejin, stresses that the renminbi will not be devalued in 1998 or in 1999.

24   Hedge fund company Long Term Capital Management receives an injection of U.S.$3.75 billion from New York Federal Reserve Bank in a rescue bid.

October

6   The People's Bank of China closes down Guangdong International Trust and Investment Corporation. The investment vehicle has incurred losses of billions of yuan on overseas investment, mainly in derivatives and properties.

8   The U. dollar slumps almost 8 percent to its lowest level in 15 months against the yen, touching a low of 111.63 and recording the largest one-day drop in 25 years.

November
11   Government-appointed Committee on Singapore's Competitiveness puts forward measures to slash manufacturers' costs by 15 percent, in a bid to boost the country's international competitiveness.
17   The SAR government merges the HK$211.4 billion Land Fund and the HK$735 billion Exchange Fund, to further enhance Hong Kong's defense against speculative attacks on its currency.

December
9   Zhu Rongji reiterates no devaluation of the renminbi.

1999
January
13   Zhu Rongji says China is pursuing active fiscal policy and appropriate monetary policy for maintaining the stability of the renminbi.
28   Yuan devaluation angrily rejected by Dai Xianglong as unnecessary.

SOURCES: *SCMP,* 31 December and 13 January 1998, and 28 January 1999; *Ta Kung Pao Daily* (TKP), 26 January 1999; and Kar-yiu Wong, 1998, pp. 3–37, 84–7.

CHAPTER 10

HKSAR's Relations with Its Chinese Sovereign

TING Wai

Introduction: The Essence of "One Country, Two Systems"

The relationship between the Hong Kong SAR and the People's Republic of China (PRC) may be a hazardous issue to discuss. First of all, it is incorrect to name the relationship as China-Hong Kong relations, as Hong Kong is already part of China. The relationship between the two is one between the central authority and a local region under a unitary state; so the two entities cannot be compared at the same level. By the same token, not even "Mainland-Hong Kong relations" is adequate, since Hong Kong is simply part of the Mainland. The proper way of naming it is "interior Mainland (*neidi*)-Hong Kong relations." For the same reason of ensuring political correctness, China-Taiwan relations is replaced by "cross-strait" relations, although the former term appears to be more precise in denoting the relations between the two discrete entities across the Taiwan Strait. For the sake of simplicity and clarity, nevertheless, we will use the term "China-Hong Kong relations" throughout the chapter. Second, this set of relations is framed in the sacrosanct principle of "one country, two systems." This principle is always presented in China as broaching no questioning, since it was the brainchild of Deng Xiaoping, which is enough guarantee in itself that Hong Kong's socio-economic status quo will continue after its return to Chinese sovereignty, at least during the first 50 years. Thus, with its political future guaranteed, its humiliation as a colony gone forever, and no foreign power ever again being able to infringe upon the territory's domestic affairs, Hong Kong need only worry about how to maintain its economic prosperity. Under the relentless support of the motherland, the future of Hong Kong should be rosy and upbeat, according to this Chinese line of thinking.

However, this rather simplistic depiction of the future of Hong Kong is not universally echoed among analysts and the general public of Hong Kong. (In this essay, I shall try to be the "devil's advocate" in articulating the worst-case scenario held by many of the Hong Kong people I know.) Even for the meaning of "one country, two systems," there are different interpretations. Some analysts argued that while Beijing puts heavy emphasis on "one country," people in Hong Kong are keener on the distinct differences of the two systems. They are more conscious in the defense of their own socio-economic system, including the freedoms and rights enjoyed since British rule, which they have taken for granted into the post-colonial era. Some even depict the "one country, two systems" as two ends of a pendulum. The pendulum may shift to the "one country" side during moments of crisis, with the likely result of tightened control by the central authority over the SAR. In trouble-free times, on the other hand, the pendulum may shift to the "two systems" end, signaling a loosening of control and a more respectful treatment of the integrity of the territory's capitalist system. However, while "one country, two systems" is the only model of Chinese unification that takes into account the historical realities of the different entities (Hong Kong, Macau, and Taiwan) involved, the two halves of the "one country, two systems" concept cannot be separated in an arbitrary way. Within a unified structure, the nature of the Chinese regime, its reform process, and the perceptions of the Chinese leadership regarding the larger changing world order will definitely impact on the future development of SAR.

This is not to say that Beijing's trustworthiness is in doubt, as regards its commitment to Hong Kong's "high degree of autonomy" guaranteed by the Basic Law. The crux of the matter is that, from the perspective of many Hong Kong residents, this commitment came from a regime founded on the principle of what Mao once called "people's democratic dictatorship," whose ruling philosophy has been demonstrated over almost half a century. To these Hong Kong people, this background will somehow overshadow the daily functioning of the SAR. Nobody questions the sincerity of the Chinese leaders, but good intentions simply is not enough, so they say.

Another problem with the concept of "one country, two systems" is the presumed "peaceful coexistence" of the capitalist and socialist systems within one nation (Buckley 1997, p.176),[1] which seems to rely on a faith in the Chinese leaders' assumption that the two systems would always remain unchanged. In fact, the professed "socialism" practiced in Mainland China has undergone significant changes since the death of Mao.

Beijing was worried about the existence of "subversive" forces in Hong Kong, which were alleged to have plans to overthrow the Communist regime. While Hong Kong people were generally worried about Beijing's possible interventionism after 1997, the latter's pledge was often couched in convoluted language. The Central Government, it said, would only be responsible for the SAR's defense and foreign relations, but would not intervene in the internal affairs of Hong Kong. However, if Hong Kong should be used as a "subversive base" against the national regime, then intervention from it would be inevitable. Pledges like this were hardly reassuring for many in Hong Kong.

In this light, whether people should trust the Chinese leaders, on whose commitment Hong Kong's autonomy depends, has become a question of secondary importance. The main concern is whether the basic texture and swing of Chinese politics will not ultimately "rub off" on Hong Kong beyond 1997. Although the SAR could continue to sway China to its ways and lifestyle, the territory by definition is expected to be politically subservient to the central authority. Under the banner of "one country, two systems," Hong Kong became an inalienable part of its Chinese sovereign, despite its unique tradition and the high degree of autonomy promised it, and is thus subject to the ultimate wills and wishes of the Chinese Communist Party (CCP). For its part, Hong Kong is in no position to interfere in Chinese (Mainland) domestic politics. The upshot is that while people in Hong Kong can be critical of the HKSAR government, they have no voice over, much less a right to meddle in, Chinese politics on the Mainland. Political parties in Hong Kong are not allowed to engage in political activities inside the Mainland, as the dominance of the CCP brooks no such challenge from any source.

Some basic tenets of Chinese Communist rule are steadfast: the CCP controls the state, while the state controls society, although such control has slackened somewhat since the open-door and reform initiative began 20 years ago. The Chinese consider that national independence, territorial integrity, and reunification are national goals of utmost importance. Having integrated into the Chinese political order, Hong Kong is supposed to uphold the same principles governing the whole country. Although enjoying a "high degree of autonomy," Hong Kong, ultimately, is not exempted from these national principles.

Some people in Hong Kong feel that if they were puzzled over the meaning of "one country, two systems" before, they are now fully aware of what the model entails. To put it in the most succinct way, they believe that if on the "political stage" all actors—be they pro-Beijing or anti-Beijing—are Hong Kongers, then the producer-director of the stage is China. In other words, Beijing's leaders set the rules of the game for the political development of this capitalist enclave. The complication, however, is that Hong Kong's unique position vis-à-vis its Chinese sovereign seems to offer for its people more elbow room than meets the eye. The city is always flaunted as the place where East meets West. Now that China accepts the game of globalization and is determined to be integrated into the global (capitalist) system, Hong Kong is uniquely playing a crucial role in furthering China's aspirations in this regard. China always treasures the international linkages that Hong Kong maintains, as the territory is a kind of bridgehead of Western capitalism in Asia. During the 1950s and 1960s, when China was isolated, Hong Kong served as a window for it to the outside world. After China opened up in the late 1970s, Hong Kong became the main source of direct capital investments. So the willingness of China to join the world's capitalist system only enhances the economic position of Hong Kong. To be more specific, Hong Kong's "bridging role" gains a new meaning from the contest and conflict between China and the West.

Having a very strong sense of cultural and national identity, Chinese leadership expresses a habitual repugnance to the "superstructure" (in the Marxist sense) of Western capitalism including its democratic ideals and human rights values. The Hong Kong people's ideological-cultural orientations and world views are basically different from those of the CCP leaders. That Hong Kong after reversion accepts a nominally submissive political status does not ipso facto imply acceptance of its sovereign's cultural and ideological values. Thus, in addition to being the meeting ground of East and West, Hong Kong is also squeezed in the middle. This can be easily demonstrated by the difficult position in which the Democratic Party (DP) and its supporters find themselves in their relations with the Chinese government. They support capitalism, the Western views of democracy and human rights, in addition to possessing a strong sense of cultural and national identity as well. They were the first to support China's resumption of sovereignty in the early 1980s; they supported mother-tongue education, and raising the "national consciousness" of the Hong Kong people through civic education. Yet, they were not accepted by the Chinese leaders in the 1990s, because they seemed to be on the side of the British Governor in his disputes with Beijing.

In brief, Hong Kong links up the Chinese economic system to the outside world, but Beijing was jittery about Hong Kong being fully absorbed and integrated into Western values. While it is inconceivable to have this capitalistic "superstructure" modified—in fact its way of life, political-legal institutions, and pluralist nature of society are all upheld by the Basic Law—China has imposed a gradualist approach in the democratization of the SAR's political institutions. The rationale is clear: the dynamics of the democratization process in Hong Kong, if carried out in gradual stages, will less likely infringe upon the supreme authority of CCP rule. What is more, if it can have its way, Beijing would like to have a strong executive branch in Hong Kong directly responsible to the Central Government, under the supervision of a preferably relatively weak legislature. (This is thwarted by the aftereffects of the reform Governor Patten pulled off before his departure, as James Hsiung has noted in his introductory and concluding chapters.) The ideal of an "executive-led" government would guarantee the adequate implementation of China's Hong Kong policy. Although the HKSAR government, conceived in the celebrated principle of "Hong Kong people ruling Hong Kong," is part of the Hong Kong "system," nobody can ignore that it is also part of the constitutional structure of the "one country" in the equation, which is China. Apart from governing the region, the SAR government's premier function is clearly not to contradict or thwart the principles and interests of the Chinese sovereign.

So the ultimate source of all the problems and skepticism in Hong Kong, now under Chinese sovereignty, is not whether the Chinese leaders can be trusted in keeping their promise as prescribed in the Sino-British Joint Declaration. Rather, it is rooted in the nature of the Chinese regime under the rule of the CCP. This chapter attempts to examine how the party-state in China might exert its influence upon the future development of Hong Kong through meddling in its civil society, as viewed from the eyes of a typically suspicious Hong Konger.

The state-society dichotomy is a useful analytical framework, and Karl Marx elaborated on this in his early works. According to Marx, during the era of feudalism, it was the state (sovereign) that "determines" or "controls" society. Capitalism is more advanced, not only because of its greatly improved productivity, but also because of the changing relationship between state and society. In fact, society has been liberated from the control of the feudalist state. The ruling class of the new bourgeois society consciously hopes to become the master of the state. Their civic consciousness can be illustrated by their actions in designing a new constitution in which check and balance among the legislative, executive, and judicial branches is well installed. Election by universal suffrage serves as a major mechanism for delegating people's power to those who serve them. Now the order is just the opposite from the feudal past: the civil society "determines" or "controls" the state. It also "participates" in state affairs and checks and balances state power. This society-over-state order means that governmental policies of the capitalist society normally have to be decided by people's wish (Rong and Yang 1989, pp.78–87, 141–77, 206–13). Although Hong Kong was a colony and its Governor was vested with enormous power, he could not become a dictator per se. As many analysts argued, the Governor had to be responsible to the British government, which was in turn elected by and responsible to the British people, although Her Majesty's Hong Kong subjects were not among them. Despite the lack of democracy, the society-over-state order assured that the freedom and rights of the people living in Hong Kong were defended, so the theory went.

In the case of China, its social-political order is often described disparagingly as "archaic" by external commentators. The party-state still controls society, although the control has been much loosened as compared with the Maoist epoch, and society has become increasingly pluralistic. Development of a civil society, however, is still under the close supervision of the state, in which the ruling CCP continues to be weary of the consequence of a strong civil society. When leaders of the proposed Chinese Democratic Party requested registration of their party in early September 1998, for instance, they were received by officials of the Departments of Civil Affairs in various cities, who laid down the conditions for the establishment of such a party. This gave the hope of further political liberalization of Chinese society. However, a week later, leaders of the said party were arrested.[2] What we try to analyze in this chapter is whether and how this state-over-society order might spill over and challenge the traditionally strong civil society of Hong Kong, which tends to have robust institutional checks on the behavior of the government.[3]

Dilemma of the Chief Executive and His Government

The major dilemma of the Chief Executive (CE) of HKSAR is difficult to resolve. On the one hand, Hong Kong people rely on him to represent their aspirations to China, to protect their interests and rights, and to resist intervention from China. In short, they want him to defend the integrity of Hong Kong's socio-political system. On the other hand, Chinese leaders rely on him to convince the Hong Kong people of Beijing's policies and wishes on major issues, or even to impose the sovereign's wills, when necessary, upon a reluctant Hong Kong. Kenneth Lieberthal (1997, p.241) made a similar observation:

> [T]he CE will face a daunting task in maintaining the confidence of both Beijing and the people of Hong Kong. He will have to work well with Beijing, as Hong Kong's long term well-being depends on that. However, . . . there will be many issues on which the CE will have to show some backbone in resisting pressures from the Mainland, and all sides will watch carefully both his style and substance in doing this.

Striking a balance is not an easy task, as there can exist an inherent contradiction between "one country" and "two systems." Though elected by an electoral college, the CE is appointed by the Central Government, and yet he hopes to become a popular leader who can secure the support of the electorate. Caught in a difficult position, the CE has to confront pressures from all sides. Above him is the Chinese sovereign, while below are the people of Hong Kong. Internally speaking, he has to lead the vast bureaucracy, the civil service, inherited from the British administration, and externally he has to secure international participation in the SAR (Wong 1997, pp.26–48).[4] The international character is a significant raison d'être for maintaining the unique status of Hong Kong as an SAR within China. If its "international" character declines, then Hong Kong is just like another big Chinese coastal city. In that case, talking about "one country, two systems" would be meaningless. Despite their significance, the external relations and international status of Hong Kong are not of concern to this essay. Our analysis focuses on three aspects: the CE's relations with China; the CE's relations with the Hong Kong people; and his relations with the civil service.

Tung Chee-hwa was elected through a complicated process. He was elected by a 400-member selection committee, the members of which were chosen by the 150-member Preparatory Committee from 5,791 Hong Kong citizens who had filed their applications in competing for the 400 slots. The complicated formality, lacking transparency, marks a sharp contrast to the democratic means of simple and straightforward methods known in the West. While nobody can claim to know the intricacy involved, it has been reported that Tung was in fact selected to become the future CE by Li Chuwen, former Deputy Director of the New China News Agency office in Hong Kong and Special Advisor to Jiang Zemin on Hong Kong affairs, as early as the mid-1980s.[5] The constitutional framework ensures that Tung be responsible to the Central Government, although it has been repeatedly stressed by spokesmen of the Chinese government, including Lu Ping and Zhang Junsheng, that no party secretary will be sent by the CCP to look over the head of the CE (Ching 1997, p.29). The message they wanted to convey is that the guarantee of SAR's full autonomy is absolute, and China is not just paying lip service.[6]

But this leaves a serious problem in the CE's rule. There is no need for the PRC government to send directives or to nakedly intervene in local matters, so long as the SAR government behaves "correctly" according to the (second-guessed) wishes of the Central Government. In other words, the CE has to find out for himself the limits of the SAR's "autonomy" and constraints of his power, and to behave prudently so as not to risk overstepping the bounds of the permissible. Beijing leaders are definitely in favor of assuring that the CE would clearly "read" their minds and then act accordingly, without creating a paradoxical situation in which a strong disapprobation from Beijing would terrify the community for an alleged interference in Hong Kong's autonomy. It seems that despite all kinds of crises and set backs in Hong Kong after the reversion, Beijing is still confident of Tung's performance, basically because he is apparently the most suitable person that they have. A study that reflects China's Hong Kong policy emphasizes the appropriate attitude expressed by Tung when dealing with China. According to the study, for Tung, coopting the DP's confrontational tactics in dealing with the CCP would not work. Instead, Tung stressed consultation instead of contestation in furthering Hong Kong's interests. "Striving by consultation" is Tung's style, which is acceptable to Chinese leaders (Guo and Qian 1998, p.77).

Tung Chee-hwa's style of appealing to the Hong Kong people is also in harmony with the CCP's "United Front" posture on mobilizing the masses. His repeated refrain "if Hong Kong is good, then China will be good; if China is good, Hong Kong will be even better," demonstrates a parochial outlook that links Hong Kong's future to the motherland, to the neglect of the institutional factors that buttress Hong Kong's successes. As the first Chinese leader in this territory, he wishes to win over the "hearts" of the Hong Kong Chinese by appealing to their nationalistic sentiments and patriotism. He believes that a "new consensus politics" can be built around a conglomeration of patriotic Chinese in Hong Kong and elsewhere. This game is radically different from the past, when colonial rule could build up its legitimacy only through the demonstration of a well-performing, competent government. Indeed, the colonial government could not win the hearts of the people any other way. After having been accustomed to (conditioned by?) judging their government by its performance since colonial times, the Hong Kong people today cannot easily be swayed by appeals to nationalistic sentiments and be freed from their worries about the SAR government's abilities to defend human rights, the rule of law, and continued economic prosperity.

Many analysts share the view that "[w]ith the demise of the Communist-Maoist orthodoxy . . . Chinese nationalism is the only major focus [sic] to galvanize mass support and to buttress the government's legitimacy, authoritativeness, and popular appeal; in short, the survival of the Beijing leadership depends on evoking strong feelings of nationalism." (M. K. Chan 1997, p.13) However, a strong nationalist feeling could be found among the students in 1989 as well as among the leaders and supporters of the DP in Hong Kong, but they are not without strong criticisms of the Chinese regime. While the party-state is bent on developing strong nationalistic feelings among the people, out of a belief that it is easier to unify patriotic people under the leadership of the CCP party, it is also true that people imbued with a strong national pride and patriotic honor may question the legitimacy of what they judge to be a non-democratic CCP leadership. Most of the

Hong Kong people tend to differentiate between the love of the motherland (nation) and that of the state *qua* government, while Chinese leaders insist that if you love your motherland, you should love the state (People's Republic) at the same time. This marked divergence of view only superficially reflects the "serious lack of national sentiments of many people resulting from ling colonial education," as alleged by pro-Beijing elements in Hong Kong. More fundamentally, it also depicts a more universalist value system of many Hong Kongers, which holds protection of human rights and democratic values to be compatible with the love of one's own country.

Thus, Tung Chee-hwa's "appeal" to Chinese characteristics is not compatible, in all cases, with the mind-set of most Hong Kong people, whose utilitarianism and pragmatism calls for the demonstration of government competence, and whose cosmopolitan outlook puts more emphasis on personal freedom and respect for human rights than any parochial appeal to nationalism.

The competence of the HKSAR government has become a major concern of the Hong Kong people, who are alarmed by its poor performance in handling the numerous crises and problems encountered during the year after its return to China. Proved inefficacy in crisis management poses a serious problem regarding the relationship between the CE and his government inherited from the former British regime. With one exception, Tung has retained all the departmental secretaries, apparently as an expedient device to assure stability of the civil service and to win the confidence of the outside world, as the role of civil service is so crucial for a purportedly "executive-led" government. Judging from the past year's experience, it seems that, despite his pledge to become a stronger leader, Tung lacks an efficient command, control, and coordination of the government bureaucracy, to borrow a catch phrase (Triple C & I)[7] from the, by now, archaic parlance of nuclear deterrence. As to "I," or intelligence/information, Tung seems to trust his personal advisors and the Executive Council (Exco), appointed by himself, more than the Civil Service. The inefficient "three-C and one-I" in his government performance leads to the following questions: Are his leadership abilities problematic? Does he succeed in securing strong support from the bureaucracy? Or does he really trust his subordinates who should be decision-makers in different policy-making bureaus of the government?

Sources close to China have revealed that there exists a certain contradiction between Tung Chee-hwa and Anson Chan, the Chief Secretary for Administration. Tung wants to become a strong leader, but Chan, a holdover from Governor Patten's cabinet, already is a strong leader who is able to command the civil service. If Tung wants to have the power to command, he can do so only by cutting Chan down to size. Conflict between these two key persons may affect the relations between Tung and the whole civil service (Guo and Qian 1998, p.77). The poor command and coordination, as witnessed in the crisis management on many issues, may be a tell-tale sign that Tung was not yet in full control. Perhaps Tung cannot trust the high-level officials. Apparently he is relying more and more on his personal advisors recruited from the traditional "leftist" camp, like Yip Kwok Wah, and the members of the Executive Council.

Yip is famous for his special linkages with the central authorities; one report even named him the most senior local member of the CCP.[8] He is in charge of a think-tank[9] that is active in studying vital policy issues of Hong Kong. He is also responsible for liaison with Taiwan. As the HKSAR government is just a local government of the PRC, governmental contracts with Taiwan are inconceivable. The Taiwan issue is not part of the portfolio of the HKSAR government. Only people authorized by the central authorities are supposed to have the credentials to deal with Taiwan. If governmental contacts are not possible, the next level of contacts should be between the two ruling parties. Thus the role of this special advisor might be an unofficial official working incognito for the CCP in Beijing.

A second group of advisors is comprised of the members of the Exco. According to a Chinese source that explains China's Hong Kong policy:

> Tung Chee-hwa mentioned that he wanted to become a strong leader, and starts to be involved in policy-making. At the same time he appointed a large group of members of Exco who are familiar with Chinese affairs and who participate in decision-making. In reality this is to build up a buffer between the central government and the high officials of the Hong Kong government, so as to harmonize the potential conflicts that may arise from contacts between the two. Indeed, in the transitional period after 1997, Hong Kong adopting this procedure of policy-making is reasonable (Guo and Qian 1998, p.141).

It has become increasingly evident that Tung wants to centralize decision-making power from the government departments to himself and his close associates. Before he assumed power in July 1997, he had assigned three members of Exco to study three major social issues: education, housing, and old-age welfare. From then on speculation has been rife about whether Tung wishes to adopt a "ministerial system." Tung's high-level associates responsible for policy issues have always insisted that they only play the role of advisors. However, if the results of the studies are eventually adopted as policy, then the policy-making power now enjoyed by the departmental secretaries will have to be removed. The Exco members would become the de facto "ministers." The crucial fact that the SAR is still in a period of transition may be the reason for the lack of command and coordination within the administration. The questions that remain unanswered are: Could the "transactional" type of leadership (Burns 1979, pp.19–20) of Tung's administration hold his followers together? Would Tung be able to really strike a balance among these different groups of high-level advisors and officials? Could he eventually become a truly strong leader?

Always seeking a balance renders Tung indecisive on many crisis issues that require firmness and determination. Hong Kong having an irresolute leadership may encounter severe difficulties in tackling the numerous problems that have resulted from the worst economic downturn it has encountered in years.

Another subject of major concern in the governance of Hong Kong is whether Tung has altered the governing philosophy and ruling style. Instead of just overseeing the whole situation and making key decisions, Tung seems to adopt a "hands-on" approach in all issue areas. With a strong wish to improve the livelihood of the people, he believes that the government should be "omnipotent" in the proper functioning of the economy and in the management of socio-economic affairs. Judging from the SAR government's performance in the first year, it is evident that the old tradition of "minimum interference" or "positive non-interventionism" from the government in Hong Kong is gradually being eroded. His deviation from a clearly defined role of the government, which used to be limited to providing infrastructure, keeping a watch on the market, and laying down the regulations (rules of the game) for the proper functioning of the market, and his drift toward a more interventionist role, has aroused serious worries in the business sector. To those institutional pillars that buttressed Hong Kong's success in the past should be added an additional variable since Tung's rise to power: a certain degree of government intervention in the economy. This changed philosophy in governance will become increasingly important in determining the future of Hong Kong. A government that tends to be interventionist, but is itself incompetent and inefficient, may bring detrimental effects to the market mechanism.

China in Hong Kong:
CCP and New China News Agency

In the first year after the reversion, praises have been heard about the Chinese hands-off stance, in that China has kept its promise not to intervene in Hong Kong's internal affairs, so that the SAR can really enjoy its high degree of autonomy. Numerous commentators have tried to explain this. They point to the crucial fact that the HKSAR is being held up as a showcase for the future reunification of Taiwan. Besides, they see Beijing as sensitive to external pressures, since Hong Kong is a cosmopolitan and international city closely connected with the interests of Western countries. Most important of all, maintaining the status quo of Hong Kong, including its economic prosperity and social stability, is simply in Chinese interests.

However, in the view of critics, the low profile of the Chinese government in Hong Kong, including its 4,000-plus soldiers who hide themselves away in their barracks, does not necessarily mean that, apart from maintaining close contacts with the CE, the PRC party-state washes its hands of all activities of Hong Kong. These critics are wary of what to them is rather simplistic thinking among some people in Hong Kong that, since China has full confidence in Tung and his government, it is bound to honor its pledge of non-intervention (Guo and Qian 1998, p.7). They doubt that the ubiquitous and "omnipotent" CCP, the ruling party of China, can simply remain out of the picture in Hong Kong for good. Ever since the 1940s, the CCP has maintained its operations in the territory. It has always been underground, keeping a low profile, but nevertheless active. After 1997, they tend to believe, the CCP approach will remain the same: underground, low-profile, but active. However, there is also one big difference. The drive to recruit new members, which began before 1997, is likely to gain more momentum after China's resumption of sovereignty. The SAR government is not directly ruled by the CCP, but the critics suspect that it will try to develop and become influential in all sectors of the community. It will try to build up its network, extending its influence among different strata of Hong Kong society, to fully ascertain the sentiments and opinions of the general public, or in other words, to acquire intelligence. This the CCP will do through its "united front" strategy, aiming to unite as many "friends" as possible, and to isolate the small batch of alleged "enemies," who, according to this view, are suspected to be subversive or otherwise detrimental to the CCP's indirect rule over Hong Kong.

In a section dealing with the State Council's Hong Kong and Macau Affairs Office, an almanac published a decade ago on the organs and organizations of the State Council clarified the functions of the Hong Kong and Macau Work Committee of the CCP. One of the major functions of the Office is:

To assist the Central Committee in handling the following tasks of the Hong Kong and Macau Work Committee:

1. United Front work of the upper class;
2. Work on the grassroots like workers and students;
3. Patriotic propaganda, cultural and educational affairs like journalism, publishing and cinema; and
4. Work relating to the Party, the Communist Youth, and the cadres.[10]

The Hong Kong and Macau Work Committee, now the Hong Kong Work Committee, leads the CCP organs and activities in the region and is housed in the New China News Agency (NCNA). The NCNA was, for a long time, the de facto representative of Beijing in Hong Kong. After the reversion, its structures and function have been modified to suit Hong Kong's new status as an SAR of China. Some of the functions, like external affairs, have shifted to another organ, the Foreign Affairs Commissioner's Office in Hong Kong. In October 1997, Jiang Enzhu, the new Director of NCNA, announced the status and functions of NCNA after Hong Kong's reintegration into China. Its tasks include:

1. The coordination of the Chinese state-owned enterprises in Hong Kong;
2. The promotion of economic, scientific, educational, and cultural exchanges between the Mainland and HKSAR;
3. The extension of liaison with all sectors of the Hong Kong community; and
4. To deal with Taiwan-Hong Kong exchange affairs.[11]

The NCNA's status has been transformed from "an organ accredited by the Central (i.e., Beijing)," to a "work organ empowered by the Central." Although the functions as well as the 600-member establishment have shrunk, it is still in charge of all liaison between the people across the China-Hong Kong border and in charge of Taiwan affairs, plus the United Front work in Hong Kong. After a major restructuring, the current 350-member establishment[12] is headed by three new Deputy Directors. One is a local woman cadre, which is very rare among the officials of NCNA. She is Chen Fengying, responsible for liaison with the grassroots, women, and youth. The second is Zou Zhekai, who was the longtime head of the United Front Department of NCNA, responsible for Taiwan affairs. His promotion clearly signifies that the Taiwan issue has become a subject of utmost importance on China's agenda. The third is Wang Fengchao, originally Deputy Director of the Hong Kong and Macau Affairs Office of the State Council. He is probably responsible for the crucial liaison between the central authorities in Beijing and the HKSAR.[13]

Apparently the NCNA's Hong Kong Office does not only head the party organs of all the state-owned enterprises in Hong Kong, but it also continues to house the Hong Kong Work Committee, which is in charge of all the party's activities in the territory. The major task of the CCP in the post-1997 era, according to speculations, will be to execute the United Front work in order to build a vast network under its leadership. In fact, many new mass organizations have been set up in the last few years for women, youth, and students, and other professions such as journalists. Leading businessmen are co-opted to the Chinese People's Political Consultative Conference, either at the national or provincial and municipal levels. Businessmen and professionals are encouraged to develop their political parties to compete for the seats in the Legco elected by functional constituencies. All these organizations are supposed to exert influence under the leadership of the CCP, which will not be easy in view of Hong Kong's pluralistic civil society. While the pro-Beijing elements are becoming better organized into a grand political coalition, what Beijing wants to see is a relatively weak organization of other political forces rooted in the strong civil society, so that they will not effectively challenge the central authorities.

State-Civil Society and the Economic Order: Chinese Enterprises in Hong Kong

In 1988, during his visit to Hong Kong to explain the process of drafting the Basic Law, Lu Ping spoke to the leaders of the Chinese state-owned enterprises in Hong Kong and asked them not to "break the rice bowl of others." The warning came repeatedly thereafter, stressing that the capitalistic "rules of the game," like fair and free competition and the rule of law, should be duly respected by the Chinese state-owned enterprises doing business in Hong Kong. Chinese leaders are well aware of the problems that arose a decade ago, but whether Hong Kong's economic environment will be affected by the increasing participation of the state-owned enterprises remains a subject not fully explored.

The regulation passed in July 1992 seeking the release of state-owned enterprises from government control did pave the way for some powerful officials to "dive into the sea" (xia hai), or leave their government positions while delving into business activities. Many of those with powerful connections, especially the offspring of top leaders, became Directors or Managers of companies set up under the auspices of a certain ministry or a local government. In principle, it can be said that they are "employed" by the government to administer the enterprises they head. They are still employees of the state, not "capitalists," because the enterprises they head are owned by the state. However, companies like these pose serious problems for the market. In the well-developed capitalist market of Hong Kong, the regime would defend the "rules of the game" of competition through a system known as the rule of law. The public sector is clearly a complication for the capitalist system, in which the government serves all the people and the legislative and judicial organs are the protectors of the integrity of the capitalist system and of the spirit of the capitalist process. But with the advent of "state capitalists," which means that political power is behind the state-owned enterprises, it makes it hard to maintain the capitalist spirit of fair competition.

It is estimated that by 1997 the total businesses of Chinese state-owned enterprises in Hong Kong accounted for about 20–25 percent of Hong Kong's economy (Allen 1997, p.206). The increasing influence of Chinese state-owned enterprises (SOE) has aroused serious attention of the business sector in Hong Kong. Local entrepreneurs are reluctant to compete with these SOE companies, for fear that if they should be regarded as "unfriendly," their future investment prospects in China would be affected. In order to protect their investments in China, they need good connections. As a result, instead of competing with Chinese enterprises, they choose to collaborate with them. In other words, they also aspire to some sort of "political support" from the state.[14]

In China, even if there is a policy since 1992 of "weaning the enterprises from the government," the very nature of being "state-owned" makes the SOE enterprises different from the private companies. Although one-third of the state-owned enterprises suffer from constant deficits, they were still subsidized and propped up by the state until 1998, because bankruptcy would mean massive unemployment, a major factor for social unrest. Moreover, only senior members of the party or the government can become managers of these SOE enterprises, thus bringing with them the "power network" they have, which is so important in Chinese society. Although the policy-makers of those "enterprises weaned from the government" insist that their enterprises must function in accordance with the market rules, these enterprises in fact bring with them the "penetration" of state power into the economy. If the state interferes in the market, unduly affecting fair competition, the market cannot function properly, because, as a prominent Chinese economist, Wu Jinglian, graphically put it, the "visible foot" disturbs the "invisible hand."[15] Borrowing the concept of "rent-seeking" from the American economist A. Krueger, Wu said that, with too much intervention and control from the state over individual or collective entrepreneurial economic activities, not only are the effects of market competition compromised, but a number of privileged people vested with power acquire the opportunity of "rent seeking," or creating benefits for themselves. Whether this can be called "bureaucratic capitalism" or "state capitalism" remains to be studied.[16] But it is obvious that the officials-turned-managers signify the concentration of both political and economic power within the hands of a few Communist cadres. With the Chinese economy steering further away from the original Stalinist system, a growing trend of socio-economic inequality between a few who are vested with both political and economic power, on the one hand, and the majority who have neither, on the other, has emerged.

In such circumstances it is almost impossible to tackle the problem of corruption in China. In capitalism, public sector and private sector are neatly delineated, with the former serving the latter. Any official in the public sector can only work on behalf of the whole society. He is not supposed to give special privileges to anybody, for everyone should be equal in the market. The state, through the executive, legislative, and judicial institutions, purports to protect the so-called capitalist rules of the game, so that fair competition and the spirit of rule of law will prevail. The structures and processes of the anti-corruption control in the West aim at preventing the illegal linkage of public and private sectors. The Independent Commission Against Corruption (ICAC) in Hong Kong, in effect, fights corruption on two fronts: (a) bribery of officials by the people in the private sector; and (b) commercial crimes including efforts that lead to unfair competition and illegal appropriation of capital within a company. However, it is very difficult to eliminate corruption in China, if and when an official and his family networks and other connections are brought to bear upon his activities in both public and private sectors at the same time. Even if an enterprise manager cannot hold a simultaneous position in government, his past experience as a government official, or his present connections in government, or his family network in the public sector all underline the fact that the private and public sectors cannot be clearly separated. Moreover, it is equally difficult, if not impossible, to get rid of improper commercial behavior that leads to unfair competition. The penetration of the state in commercial activities, as demonstrated by officials turning into managers, and the exchange of interests within the complex networks of relationship, are the basic features of the political-economic system of China in transition. Corruption, it can be said, is "built in the system." Not only may the party-state in effect render the market unworkable, but the cadres are making use of state power to further their private ends and enjoy rent-seeking benefits. The worst thing that could happen is the straight appropriation of state capital.

So, how is Hong Kong affected by this unique phenomenon in the face of the transformation of "socialism" in China? In March 1997, when the Hong Kong stock market was skyrocketing, analysts in four stockbrokerages reported on the poor performance of some Chinese enterprises listed locally, but they were discouraged from doing so, and two of them were subsequently fired.[17]

What is at stake is the detrimental effects to competition in the Hong Kong market. It has been noted, especially by Western analysts, that "[t]he worst is that Mainland state enterprises will force themselves into strategic sectors which they covet . . . , the danger is that China will take an increasingly proprietary attitude towards it and squeeze out existing interests" (Allen 1997, p.216). Examples have been noted by analysts, and the most remarkable one is the Chinese acquisition of Dragonair. Aviation is a highly lucrative business that is still a state monopoly in China; and political considerations certainly prevail over economic interests in that, as a British writer (Allen 1997, p.225) notes, Beijing does not like the idea of having a British-owned company operating in HKSAR and dominating the market share on the flights between Hong Kong and China. The peculiar features of Chinese state enterprises can be demonstrated by a simple statement of the Managing Director of Cathay Pacific at that time, " . . . Chinese National Aviation Corporation is the commercial arm of the [Chinese] regulatory authority. How do you compete with somebody who is a body of the regulatory authority?" (Allen 1997, p.225). The result of course was cooperation rather than competition, and Cathay surrounded the controlling stake. People were worried about further deterioration or "erosion" of the market with the increasing participation of Chinese enterprises. The market cannot work properly with the intervention of political power. The situation could have worsened after the proposed bill on fair competition was scrapped by the new HKSAR government, using the pretext that it did not suit Hong Kong's interests, despite years of efforts by the Consumer Council to study and find ways to liberalize the many monopolistic or oligopolistic service sectors. The 1996 report by the Consumer Council received favorable response from the then Legco in 1997, but owing to a tight agenda, Legco was unable to discuss and pass the bill before the 1 July 1997 handover date. A peculiarity of Chinese state enterprises is that they are subject only to soft-budget constraints (Kornai 1992, pp.140–5) and are not necessarily restrained to profit-seeking activities. An implication is that the "state-owned-enterprises could outbid Hong Kong firms in their acquisition activities and thus may adopt an overly aggressive investment stance." (Huang 1997, pp.103–6) How the Chinese working style could affect the proper functioning of the market, and whether Hong Kong could maintain strong regulatory institutions and clear-set regulations to protect the market, will always remain a subject of concern (H. L. Chan 1995, pp.941–54).

After Zhu Rongji became the Prime Minister of China in March 1998, restructuring of the state-owned enterprises received his top priority. As a result, only 512 large conglomerates will stay state-owned. Tighter and more efficient administrative and budget control is being exercised by the State Council. Some notable state enterprises listed in the Hong Kong Stock Exchange in recent years, which are under the control of the offspring of high officials, are now undergoing major reshuffle. These include the famous OXFEM Holdings, the board chairman of which was Wu Jiancheng, the son-in-law of Deng Xiaoping, and the China Venturetech Investments Corporation, whose Deputy General Manager was Chen Weili, the daughter of Chen Yun. Some of the high officials of the former company were arrested in April and May 1998, personnel changes were carried out, and Wu was sent back to Beijing to become the Deputy Director of the newly-created Bureau of Metallurgical Industry.[18] The latter company was simply closed down because of its great losses in the financial market.[19] The strenuous efforts made by Zhu in reforming the state enterprises to abide by the rules of the market mechanism are by all means helpful in regulating the behavior of Chinese enterprises in the highly open and transparent market of Hong Kong.

State-Civil Society and the Social Order: Media Ownership

The mass media is an important component of the public sphere. It not only informs the public, but it is where debates about policy issues are aired beyond the interference of the state. Privately owned media is regarded as a benchmark of liberty. However, the market would tend to shift the media toward profitability considerations, thus publicly owned (not necessarily state-owned) media, financed by taxpayers but not directly controlled by the state, fulfill the proper functions for public education and debate. China under its one-party rule has a radically different media system. Its function is to propagate official views of the party on policy issues. Thus, a strong control is still exercised over the media and publishing companies, despite the demand for a more pluralistic society, which should have a more heterogeneous and enterprising media system.

Hong Kong is long famous for its splendid media scene, the very large readership of newspapers, and the fierce competition among the media in a relatively small market of 6.3 million people. Yet, although the media is a special feature of a vital capitalist system and a symbol of freedom of speech, many people fear that living under "one country" would not exempt Hong Kong from the sway of the state-media nexus of China.

In a large sense, it appears, the market mechanism still plays a predominant role in post-handover Hong Kong. Those few newspapers that folded were closed down for financial reasons after 1997, as happened before, leaving the two big tabloid dailies, the *Oriental Daily* and *Apple Daily* to enjoy the lion's share of the market. But significant changes in media ownership have been noticed in the past few years. Media ownership is always a highly sensitive issue in all democratic societies, as the media is regarded as the "fourth estate" of power because of its power over public opinion. For instance, in Hong Kong "foreigners" are strictly forbidden to control the majority of the stake in any newspaper. However, the post-reversion period has witnessed the change of ownership in many media organizations. While the original owners were of various social and political backgrounds, the new owners are all leading businessmen with known pro-Beijing leanings and having major investments in China. Thus, critics who suspect an exchange of interests between business elites and state power, often question the independence of some newspapers. It is not easy to find out the intricate and delicate relationship between the state and those businessmen (Ting 1996, pp.49–67). But, considering the traditionally strong concern of the Chinese party-state for media control, some critics suspect the playing out of the tactic of *Yi Shang Yang Pao* (using business earnings to sustain newspapers) when newspapers change hands. This may connote an indirect and subtle control of the state over the media industry, say the critics. In relatively "peaceful" times, in this view, the media can perform as usual and the editorial policy need not be changed with the change of ownership, but in times of crisis, China can ensure that the media be under some sort of control.

The latest changes in media ownership concern two major companies, Asia Television (ATV), and the Sing Tao Group. The new board Chairman of ATV is Wong Po-Yan, a long time friend of Beijing, but the main shareholders are recent immigrants from China, whose companies are suspected of having government connections in Beijing.[20] The Dragon Viceroy Limited, partly owned by Liu Changle, who reportedly was a former officer of the Chinese Army, acquired stakes of 46 percent of ATV. Another owner of the company is Feng Xiaoping, a real estate developer who immigrated from Guangdong in the mid-1980s. Another company, Rankon Limited, holds stakes of 5 percent, but the majority owner of this company, Wu Zhen, a Chinese emigrant in the United States for some years, emerged as the Managing Director of ATV. It is also reported that Cha Chi Ming, another pro-Beijing businessman, was interested in the Sing Tao Holdings, an Aw family business that has been influential throughout Southeast Asia for decades. Again, he might serve only as the board Chairman, while the majority of stakes belong to other Chinese enterprises.[21] However, Cha eventually declined to buy the group, possibly because of the poor financial situation of Sing Tao. The alleged intrusion of the Chinese party-state in the form of censorship, however, is difficult to prove. But since the handover, one often hears of self-censorship by the media, resulting in the eclipse of professional ethics by concerns for political corrections and macro-economic interests.[22]

Concluding Remarks:
Between State and the Individual

When discussing the relationship between the state and civil society of Hong Kong, the legal-constitutional order cannot be ignored. The legal framework that protects the sovereign in managing the Hong Kong system is not treated in detail here since there is another relevant chapter in this book. It is worth mentioning that the Preparatory Committee was determined to abolish some laws prior to 1 July 1997. After the Bill of Rights was passed in 1991, legislative amendments have to be made to numerous laws in compliance with the Bill of Rights. These include laws that originally served to safeguard the colonial administration by enabling it to restrict Hong Kong people's rights to political activities. Amendments were made to the following laws, among others: the Societies Amendment Ordinance (1992), the Television Amendment Ordinance (1993), the Telecommunication Amendment Ordinance (1993), the Broadcasting Authorities Amendment Ordinance (1993), the Public Order Amendment Ordinance (1995), and the Emergency Regulations Order (1995). The Preliminary Working Committee of the Preparatory Committee for the establishment of the HKSAR decided in 1995 that provisions of the Bill of Rights that gave it an overriding status, as well as the above-mentioned amended laws, should be repealed. The older versions of the laws should then be reinstated. Owing to rigorous opposition of the people of Hong Kong, especially the legal profession, the National People's Congress of China eventually repealed only the most notable ones of those amendments in February 1997, including Article 2(3), 3, and 4 of the Bill of Rights, Article 3(2) of the Personal Data (Privacy) Ordinance on the overriding status of the Bill, the Societies Amendment Ordinance (1992), and the Public Order Amendment Ordinance (1995).[23] Whether the older versions of the laws would be reinstated was left to the future Legco to decide. Legal scholars, pointing to Article 39 of the Basic Law, argued that the so-called "overriding" status of the Bill of Rights is not inconsistent with that law. The reinstatement of the old draconian laws, whose original function was to fortify colonial rule, is thus considered inappropriate.

Nonetheless, the Public Order Ordinance and Societies Ordinance were revised and eventually passed on 14 June 1997 by the Provisional Legco. While people are concerned with the further restriction of their civil liberties, the SAR government has given more emphasis to the dignity and integrity of the state, though this is camouflaged by the statement that a balance should be struck between personal freedom and national interests. The Provisional Legco also passed the Adaptation of Law (Interpretative Provisions) Bill in April 1998 to replace the word "Crown" with "State" in all the preexisting laws from colonial times. Among the 600 laws of Hong Kong, only 52 are restrictive or applicable to the "state,"[24] which means state organs under the Central Government including the NCNA, will be exempt from the remaining laws in Hong Kong. The exemption had been granted to the Crown in the past because Hong Kong was a colony. But Article 22 of the Basic Law, which stipulates that all offices set up by the Chinese government in HKSAR shall abide by the local laws, seems to present a problem for the exemption. With hindsight, the drafters of the Basic Law seem to have overlooked this problem.

All the controversial legislation made before and after the reversion, as described above, touches on one key issue: the relationship between the state and the individual. The state is keen on achieving its objectives such as national independence and territorial integrity, in addition to safeguarding the supremacy of the CCP. But individuals in Hong Kong, who have a more outward-looking attitude due to the international character of the city and are ready to accept a more universalist interpretation of human rights, may find it difficult to comprehend why those issues like Tibet or Taiwan are so sensitive to Chinese leaders. The HKSAR government, or more precisely, the Chief Executive, is just caught in the middle. As we have explained before, he has to persuade the Central Government to accept the popular attitudes and mentality of the Hong Kong people while, at the same time, he has to rigorously defend the positions and implement the policies of the central authorities. What he needs to do is not only strike a balance between national interests and individual liberty, but pursue the almost impossible task of reconciling the views of the state and the Hong Kong people. After the recent election of the first Legco in the SAR, another controversial issue will be a real test of the political wisdom of Chief Executive Tung: that is, the enactment of a new law to implement the celebrated Article 23 of the Basic Law, which requires the SAR to enact its own law governing subversion, among other crimes. For instance, how does one define "subversion," a crime that simply does not exist in the common law tradition.

A Chief Executive who can appease the capitalists, stabilize the civil servants, and is resolute in defending the Hong Kong policy of the sovereign perfectly fulfills the expectations of the Chinese sovereign. But, he passes only half the test. The other half of the test is whether he can win the hearts of the Hong Kong people. This he can do only by demonstrating good performance of the SAR government he heads, which will depend on his political wisdom and vision. Judging from his first year experience, the road ahead for Tung will be rough and thorny. But, it remains true, the soundness of the Mainland-Hong Kong relationship depends very much on the adeptness of the Chief Executive.

Notes

1. Buckley mentioned the capitalist-socialist alliance formed between capitalists in Hong Kong and Beijing in determining the SAR's policies.
2. See the report in *Ming Pao,* 20 September 1998, p.B8, and in *Hong Kong Economic Journal,* 21 September 1998, p.6.

3. A recent study on the civil society of Hong Kong is Ip Po-keung, "Development of Civil Society in Hong Kong: Constraints, Problems and Risks," in Li Pang-kwong, ed., *Political Order and Power Transition in Hong Kong,* Hong Kong: Chinese University Press, 1997, pp.159–86.

4. Wong Ka Ying has even elaborated on seven areas of constraints faced by Tung Chee-hwa. See Timothy K. Y. Wong, "Constraints on Tung Chee-hwa's Power and His Governance of Hong Kong," *Issues & Studies* 33, no. 8 (August 1997): 26–48.

5. *The European,* 26 June - 2 July 1997, p.3.

6. Of course it does not preclude the fact that the CCP will still develop in Hong Kong. Please see the analysis below.

7. C3/I represents control, command, and coordination, plus intelligence, all ingredients important in the performance of any system of nuclear weaponry. These four categories are used here to assess Tung's performance.

8. *The Open Magazine,* February 1997, pp.50–5.

9. Hong Kong Policy Research Institute, founded in 1996.

10. *Introducing the Organs and Organizations of China's State Council,* Beijing: Worker's Press, 1988, p.341.

11. See *Ming Pao,* 31 October 1997, p.A8.

12. *Cheng Ming,* September 1998, p.22.

13. *The Open Magazine,* September 1998, pp.64–5.

14. I have analyzed the effects of Chinese enterprises on the economy of Hong Kong in Ting Wai, 1995: 22–27. (In French)

15. See *Corruption: Exchange of Power and Money,* edited by the journal *Comparative Economic and Social Institutions,* Beijing: Chinese Economic Press, 1993, pp.4–5.

16. In Taiwan some economists use the term "party-state capitalism" to characterize the economy of Taiwan after 1949, where the Kuomintang-controlled enterprises monopolize the market of basic necessities and all the important products.

17. *Hong Kong Economic Journal,* 4 March 1997, p.2.

18. *Hong Kong Economic Journal,* 30 June 1998, p.4.

19. *Hong Kong Economic Journal,* 23 June 1998, p.4.

20. *Hong Kong Economic Journal,* 27 May 1998, p.3.

21. *Hong Kong Economic Journal,* 15 May 1998, p.3; 21 May 1998, p.4.

22. See *Questionable Beginnings: Freedom of Expression in Hong Kong One Year after the Handover to China,* Joint Report of the Hong Kong Journalist Association and Article 19, June 1998.

23. *Tai Kung Pao,* 24 February 1997, p.B3.

24. *Hong Kong Economic Journal,* 21 April 1998, p.8.

References

Allen, J. 1997. *See Red: Chinese Uncompromising Takeover of Hong Kong.* Singapore: Butterworth-Heinemann Asia.

Buckley, R. 1997. *Hong Kong: The Road to 1997.* Cambridge: Cambridge University Press.

Burns, John M. 1979. *Leadership.* New York: Harper & Row.

Chan, Hing-lin. 1995. "Chinese Investment in Hong Kong: Issues and Problems." *Asian Survey* 35, no. 10: 941–54.

Chan, M. K. 1997. "The Politics of Hong Kong's Imperfect Transition: Dimensions of the China Factor." *The Challenge of Hong Kong Reintegration with China.* Edited by M. K. Chan. Hong Kong: Hong Kong University Press.

Ching, Frank. 1997. "China-Hong Kong Relations." *The Other Hong Kong Report 1997.* Hong Kong: Chinese University Press. (In Chinese)

Cohen, W. I., and L. Zhao, eds. 1997. *Hong Kong under Chinese Rule: The Economic and Political Implications of Reversion.* Cambridge: Cambridge University Press.

JCESI. 1993. *Corruption: Exchange of Power and Money.* Edited by the journal *Comparative Economic and Social Institutions.* Beijing: Chinese Economic Press. (In Chinese)

Guo, S. P., and X. J. Qian. 1998. *The Post-1997 Relationship between China and Hong Kong SAR.* Hong Kong: Pacific Century Press. (In Chinese)

Huang, Y. S. 1997. "The Economic and Political Integration of Hong Kong." *Hong Kong under Chinese Rule.* Edited by W. I. Cohen and L. Zhao. Cambridge: Cambridge University Press.

State Council. 1998. *Introducing the Organs and Organizations of China State Council.* Beijing: Workers' Press. (In Chinese)

Ip, P. K. 1997. "Development of Civil Society of Hong Kong: Constraints, Problems and Risks." *Political and Power Transition in Hong Kong.* Edited by Hong Kong: Chinese University Press.

Kornai, J. 1992. *The Socialist System.* Princeton: Princeton University Press.

Li, P. K., ed. 1997. *Political Order and Power Transition in Hong Kong.* Hong Kong: Chinese University Press.

Lieberthal, K. 1997. "Post July–1997 Challenges," in *Hong Kong under Chinese Rule: The Economic and Political Implications of Reversion.* Edited by W. I. Cohen and L. Zhao. Cambridge: Cambridge University Press.

HKJA. 1998. *Questionable Beginnings: Freedom of Expression in Hong Kong One Year after the Handover to China.* Joint Report of the Hong Kong Journalists Association.

Rong, J., and F. C. Yang. 1989. *On Democracy.* Shanghai: People's Press. (In Chinese)

Ting, Wai. 1995. "Les entreprises chinoises à Hong Kong." *Perspectives Chinoises* 31 (September-October): 22–27. (In French)

Ting, Wai. 1997. "The Orientations and Challenges of Hong Kong Mass Media in the Face of 1997 Impact." *Mainland China Studies* 39, no. 1 (January): 49–67. (In Chinese)

Wong, T. K. Y. 1997. "Constraints on Tung Chee-hwa's Power and His Governance of Hong Kong." *Issues and Studies* 33, no. 8 (August): 26–48.

CHAPTER 11

Hong Kong's Reversion and Its Impact on Macau

Bolong Liu

Introduction

Macau, a small enclave of about half a million people, has been ruled by Portugal for over 400 years. Macau will be returned to China on 20 December 1999, two and half years after Hong Kong's reversion. In consideration of Macau's population and economic scale, and the fact that Macau can follow the example of Hong Kong's reversion to China, one may come to the view that Macau's return would be an easy task. However, the reality is that it may not be as smooth sailing as most people might imagine. Over one year after Hong Kong's reversion, Macau's economy is falling further into recession under the crush of the Asian financial crisis, and Hong Kong's downturn also impacts on Macau, whose economy is dependent on Hong Kong's. Five years of continuing recession has eroded people's faith in the present administration, and frustration can be witnessed everywhere in Macau.[1] Besides, owing to the influx of more triads from Hong Kong, Taiwan, and even Mainland China, street killings and gangster wars have become daily routines, making frequent headlines in Hong Kong newspapers. The decline in the number of holidaymakers to Macau further aggravated the recession in tourism, affecting the property market and hotel and restaurant businesses, Macau's staple sources of revenue as a tourist city. Macau has recorded negative GDP growth of –0.5 percent and –0.1 percent in 1996 and 1997 respectively.[2]

People in Macau are watching with rapt attention the development in Hong Kong after the handover, because they believe that Hong Kong's success or failure will herald Macau's future after 1999. If Hong Kong can keep its prosperity and order after the transition, so will Macau, as the two areas share similar systems and have a local Chinese elite of approximate quality and education. One of Macau's Chief Executive hopefuls, Edmund Ho, once said that Hong Kong today is Macau tomorrow," reflecting a common sentiment of the Macau people.[3] The main Macau Chinese newspapers also echoed the same theme. *Citizen Daily* has this to say: "local political observers believe that Hong Kong performance at the initial period will be a crystal ball' for Macau's future."[4]

In Macau, people usually feel they are lucky that the enclave's return to China comes after Hong Kong's in time. Macau can watch and learn, before making final decisions. What has happened in Hong Kong will have a direct and profound impact on Macau, in respect of people's faith in the one country, two systems" model, as it has been first tried in Hong Kong. Macau has never been a recognized colony of Portugal in history, and after the 1974 Portuguese revolution, the Portuguese government offered to return the enclave to China immediately, only to be declined by China. This is one of the factors accounting for Macau's eventual return to China, two and half years after Hong Kong. According to Huang Wenfang, a political commentator who knows the inside story of Chinese negotiations with the U.K. and Portugal, Beijing decided to put Macau behind Hong Kong on the retrocession timetable as a precaution, because the British could easily monopolize the Macau economy to the detriment of China, if Hong Kong should still be in British hands.[5] This way, so the reasoning goes, Macau people can witness what has happened in Hong Kong and decide on their own what they should be doing, to remain in the enclave or go somewhere else.[6] However, we find a paradox here again. Owing to its economic downturn and mounting public disorder, most of the local people in Macau are getting frustrated and are calling for an earlier return to China, instead.[7] They know, of course, this is a sheer wish and China will go by the timetable set in the Sino-Portuguese Agreement, which is an international agreement binding on China.

This chapter intends to discuss many issues related to Macau's views, including its evaluation of Hong Kong's experience after reversion. It will briefly introduce Macau's status quo and its main problems pending its own handover, and then it will discuss the general impact of Hong Kong's post-handover experience on Macau, followed by a detailed analysis of its possible effects, both positive and negative. Finally, the discussion will turn to what lessons Macau can draw from the administrative experiences of the Hong Kong SAR, to guide the future Macau SAR, so that its government can better cope with the challenging task ahead.

As an international metropolis and financial center, Hong Kong has drawn wide media attention and numerous commentaries the world over.

The same is true with its first anniversary after reversion. But, almost none of the commentaries took in view the Macau angle. So, a review from Macau's perspective, such as this one, could conceivably offer something both new and relevant to Macau's own forthcoming retrocession to Chinese sovereignty.

Macau's Status

Macau and Hong Kong have similar histories. Both were Western colonies eventually returned to China, from which they had been snatched in what was by now a closed chapter of history. In their post-handover reincarnation, they will have similar political and economic systems under the one country, two systems" formula. The future Macau SAR government will follow the example of the Hong Kong SAR, in both structure and policy. The main difference lies in the fact that Macau is dwarfed by Hong Kong, which is a giant by comparison, not only in territory and population, but also on the financial and economic scale. However, Macau has a much longer colonial history, as it has been ruled by Portugal for about 400 years. Macau takes pride in its cultural heritage as a unique place where European culture meets Chinese culture. Since the Qing Dynasty, Macau has had 32 Governors, overtaking Hong Kong's in number. Tourists can easily spot European-style buildings and streets. Macau had the glory of having established a first Asian university three centuries ago, which contributed to the spread of Christianity in China and Japan.[8] Among its population there is a large component of Macanese, a mixture of Portuguese and Chinese ancestry, which is unique in the world. Macau will remain a historical city even after its reversion.

The Sino-Portuguese Joint Declaration was signed in 1987, under which Macau will be returned to China in 1999; and the Basic Law of Macau was drafted and promulgated in 1995. Macau has undergone intense preparation for the transition. The Chinese and Portuguese governments reached a consensus on three main issues important for a smooth transition: localization of the civil service, localization of law, and the use of Chinese as the official language. The top posts of the civil service have always been occupied by Portuguese expatriates throughout colonial history, and the Macanese have mainly found government jobs at the middle level. While Chinese account for 95 percent of the population, their representation in the civil service is disproportionate to their ratio in the total population. Macau's laws are written in Portuguese, and judicial cases were deliberated and written in Portuguese only. There is an urgency for the translation of Macau's laws into Chinese and the training of more bilingual lawyers and judges. The Basic Law of the Macau SAR stipulates that Chinese will become an official language besides Portuguese. More

Chinese-speaking civil servants and translators will have to be recruited. Ten years have passed since the signing of the Sino-Portuguese Agreement, but the process of localization is far from satisfactory. The civil service remains undertrained and inefficient, which is considered as one of the main problems for the slow progress in attracting foreign investment.[9] Macau's legal tradition was introduced from Portugal but adapted to the local Chinese customs. But many of the laws on the book were obsolete and need to be modernized. For example, the present tax system was introduced in 1987, and modified in 1990, but cannot meet the needs of the changed economic environment in Macau today.[10] Owing to the lack of qualified translators and bilingual officials in the civil service, the Chinese language is yet to be widely introduced in many government offices. For foreign businessmen who come with the intent to invest in Macau, all forms have to be filled out in Portuguese. Chances are many will be forced to give up because of language difficulties. So far, the most worrisome task is localization of the top government posts—at the Undersecretary and Director levels. If inexperienced local candidates come to these posts overnight upon handover, their lack of experience will most likely put the local community in peril. Chinese officials are unsatisfied with the slow pace of localization and urge the Macau government to promote more Chinese-language officials to the senior levels of administration.

On the issue of Macau's international status and external links, one measure is its participation in international organizations. Both Chinese and Portuguese governments are content with the progress on this score. China stresses the importance of this linkage and is working very hard to help Macau continue its membership in many important international organizations, including WTO, UNESCO, Asian Pacific Economic and Social Commission, etc. Both sides are negotiating on Macau's continued participation in more than 200 international treaties. However, in view of the complexity of international law, Macau may have to undergo trying tests in this respect, if its international status is not readily recognized worldwide. As Professor James Hsiung has noted in a similar situation, Hong Kong has encountered problems of its own international status recognized in international law in Europe and America.[11] Things can only be more difficult for Macau. As a whole, Macau has to step-up its preparation for the transition, to build up a solid core of Chinese-speaking civil servants in public administration, and to speed up the localization and modernization of the laws. But, how to achieve recognition of its independent status by the world community is something not entirely within its reach.

The Chinese side has established a set timetable for the transition. The Macau Preparatory Committee was set up in May 1998, to probe into ways of electing the first Chief Executive, the Legislative Council members, and the main officials of the SAR government. As Macau can borrow from Hong Kong's experience, the task facing the Macau Preparatory Committee was not so insurmountable. Many Mainland members have had experience serving in the Hong Kong Preparatory Committee, and their leadership is useful for the smooth transition in Macau.[12]

However, some external factors are hard for Macau to cope with. First, the Asian financial crisis has dealt a heavy blow to a very fragile local economy that has undergone a period of recession from 1993, when China adopted a stringent policy of economic adjustment, resulting in the withdrawal of large amounts of capital from Macau. Over 30,000 vacant flats are awaiting buyers in Macau. As recession has beset Hong Kong and Japan, fewer tourists from these two important sources came to Macau, causing a precipitous decline in tourism, which in turn has upset the general economy. The official unemployment rate reached 3.3 percent for the first quarter of 1998, and forecasts were for a further decline unless the Asian financial crisis is over. Over 10,000 people had to leave their families behind and traveled to Taiwan in search of jobs. Worst of all, the triad gangs have come to Macau to squeeze out the final profits before the reversion, and the continuous turf fights among them make the world's headlines, further hurting Macau's business and tourism.

In sum, the situation is very complicated as Macau is beset by a combination of problems, including its low speed of localization, long-term economic slowdown, reduction in foreign investment, and a worsening in welfare for its citizens. The Portuguese and Chinese governments are working very hard to transform Macau into a city playing the role of a gateway between China and the West, especially the Latin language countries in Europe. However, unless effective measures are introduced to improve the economic situation and to enlarge its trade, it is very difficult to foresee a meaningful role played by Macau in this respect.

Macau's Perspective on Hong Kong's Transition

Macau is following in Hong Kong's footsteps in its reunion with China. For this reason, the Macau media have had extensive coverage of Hong Kong events after the handover. In fact, Hong Kong newspapers have a bigger circulation than the main Macau newspapers in their own hometown. People in Macau are used to watching Hong Kong TV for news and entertainment. They are impressed with the Hong Kong public opinion polls, as there is no similar survey in Macau. Commentaries and in-depth analyses are rare in Macau newspapers. Besides, Macau people have shown a tendency of "subject political culture," in which they closely follow the current events in Hong Kong but express very few opinions in public.[13] The main newspapers are pro-China and tend to praise the excellent performance of the Hong Kong SAR government. However, there are several small-circulation newspapers that are independent and usually are not one-sided.

Generally speaking, the Macau media show a general consensus that China is sincere about "one country, two systems," has not intruded into the internal affairs of the Hong Kong SAR, and has adhered to the Basic Law. The leaders of the PRC government, including President Jiang Zemin and Premier Zhu Rongji, have reiterated on many occasions that China will not interfere in issues falling within the Hong Kong SAR's jurisdiction under the Basic Law. As President Jiang Zemin pledged at the handover ceremony, the Hong Kong SAR shall be vested, in accordance with the Basic Law, with executive power, legislative power, and independent judicial power, including that of final adjudication."[14] After the Chinese government resumed the exercise of sovereignty over it, as the message goes, Hong Kong's capitalist system and way of life will remain unchanged, while the sovereign continues with its socialist system."[15] China's new Premier, Zhu Rongji, also gave his full support for the Hong Kong SAR to fend off the Asian financial turmoil.[16] The record of events in Hong Kong since the handover, as seen through the eyes of Macau, has proven that this has been the case.

One serious test came in 1998, when a Hong Kong member of the Chinese People's Political Consultative Conference, Xu Simin, while attending a CPPCC meeting in Beijing, openly attacked the RTHK (Radio and Television Hong Kong) for slandering Hong Kong and creating a false negative image. He urged the Central Government to intervene and rectify the editorial policy of RTHK. This event aroused great anxiety and anger among Hong Kong media and public, as they feared it would signal the loss of the freedom of speech and the press in the territory. To the amazement and surprise of the Hong Kong people, however, the Chinese leaders immediately made it clear they would not have anything to do with the accusation, as it was a domestic issue of the Hong Kong SAR. The SAR government, they added, had the sole authority over how the RTHK, which is subsidized by the SAR government, should be run. Li Ruihuan, Chairman of the CPPCC, said: it is not fit [for Beijing's intervention] and if not handled properly, it would undermine the enforcement of the one country, two systems' policy. "[17]

Again, in May 1998, when a Television Broadcasts Limited (TVB) journalist was covering Premier Zhu's official visit to France, he was sternly scolded by an official of the Chinese embassy for asking an offensive" question. The embarrassing moment was shown on local television and aroused protests in Hong Kong. Premier Zhu soon offered his apology by commenting that the official's attitude was inappropriate. His apology was accepted in Hong Kong with satisfaction, as Zhu is the only Chinese leader to apologize in public for such a sensitive issue.[18] Since the handover, the Xin

Hua (New China) News Agency, which played a prominent role in rallying the local pro-China groups in the campaign to oppose political reforms of the last British Governor, has changed its posture and kept a very low profile. Although Jiang Enzhu, the NCNA Director, and Ma Yuzheng, who is the resident Foreign Relations Commissioner, are two heavyweights in China's foreign policy establishment, they have managed to remain in their offices in Hong Kong, trying their best not to act or appear like a tai shang huang," or pro-consul sent from Beijing.[19]

A most interesting development involved the status and role of the Democratic Party in Hong Kong. The Democrats were forced out when Beijing, immediately after the handover, dissolved the Legco elected in 1995 under Governor Patten's reform, which gave them their newly found political prominence, and instituted a Provisional Legislative Council that lasted until a new elected Legco was put in place in May the following year. As China never allows the existence of political opposition on the Mainland, many people in Macau had predicted that the Democratic Party would disappear in Hong Kong after the handover. However, over one year later, the Democrats were doing fine. They won 16 of the total 60 seats in the May 1998 election for the Legco, including 13 of the 21 seats directly elected by popular vote. The Macau people now realized that China is not out to destroy the Democrats.[20] They were also impressed with the recordbreaking turnout of Hong Kong voters, despite a pouring rain, on election day, 24 May 1998.[21] The Chinese leadership, all indications are, has lived up to its pledge to allow a high degree of autonomy" to the Hong Kong people and to let them decide on the speed of democratization after 2,007.[22] The usual gathering on 4 June 1998 of large crowds demonstrating against Beijing, demonstrates that Hong Kong is living in a system very different from China's.

The mature way by which party politics has played out in Hong Kong has received favorable comments from the Macau press. One commentary had this to say: the emergence of party politics has demonstrated a mature balance of political forces necessary for democracy to appear in Hong Kong society. The Central Government does not interfere, nor takes credit for this. One notices that Beijing has adopted a hands-off policy towards Hong Kong, which shows that the one country, two systems' is not a tentative scheme. Macau, by contrast, lacks an open, transparent and clean political system. Bribery in elections is rampant; ballots can be exchanged for money, and local nepotism has become a fact of life."[23]

Some people in Macau have advocated the inauguration of a party system, for the purpose of electing a Chief Executive and members of the Legislative Council. During the Asian financial crisis, Beijing has stated on many occasions that China will help Hong Kong defend the Hong Kong dollar peg to the U.S. dollar and, for that purpose, has pledged so many times not to devalue the renminbi. Premier Zhu insisted that China will protect the Hong Kong dollar at all costs. At his first press conference as the new premier, Mr. Zhu said: "if the Hong Kong SAR Government were to need support from the central government . . . , no sooner does the Hong Kong SAR Government make a request than the central government will act, sparing no effort nor cost, to take steps to maintain the prosperity and stability of Hong Kong, including the defense of the linkage of the Hong Kong dollar to the U.S. dollar."[24] The announced policy as such not only boosted confidence in Hong Kong, but also gave a positive image of Beijing in the eyes of the Macau people, who had reasons to be grateful, because the same policy also served to safeguard their own currency.[25] In addition, they feel proud that China could withstand the challenge and escape unscathed by the regional financial storm. Many believe that Macau's reversion to China could give the enclave a respite and open the gate to the pouring in of Chinese capital to lift the Macau economy out of its long recession.

On social issues, China has promised not to interfere in the normal life of the Hong Kong residents. The HKSAR has been given total autonomy in formulating its own policies, including housing, education, industrial, and other policies. The PLA soldiers in Hong Kong have kept a low profile and are confined to their barracks even during weekends. Universities continue to operate as they have always done, enjoying the traditional autonomy allowed under the principle of academic freedom, except for budgetary constraints in the case of public-funded tertiary institutions. Professor Edward Chan, a prominent economist, President of the Lingnan University, who had been denied membership on the Hong Kong Preparatory Committee for his unflattering comments on Deng Xiaoping, remains at the helm of Lingnan, which is cleared for elevation to a full-fledged public-funded university.

As for the controversial SAR policy of requiring some schools to convert their medium of instruction from English to Chinese, complaints have been heard from many quarters including the parents. Some parents who can afford the cost have opted to send their children to private English schools. But the decision on the controversial policy was made by the SAR government, with no influence from Beijing. Ironically, China has no power to suspend or reverse the implementation of this policy shift.

As a matter of fact, what concerns Macau's civil servants and the public most is the Hong Kong SAR government's decision to retain all high-level civil servants at their original posts. Even those who supported the political reform of the last British Governor, Christopher Pattern, were allowed to work as usual without discrimination.[26] The resultant stability has boosted the confidence of the local Macanese who have been given chances by Portugal to return to their motherland for jobs at the same levels.[27] Most of them have chosen to stay in Macau as they know their future is secure because of the precedent set in Hong Kong. They are more familiar with the local culture and environment, and in fact the fringe benefits for the Macau civil service are much better than those in Portugal.

Possible Negative Impact

As the major newspapers in Macau are pro-China, they are in the habit of routinely lauding the achievement of the Hong Kong SAR after its handover from the British. While on 1 July 1998, the first anniversary of the HKSAR, the Hong Kong press voiced harsh criticisms on the performance of the government, major newspapers in Macau, such as the *Macau Daily* and the *Overseas Chinese Daily*, all published congratulatory commentaries on the success of the Hong Kong SAR without mentioning a word about the difficulties and problems confronting its government. One can only get a glimpse of a balanced picture of both success and problems from the handful of small-circulation newspapers, which usually are more independent in their editorial policy.

To be objective, one has to be apprised of the distracting problems besetting the Hong Kong SAR government, such as the bird flu and the poisonous red tide," when considering the devastating effect brought on by the Asian financial crisis. Macau is a mini-economy. As the saying goes, if Hong Kong catches cold, Macau falls seriously ill. As a result of the financial crisis, Hong Kong has entered a period of economic slowdown and stagnation. Property values have been halved, and stock prices have been tumbling down. Hong Kong's economy recorded an astonishing −2 percent growth rate and 5.2 percent unemployment rate by the end of 1998, the highest in ten years. Hong Kong's economic meltdown has affected Macau seriously. Its gambling industry has witnessed a fall in income for the first time in ten years.[28] The comparatively lower property prices in Macau are no attraction to the Hong Kong people. According to statistics, Macau's property market is over-supplied with over 30,000 vacant flats.[29] Even if the Macau government is promoting a policy of immigration by investment," the response thus far has been lukewarm.[30] The decline in the number of Hong Kong buyers in Macau's local property market is causing the Macau economy to slip into further recession. Although no one in Macau blames Hong Kong for its own economic difficulties, the fact remains that Hong Kong's stagnation has direct negative impact on Macau. As the *Macau Citizen Daily* commented some time ago: "just half a year after Hong Kong's reversion, how could it be possible that things changed so quickly? Was it because of bad Feng Sui? Or problems with the SAR Administration? Or external factors?

Macau will be returned to China in two years' time. Macau has to draw the relevant lessons from Hong Kong and should be extremely careful with its turn of reversion."[31]

Besides, Macau has experienced a continuous spate of gang killings, arson, kidnappings, and other crimes, to an extent unheard of in it's history. The negative image evoked by wide reports of these crimes, as seen on Hong Kong TV and in the print media, like *Time* magazine, has frightened away potential visitors to Macau, causing damage to the gambling, hotel, and restaurant businesses. The Macau government has been blaming the influx of outside gangsters, especially Hong Kong triads, for causing troubles in Macau and further complicating crime fighting by police. In fact, they are blaming Hong Kong gangsters for the huge backlog of unsolved crimes in Macau.[32] During the past year, Hong Kong has maintained a very good record of public order, with reduced cases of robbery and killing. Some Hong Kong gangsters have fled to Macau and instigated turf wars against local triads, thus bringing more violence to this small enclave.

During the months after reversion, Hong Kong has suffered a series of unexpected incidents that put the world's limelight on the territory, calling its reputation in doubt. For example, Hong Kong had to slaughter one million chickens in its bid to combat the bird flu virus. For the following three months, Hong Kong people were without chicken on the dining table. The oddity gave a detrimental blow to its sliding tourism. Government indecisiveness in handling many crises was also a cause of worry. The bird flu was followed by a series of food-related problems, affecting beef, deepwater fish, and pig intestines, a special gourmet item for the Cantonese. Many of the Hong Kong press have criticized the Hong Kong SAR government for incompetence and hesitation. One isolated incident may be bad luck, but a continued flow of incidents and mismanagement by the government can only be evidence of administrative incompetence. Take the "red tide" incident as an example, when large numbers of deepwater fish died in the fishing farms along the coastal lines of Hong Kong following the tide's onslaught. There were repeated reports in China that red tide was attacking coastal lines of Fujian and Guangxi provinces, which had already taken a heavy toll on fish farming in 1997.[33] Why could the Hong Kong government not have taken precautions? Why was Hong Kong caught by surprise when the red tide hit the region?

To bystanders elsewhere, the Hong Kong SAR government showed a certain degree of incompetence and disorder under the onslaught of the series of crises. In the Asian financial crisis, Hong Kong SAR financial officials seemed to know only to raise interest rates to protect the Hong Kong dollar, but were very reluctant to confer with academics in search of useful ways to mitigate the economic impact, while protecting the Hong Kong dollar. The

Chief Executive, Tung Chee-hwa, had set a housing target for Hong Kong at the beginning of his administration, when property price hit an all-time high. After the Asian financial crisis hit Hong Kong and the local interest rates suddenly skyrocketed, as a way of defending the Hong Kong dollar, Tung and his government officials still adhered to their original pledge of supplying 85,000 new flats a year, and refused to revise it. The result was the tumbling down of property prices, which endangered the property-dependent banking system. As Christine Lok, a prominent Democrat and Chairperson of the Citizen Party in the Legco, commented: that is a good idea under sunny skies may be a disaster in a typhoon. For example, aggressively promoting a policy of selling public housing flats at steeply subsidized prices may be justifiable in very good times as a measure to lower property prices and redistribute wealth, but to continue to push it when property prices are sagging is sheer folly."[34]

Now, the decline of the people's confidence in the Hong Kong economy, stemming from their low expectations of what the SAR government can do to fix it, does not augur well for Hong Kong's future. Even the pro-China politician Tsang Yok Sing had this to say: Tung Chee-hwa is not a trained politician, and neither are his senior aides. They do not possess the political skills required for handling crisis situations in a way that inspires public confidence, and as the past year has been fraught with crisis, people have reasons to criticize the SAR administration."[35]

To the future Macau SAR government, the message is clear: Macau has to select a Chief Executive who is not only experienced and honest (Mr. Tung is an honest man), but who is aggressive in managing the economy. He or she should be a trailblazer, ready to cope with sudden changes in the international, as well domestic, environment. This leader should be preferably a relatively young person, and should have the courage to break old shackles and be responsible and creative. He or she should not be a person of conservative bent, with no vision for the future. In consideration of the present economic and social difficulties, Macau needs a capable leader to lead the new Macau SAR into a new era. The enclave has a limited population, a mini-economy, with a poorly-trained civil service, and almost no foreign currency reserves.[36] It has entered a period of economic depression and social security disorder that cries out for a bold and capable Chief Executive who can lift Macau out of this deep morass. Macau can not afford to stick to the past, sit, and do nothing. In all fairness, we ought to give Hong Kong SAR more time to accumulate experience and test its ability, but if this crisis endures two or three years without let-up, Hong Kong could be in for resultant chaos and even riots.[37] Macau, on the other hand, cannot afford any testing time. A quick and effective turnaround of the economy is a urgent necessity. An honest, aggressive, and responsive Chief Executive, who can lead the Macau SAR government accordingly, is a prerequisite for the success of "one country, two systems" in Macau. Such is the ultimate lesson learned from the experience of the Hong Kong SAR, which was inaugurated two and a half years ahead of the Macau SAR.

Reviewing its process of selecting the members of government advisory committees and the main decision-making body—the Executive Council—one has the distinct impression that the Hong Kong SAR government followed a model of businessmen ruling Hong Kong. The Chief Executive and his main advisors in the Executive Council are mainly composed of business executives. Surprisingly, no academics were recruited into the Executive Council, or consulted in combating the Asian financial crisis. The pattern raises a question of whether a team made up entirely of former business executives has the wherewithal to cope with political governance. In public administration theories taught in academe, students are always told that public administration has its own characteristics and cannot be dealt with by the same principles as in business.[38] Successful business management does not guarantee equal success in public service. Businesspeople can manage public administration only if they have accumulated enough experience in public management and understand the differences between managing private organizations and public institutions. The composition of the HKSAR Executive Council shows that Hong Kong relies too heavily on businessmen for policy formulation and decision-making, to the neglect of the value of academics, especially those with a Ph.D. education. In the present age of high-tech and complicated financial management, leadership buttressed by a good quality, educational background is needed in managing the economy and finance. Macau has to learn this lesson from the experience of the Hong Kong SAR. Its future Chief Executive must heed the need for the recruitment of more professionals into its decision-making and consultative organizations, in order to guarantee good quality in its policy-making and government functioning.

Macau Pulse on one occasion reprinted an article lifted from the Hong Kong media, titled "Emphasize Professionals; Open Recruitment Channels," in which Hong Kong's civil service recruitment system was criticized for its exclusion of professionals.[39] Macau has had no ethnic Chinese at the Undersecretary level, and the pace of localization at senior levels is still limited. With little or no experience in public management, one can imagine the difficulties the future Chief Executive and his cabinet will face during the initial period of the Macau SAR government. They, all the more, have to rely on professional consultants in formulating policies and in pulling off successful management of the economy. The future Macau SAR government should incorporate more academics into its administrative structure and consultative organizations. Of course, the good quality and honest character of these academics will be considered essential.

Conclusion

In sum, reviewing the Hong Kong SAR's performance after its first anniversary, most people in Macau share the consensus that its "one country, two systems" grand design is working satisfactorily. China has amply demonstrated its determination to adhere to the Basic Law and respect the autonomy of the Hong Kong SAR. Despite some skirmishes, such as over the electoral system change, repeal of the Labor Law, and the controversial status of the New China News Agency, one has to recognize that Hong Kong still has its own political and economic systems intact. China has shown restraint in Hong Kong by allowing it freedom of the press, keeping a gag rule over its agencies in Hong Kong (NCNA, the Foreign Relations Commissioner's Office, and the PLA unit), and giving a free hand to the SAR government in coping with the Asian financial crisis. Even the U.S. State Department had to come to the only logical conclusion that Hong Kong has its autonomy for real.[40] Macau's major newspapers are owned by Chinese capital and may therefore tend to whitewash" for Beijing the conditions in the Hong Kong SAR or the Macau people's goodwill toward it. But, in case of doubt, one could always check with taxi drivers, patrons at the tea houses, or the better informed professionals about the Macau people's appreciation of Chinese sincerity in one country, two systems" and their confidence that the same model is going to work in their enclave after 20 December 1999, when it is due to be returned to Chinese sovereignty under the same formula as Hong Kong was returned from the British.

However, the motto "Hong Kongers governing Hong Kong," it seems, has yet to be tested and observed. The past year, as noted above, saw many unexpected incidents occurring in Hong Kong, and how successfully the SAR government, staffed almost entirely with Hong Kongers and headed by a Hong Konger Chief Executive, proved that Hong Kongers could run their own affairs remains a question. When the new Chek Lap Kok Airport opened on 7 July 1998, the unprecedented disorder and mismanagement, resulting in flight delays, missing luggage, and huge economic loss, shocked the world. In Macau, where the economy is badly in need of help, people are looking to China for a bail-out.

As two former Western colonies, Hong Kong and Macau will become the only two SARs in China, and they share a lot in common. The future of Macau lies in a scenario of closer economic cooperation with the neighboring areas, especially the China Mainland, Hong Kong, and Taiwan. In view of its high degree of dependence on Hong Kong in matters relating to trade, finance, currency, and tourism, Macau has to design a more intimate relationship with Hong Kong in the future. A Hong Kong–Macau bridge project is widely supported by Macau business and economic circles, because it will attract Hong Kong retirees, investment, and tourism.[41] Yeung Juenwen, a Macau Chief Executive candidate, believes that the enclave could become an ideal place for so many small international enterprises who cannot afford the high land and labor cost in Hong Kong. They could use Macau as a bridgehead to do business with and invest in China.[42] The idea, if it materializes, might turn Macau around and launch it into a new era after its retrocession to China.

Notes

1. Macau's economic recession began in 1993 when China adopted a stringent policy and withdrew capital from Macau. From then on, Macau has undergone five years of continuous economic recession.

2. Director of Statistics and Census, *Monthly Bulletin of Statistics,* April 1998, p.10.

3. See Edmund Ho, "Hong Kong's success in reversion will create opportunities," *Macau Daily,* 7 October 1997, p.1.

4. "Today's Hong Kong, Tomorrow's Macau," *Citizen Daily,* 28 January 1998, p.1.

5. Huang Wenfang, "For Macau, Returning is easy, Prosperity is difficult," *Ming Pao Daily,* 12 May 1998, p.E9.

6. The Portuguese government allows all Macau-born citizens to have access to Portuguese passports. Also, Macau civil servants will be allowed to return and work in the Portuguese civil service if they so wish.

7. The Vice-Chairlady of the Preparatory Committee, Ms. Tsao Chezheng, told the Chinese President, Jiang Zemin, that ten years ago, when the Sino-Portuguese Agreement was signed, many people were worried about returning and resorted to immigration. But recent years have witnessed a smooth transition in Hong Kong and the satisfactory functioning of the SAR government. The same people now have the desire that Macau be returned earlier, as a result of their dissatisfaction with the status quo in Macau. See "President Jiang meets Tsao Guangbiao and his family," *Overseas Chinese Daily,* 9 September 1997, p.1.

8. The Saint Paul College was set up at the end of the sixteenth century and trained many prominent Jesuits to spread Christianity and science in China, but these Jesuits also introduced Chinese culture to the West in return.

9. If one wants to start a business in Macau, he or she has to undergo incredibly complex processes that normally take six to eight months to accomplish. See Xu Qingfeng, "Macau Investment Environment Must Improve," *Macau Daily,* 28 June 1998, p.A9.

10. Ibid.

11. For details, see James C. Hsiung, "The Paradox of Hong Kong as a Non-sovereign International Actor," paper presented at the conference, "Hong Kong: A Year After Reversion to China," Lingnan College, Hong Kong, 2–3 June 1998, and an update in Chapter 7 of this book.

12. Among the 40 Mainland members of the Macau SAR Preparatory Committee, 20 come from different ministries of the Central Government, among which many were former Hong Kong SAR Preparatory Committee members. Three out of four Mainland law professors inside Macau Preparatory Committee were also Hong Kong SAR Preparatory Committee members.

13. Macau people show characteristics of "subject and parochial political cultures." For detail, see H.Yee and B. L. Liu, *Macau Political Cultures* (Macau: Macau Foundation, 1994).

14. "Speech by President Jiang Zemin at the Hong Kong Handover Ceremony Jointly held by the Chinese and British Governments," *Beijing Review*, 14–20 July 1997, pp.12–5.

15. "Speech by President Jiang Zemin at the Party Celebrating the Establishment of the Hong Kong SAR of the People's Republic China," *Beijing Review*, 14–20 July 1997, pp.16–17.

16. *Beijing Review*, 6–12April 1998, pp.39–41.

17. "Xu's attack on RTHK dismissed," *South China Morning Post*, 8 March 1998, p 4.

18. Hong Kong TV Jade morning news, 30 April 1998.

19. Mr. Ma Yuzheng is the Commissioner of the Office of the Ministry of Foreign Affairs of the People's Republic of China in Hong Kong; Mr. Jiang Enzhu is the Director of Xinhua News Agency in Hong Kong.

20. "For the time being, Hong Kong people are still speaking freely, enjoying a free press, and having justice dispensed fairly." Martin Lee, "Beijing new colonial ruler," *South China Morning Post*, 1 July 1998, p.19.

21. On 24 May 1998, 53 percent of the voters come out to vote despite the heavy rain. See *Ming Pao Daily*, 25 May 1998, p.A1.

22. According to the Basic Law, the Hong Kong people will decide on the speed of democratization in 2007.

23. *Overseas Chinese Daily*, 23 December 1997, p.1.

24. "New Premier charts development course," *Beijing Review*, 6–12 April 1998, p.40.

25. Macau's currency is linked to the Hong Kong dollar at a rate of 103:100.

26. Take, for example, Mr. Peter Lai, Secretary of Security, who worked as Deputy Secretary for Constitutional Affairs and played a key role in the failed 17 rounds of talks with Beijing over the Patten political reforms. He had worked in the Hong Kong SAR government as Secretary for Security without political pressure until he tendered his resignation recently. See *South China Morning Post*, 28 July 1998, p.1.

27. The Portuguese government has allowed all Macau civil servants to work in the Portuguese civil service as long as they master Portuguese.

28. In the first half of 1998, the gambling industry recorded nearly 10 percent drop in its income; government taxes on gambling decreased by 8 percent. *Macau Daily*, 3 July 1998. p.1.

29. Li Nanchuen, "Macau Economy and Property Market are Interdependent," *Overseas Chinese Daily*, 24 December 1997, p.2.

30. In Macau, expatriates can invest over one million MOP (Macau pataca, a currency unit) on property, then get Macau residence. However, the scheme has proved to be too complicated in application and achieved limited effect.

31. *Citizen Daily,* 25 June 1998, p.1.

32. The Portuguese Prime Minister made it clear that triads in Macau are from neighboring areas. See *Macau Daily,* 20 April 1998, p.1.

33. According to officials from the China State Ocean Affairs Bureau, there were eight cases of red tide along the Chinese coastal line, costing losses of about 30 billion RMB (renminbi, the Chinese currency unit). "Ocean disasters worst in 50 years," *Ming Pao Daily,* 22 March 1998, p.A13.

34. Christine Lok, "First lesson is autonomy a painful affair," *South China Morning Post,* 3 July 1998, p.21.

35. Tsang Yok-sing, "A case of false prophesies," *South China Morning Post,* 3 July 1998, p.2

36. The Macau government has no reserves, but China, through its land fund, has accumulated eight billion MOP for Macau. See *Macau Daily,* 4 February 1998, p.2.

37. Political analysts in Hong Kong have warned that if the economic crisis cannot be solved, anti-government riots will break out. See "Retrospects on the first anniversary," *Ming Pao Daily,* 29 June 1998, p.A6.

38. The relevant theories can be found in Harold F. Gortner, et al., *Organization Theory: A Public Perspective* (Los Angeles, CA: Brooks/Cole Publishing, 1989), pp.18–39.

39. "Emphasizing Professionals, Opening Recruitment Channels," *Macau Pulse,* 2 January 1998, p.3.

40. "The U.S. State Department, in its 1 April 1998 report, said that within the nine months after the British departure, there were clear signs that Hong Kong was acting autonomously," *South China Morning Post,* 2 July 1998, Business Section, p.3.

41. Macau people do not favor "Lin Tin Yang Bridge," designed by Zhuhai to link Zhuhai to Hong Kong, considering it no benefit to the Macau economy.

42. Mr. Yeung Juenwen, Chairman of Macau Productivity and Technical Transfer Center, said: "Macau can help Hong Kong to raise its productivity level in many aspects. The low cost in Macau could attract those small and medium enterprises who could not afford high cost in Hong Kong and for those transitional corporations who build up their logistic support in Macau," *Macau Daily,* 2 July 1997, p.2.

PART III

Conclusions

CHAPTER 12

The Hong Kong SAR:
Prisoner of Legacy
or History's Bellwether?

James C. Hsiung

This book is an update on the story of Hong Kong after its retrocession to China, ending over a century and a half of British colonial rule. Throughout the volume, we have noted consistent wide discrepancies between earlier predictions, both good and bad, and subsequent outcomes in Hong Kong's transition into a Chinese special administrative region (SAR). On the domestic political scene, Hong Kong has disappointed the prophets of doom for not having turned out nearly as woefully as it was supposed to. Nor, on the other hand, has the transition turned out to be as nearly without a hitch, on the external front, as was forecast by some of the most respected jurists. The latter had held out for an uneventful continuity in Hong Kong's international status as a non-sovereign actor and its unquestioned acceptance as such by the international judicial community. As we have seen from the selected cases examined in this book, the outcome was not so clear-cut. The least anticipated development of all was, of course, the economic downturn befalling the territory almost immediately after the handover. By coincidence, the financial crisis hit the Asian Pacific region on the day after the British left their former colony. In brief, a "super paradox" syndrome epitomized Hong Kong's historic transition.

The most significant outcome, as we have seen, is that internally the territory has, relative to the downcast predictions, fared surprisingly well in terms of the normal state of people's lives and the continuance of the rights and freedoms they enjoy. Disputes over such things as the legal status of the

New China News Agency and the speed of the SAR's democratization, more especially whether the Legco should be entirely elected from geographical constituencies, thus doing away with the functional constituencies, were relatively minor and not fatal to the SAR's functioning (Wong 1998, p.17). In the most critical chapter in this book (Chapter 10), which presents the worst-case scenario as seen by the typically most suspicious Hong Konger, the author has raised fears about whether China's own party-state order would "rub off" on Hong Kong but stops short of chastising Beijing for its interference, simply because there has been none. The only time Beijing openly criticized the SAR was over the *Chan Kam-nga* case decided by its Court of Final Appeal, but, as we will see below, it was a storm in a teacup. In fact, a noted publicist Anthony Neoh, a member of the HKSAR Basic Law Committee and a contributor to this volume, called it in an op-ed commentary "a case of legal misunderstanding."[1] Although the economic slowdown was a crushing blow, everybody knew it was not going to stay forever and, moreover, it was not a misfortune inflicted by Hong Kong's return to Chinese sovereignty. Despite the change of guards, and of the emblems, little seemed to have changed for the people in the street. Demonstrations for all causes, averaging over three a day during the first year, some directly challenging the Beijing government on human rights in China proper, also survived the handover.

Even at the height of the economic slowdown, as polls revealed, the public's confidence in the SAR's political lot was unflinchingly strong. If there were doubts in the public's collective mind, they were mainly over the Chief Executive's ability to cope with and reverse the territory's sudden economic setback, not over Beijing meddling in Hong Kong's internal affairs. In the views of most commentators, the long-range prospects for the HKSAR's future seem to be positive. An indicator was the unequivocal conclusion by the Sino-British Joint Liaison Group, issued at the end of its session in Beijing in December 1998. This body, created per the 1984 Sino-U.K. agreement on the return of Hong Kong, has a mandate to oversee the territory's transition through the year 2000. Its report not only acknowledged the Joint Group's confidence in the future of the territory, but aired its unqualified confidence that the good post-handover developments in Hong Kong were a good omen for bright UK-China relations well into the next century.[2] Gone was the rancor of the previous Pattenesque Sino-British disputes that rattled the world's media. Positive assessments even came from a remote source. A European Commission report, released on 11 January 1999, concluded that "basic rights, freedoms, and autonomy have been broadly upheld."[3]

As has been noted throughout the book, the normalcy prevailing in the domestic political sphere of post-handover Hong Kong was in large measure attributable to Beijing's surprising hands-off policy. We shall not rehash the consequential, but futile, attempts by some former critics to explain away what had gone wrong with their unfulfilled prophecies of doom, nor their endeavors to divert attention to a red-herring, i.e., the alleged link between the British departure and the sudden economic setback hitting the territory, since it is part of a regional crisis hitting all Asia Pacific.

We have noted that, even at the height of the economic crisis, the Hong Kong public's confidence in the viability of the "one country, two systems" governance structure, introduced following the territory's return to China, proved unswerving. The Hong Kong Institute of Asia-Pacific Studies, of the Chinese University of Hong Kong, has been tracking public opinion trends on this and other issues on a monthly basis. According to its statistics, made available to me,[4] the percentage of the people that expressed greater confidence in the "one country, two systems" model increased over time, although the exact ratio may fluctuate in the monthly surveys. In July 1997, for example, 18.1 percent indicated an increase in their confidence in the new political order and 62.6 percent said their attitude remained unchanged; 13.0 percent said they had less confidence than before. By September, only 5.0 percent said they had less confidence, while 54.8 percent and 36.1 percent, respectively, said that their attitudes remained the same and changed to the better. While those who expressed an increase in their confidence dropped to 19.6 percent in January 1998, the percentage soon went back up. For the rest of 1998, some 25–31 percent of those polled in the monthly survey indicated they had more confidence in the "one country, two systems" regime of the SAR than before. By contrast, the proportions of respondents whose attitude toward the "one country, two systems" model did not improve, or who in fact indicated a decline in confidence, went down consistently. Taken as a whole, the public's confidence indicators as such, from the monthly polls taken between July 1997 and the end of 1998, showed a sustained pattern likely to continue beyond the first one and a half years after the reversion.

On the economic front, the public's confidence rating went up after the 1 July 1997 reversion date, and it promptly reached its first peak in September 1997. In that month, 41.4 percent of those polled believed Hong Kong's economic prospect was better, while 35.9 percent thought it was as good as before. Only 14.9 percent thought it was worse. The confidence index then dropped to a low point in May 1998, when only 9.9 percent believed the economy was doing as well as before the handover. It soon recovered to 33.7 percent the following month and reached 41.3 percent by year's end, however.

As regards the Hong Kong SAR's internal political situation, the public's confidence remained relatively strong throughout the months following the handover. It hit a noticeable high in September 1997, when 21.9 percent of those polled indicated that the political conditions were better than during the colonial times and 38.1 percent thought they were the same, while 25.4 percent thought they were worse than before. After a momentary drop to 15.9 percent in March 1998, however, the confidence rating on the SAR's political health went back up to a 17–22 percent level until November, when the proportion of respondents who believed the situation was better than before climbed to 24.0 percent. While 49.9 percent considered it about the same as before the handover, those who believed it was worse off dropped to 15.2 percent. Thus a consistent pattern emerged, in which the number of people who thought the political situation was worse than during colonial rule increasingly declined, and the combined total of those who either thought it was better or about the same as before went up. It seems also to be a pattern most likely to continue into the foreseeable future.

Public ratings of the SAR government performance are also interesting to watch. Despite the relatively consistent support of the respondents for the "one country, two systems" governance structure installed in the SAR, as just noted, their satisfaction rate with the SAR government itself fluctuated from the 54.3 percent record high in September 1997, or two months after the reversion, to a low of 14.7 percent in May 1998. By the end of 1998, however, it steadied at the 22–25 percent range in the monthly pollings. The Institute's pollsters kept separate scores for the public's satisfaction with the Chief Executive, as distinct from the SAR government he heads. The polling results showed, for instance, that during May 1998, when the public gave the SAR government a paltry 14.7 percent satisfaction score, the Chief Executive received a 56.2 percent approval rating. Looking at the Chief Executive's ratings over time since July 1997, one finds they were relatively stable, hovering within the 56–68 percent range, as contrasted with the SAR government's more widely fluctuating scores between 14 percent and 54 percent, over the months. The Institute's polling consistently surveyed the public for its views on the Chief Executive's qualities in ten categories, such as integrity, general competence, trustworthiness, ability to balance the SAR interests against those of the Chinese sovereign, etc. In a survey stretching over two years since Tung Chee-hwa was first selected in December 1996, the results showed he constantly received decent marks except in two areas: "promotion of democracy," and later "improving Hong Kong's economy." On the former he received a low mark of 45.7 shortly before the 24 Mary 1998 election, but higher marks were reported after the election. His lowest score for all ten categories was over the question of whether he had the ability to reverse Hong Kong's economic fortunes. On this tangent, he received a 45.2, or below the 50 passing grade, for the month of September, 1998. Over all, Tung received higher scores early on until October 1997, but continued to receive poorer grades afterwards, or after the onslaught of the regional financial crisis began to show its effects in Hong Kong.

Putting things together, we can safely conclude that the Hong Kong public's dissatisfaction has been more pronounced with the SAR government than with the Chief Executive and that, on balance, the disgruntlement has been closely correlated with their perceived inability to contain the financial turbulence that has disrupted the territory's economy. From the consistently high ratings the respondents gave to the viability of "one country, two systems," it is also safe to conclude that the public is satisfied with Beijing's deliberate low profile and hands-off stance toward the SAR, which has quieted earlier apprehensions among many in Hong Kong.

This is a point on which there is an overwhelming consensus, as attested to by the various chapters above, in Part I: The Domestic Scene. It was also corroborated by the Speaker's Task Force on Hong Kong, headed by Congressman Doug Bereuter (R-Nebraska), in a report dated August 4, 1998 to the Speaker of the House, U.S. Congress.[5] While the report covered a wide range of issues including the SAR's 24 May 1998 election, rule of law, economic developments, etc., it began with an acknowledgment that Beijing, despite all the prior apprehensions of the outside world, proved itself capable of resisting "the temptation to meddle in Hong Kong's internal affairs" after all.

In this concluding chapter, we shall resist the temptation to rehash what has already been said in the various chapters above, except in noting some specific items that truly merit reiteration. Even when we do so, the reason is not repetition for emphasis so much as for an ontological reason, namely, how a noted peculiarity from the HKSAR experience can or cannot be compared with other bodies of human experience. One example is: what does Hong Kong's "one country, two systems" model tell us about governance? Conversely, how can our existing knowledge of governance elsewhere help us come to grips with the essence of the SAR's political developments.

Along the way, we need to look at what the Hong Kong SAR has become and what unique problems it has run into, especially if they present a theoretical "puzzle," in the sense that Thomas Kuhn (1970, p.92) used it, to those concerned with human governance. Not the least of these, for example, is the question of what need be present as a precondition for the continued success of the Hong Kong SAR living in the shadow of its Chinese sovereign, the initial normal relations marked by a camaraderie of unexpected proportions notwithstanding.

Colonial Legacy: Asset or Albatross?

In a broader perspective, it has to be noted that the reversion of Hong Kong came at the end of the world's long process of decolonization, which began under the impetus of what in retrospect was probably the most decisive force released by the United Nations, in the form of a resolution adopted by the General Assembly in 1960. The Declaration on Granting of Independence (GA Res. 1514 (XVI)) converted self-determination as a "principle" noted in the United Nations Charter into a human right of all subjugated peoples. It also transformed the ideal of "self-government" as a long-range goal in the Charter for all dependent peoples into an immediate goal of independence, to be achieved expeditiously through decolonization. The powerful sweep of this coup became evident only in retrospect. By the early 1990s, or a short 30 years later, when the process was near complete, over 80 former dependencies, including trust territories and colonies, had achieved independence. The face of the world has changed (cf. Hsiung 1997, ch. 7, esp. pp.134f). Hong Kong, while coming at the tail end of this historic process, was unique in two senses.

First, in exiting from colonial rule, Hong Kong did not become an independent state. In fact, it is the only former colony of a Western power that was retroceded to its former motherland thus far, although Macau is to follow suit in less than two years. As such, it represents a rare instance of decolonization without self-determination. Second, unlike other former colonies, most people in Hong Kong are not known to harbor any serious resentment to their former colonial masters. In a public talk, Alan Lung (1998, p.1), chairman of the Hong Kong Democracy Foundation, probably was speaking for most Hong Kongers, particularly the British-educated local elite, when he said: "I could still remember on the first day I reported to work at a new job in Singapore in the mid-1980s, Mr. Goh Geng Swee [then Finance Minister, later to become First Deputy Prime Minister]—decided to kick Jardine Fleming out of Singapore because he did not like what the [British] company did in his country." But, Lung continued, "here in Hong Kong, the cause of bashing our former British colonial master was taken up by 'patriotic' fellow citizens who tried to blame the economy, the education system, our housing shortage and, incredibly, the foul-up at the Airport on the departed British administration. The cause did not go very far because *the vast majority of us do not seem to hate the British*" (emphasis added).

Certainly, the resentment of the Singaporeans was more typical of the sentiments in most other former colonies toward their erstwhile colonial rulers. Hong Kong, in this sense, pretty much stands alone. I can think of three possible reasons for it. First, Hong Kong exited from colonial rule not after a struggle with its colonial ruler, as was true in almost all other cases, some involving real bloody fights, even armed rebellions. Rather, the end of colonial rule in Hong Kong was the result of a deal between its colonial ruler and its past motherland and future sovereign, without the participation of its population who, under the circumstances, had legitimate and understandable concerns and fears. Secondly, although the ingenious "one country, two systems" solution helped ease many people's fears, it was true that even on the day of Hong Kong's handover, some inexplicable uneasiness was evident in the air, not to die down until it was proven that Beijing refused to meddle in the internal affairs of the new SAR. The anticipatory suspense in the meantime may have diverted the Hong Kong people's attention away from the stigma of colonial rule and what it stood for. And, thirdly, the British, after signing the 1984 agreement on Hong Kong's return to China, embarked upon a systematic campaign, though belatedly, to introduce democracy in the territory. Thus, it is only natural if most people in Hong Kong should tend intuitively to judge the British colonial rule by its most recent record (1985–97), to the neglect of what had prevailed in all the previous 148 years since 1842. In fact, a case can be made that the belated British democratizing reforms after the 1984 agreement were specifically intended, in the words of Liu Shuyong (1997, p.586), to "prolong British influence in Hong Kong beyond 1997." Hence, not surprisingly, the Hong Kong people, as Lung articulated it, do not harbor any resentment toward their former British colonial masters.

In this connection, it is necessary for us to examine schematically what was the legacy that the departing British left behind, in order that we can assess how the legacy has helped or, conversely, unduly aggravated the SAR's tasks of self-governance during the transition after 1 July 1997. The discussion here is intended neither to commend, nor to condemn, but is aimed at a balanced explication of the true reality that the SAR has inherited, within the parameters of which the SAR government has to function.

As the experience of Hong Kong has shown in combating the regional financial crisis of 1997–98, its comparative success can be said to be ascribable to the virtue of certain tested principles inherited from the past, including adherence to the rule of law and the rules of the market, the free flow of information, capital, and goods, and a clean, efficient, and non-interventionist government (see Boucher, 1988, p.1). In all fairness, these are part of the British legacy in Hong Kong, which in its entirety is often summarized largely in four areas: the rule of law, the civil service, economic freedom, and, belatedly, efforts at democratization. Many writers have acknowledged these contributions of the British legacy, but to my knowledge only Ming K. Chan (1997) has, with undoubted courage, taken the pains to point out that this legacy, while noteworthy, has warts that the SAR has to live with. Earlier, in my Introduction, I have noted that the "bequest" of Governor Christopher Patten, such as a carefully insulated bureaucracy and a precipitously elevated Legco, has presented problems of governance to the SAR government. We shall not repeat what has already been said there and in the chapters in Part I on this point, except to note, in hindsight, that in their last years in Hong

Kong the British rulers created, by their unquestionably well-intentioned reform efforts, a "revolution of rising expectations" that can hardly be readily met by the subsequent SAR government after its inauguration in July 1997. To make a long story short, the persistent high ratings of dissatisfaction with the Chief Executive, as well as with the SAR government, on the question of "promoting democracy" is an indication of this very problem. One issue that has often been named as a disappointment, and even a "retrogression on democracy," was the mode of partial election—in combination with the role of a selection committee—used in the 24 May 1998 election for the Legco.[6] But, it should be noted, the current election method, which was first drafted in 1985 under Governor Wilson and found its way into the Basic Law, was used by Governor Patten in the 1995 election.[7] And, besides, it was a carried-over stop-gap measure for now, as the Basic Law requires that universal suffrage be achieved by 2007.

People who thirst for instant democracy might, out of genuine concern, object to this delay and the gradualist approach responsible for it. But a more equitable comparison, it seems to me, should be made with what had prevailed under the long British colonial rule before 1984, not with the last-minute reform period in the ensuing 13 years. Throughout the colonial period before 1985, the Legco members were always appointed by the Governor, who in turn was appointed by London. Martin Lee, leader of the Democratic Party in Legco, went on television to hold up Taiwan's successful democratization as an example for Hong Kong to follow.[8] But, a question he did not, but should, ask was: Where were Hong Kong's democratic champions when their Taiwan counterparts began to fight for freedom and democracy, as they did at considerable cost during the darkest hours of Taiwan's authoritarian rule? The fact behind the non-answer is that, until 1985, no people in their right mind in Hong Kong would dare to think of agitating for democracy under British colonial rule. And, the fact that the first ever political party to appear, the ADPL, did not come into being until 1986, or two years after the signing of the Sino-U.K. Joint Declaration on Hong Kong's return, is instructive on this point. Quite revealingly, when this first political party arrived, it chose for itself a non-descript name, "Association for the Development of People's Livelihood," as if to soften its political impact or possible threat to the colonial political order.

In the last years of British rule, it should be noted, the local democrats and the last British Governor, Christopher Patten, were actually fighting on the same side, against a common adversary perceived to be of threat to the Hong Kongers' hankering for democracy. The Hong Kong democrats' fight, as such, was not staged against the British colonial rulers, but against the territory's former motherland and future sovereign. This, parenthetically, may be an additional reason why, as Lung noted above, the people in Hong

Kong, especially the elite, do not harbor resentment against the British. The latter, in effect, posed as a benign, kind uncle shielding the local democracy-seekers against a prospective bully. The irony is that the would-be bully turned out to be a surprisingly harmless patron, at least thus far, and the British uncle left a mixed bequest—increasingly found to be less benign in retrospect than was previously assumed—that the SAR has to live with and, in some instances, extricate itself from (see below).

The point here is that while credit has been given, justifiably I think, to such British legacy as the rule of law and so on, due recognition should equally be given to how the Chinese have adapted to the entire "givens" inherited from the British. It is the Chinese, including the SAR government and the remote Central Government in Beijing, who have had to make everything work within the confines of an inherited status quo from the departed British rulers, who have left behind, among other things, a "revolution of rising expectations" they helped create. This realization is all the more crucial when considering that it was not until after July 1997 that the local democrats for the first time ever have an uninhibited chance to fight their government in demand for democracy. And, to put it nakedly, this is a luxury they can now afford precisely because the British are no longer running Hong Kong. Their fight, lofty and legitimate in itself, bears witness, paradoxically, to the fact that the "one country, two systems" governance structure is working in the SAR—a fact gradually acknowledged in many quarters. At a minimum, it should not be viewed as a reflection of the SAR's failure, as certain critics with an ax to grind might be tempted to do.

Harking back to the point made by Ming K. Chan (1997) above, that the legacy from the past may, like roses, have hidden pricks and thorns, let us ask if it is true and, if so, where do we find them. Here, I would like to offer an alternative, *revisionist,* view on how to appraise certain parts of the British legacy as a problem or burden, as distinct from a uniformally assumed asset.

A Revisionist View

Here, I would stay away from the immediate bequest of Governor Patten, whose reform, shifting the gravity of government from the executive to the legislature and placing certain high officials (chosen for their political convictions) in holdover positions, created fortuitous problems for the SAR, as already noted in the chapters in Part I and my Introduction above. Instead, I would call attention to the bequest of a different sort, representing long-term neglect from colonial policy. I am referring to the egregious absence of a whole sleuth of social entitlement programs ranging from comprehensive pension, welfare, social security (as we understand it in the United States),[9] to unemployment insurance—in short, the hallmark of a modern society since the Great Depression of the 1930s. Public welfare programs were unheard of in Hong Kong before 1965 (Zhou 1992, p.42). The lacuna, which is particularly shocking to a New Yorker, was due to colonial policy, since undeniably the colonial government's first priority was to look after the interests of the British expatriates. So there was a pension system exclusively for the government bureaucracy, staffed in the past mostly by expatriates. It is plain that the expatriates would not need any of the other programs, certainly not unemployment insurance. Hence, unemployment insurance was not only unknown, but, worse, it was stigmatized as something that would only pamper the lazybones. The sad thing is that this lack of compassion perpetuated by colonial policy as such has rubbed off on the British-educated local elite even into the post-colonial era.

For the same reason, the colonial government showed little concern for the inordinate income distribution inequities and, for that matter, the very problem of abject poverty amidst affluence, throughout the British rule in Hong Kong. Statistics presented by Lui (1997, p.60: table 3.5) demonstrate a consistent pattern that, for instance, in the period of 1976 through 1991, the top 10 percent of the population earned eight times as much income as the bottom 10 percent. And, from 1 percent to 4.6 percent of the population during the same period had zero income. The gap was widening instead of narrowing. Over the ten-year period of 1986–96, the top 20 percent of wage earners received a hefty 60 percent increase in income. But, the bottom 20 percent of all wage earners had only a 20 percent pay rise, to an average monthly income of HK$5,500 (roughly U.S.$737).[10] In the ten years to 1996, the number of Hong Kong people living below the poverty line (set at monthly wages lower than HK$4,500 in 1996 HK dollars) increased 80 percent to a total of 850,000 among the territory's 6.5 million population.[11] Thousands of people, known as cage dwellers, lived in rented cubicles euphemistically called bed-space rooming. According to all reports, the colonial government remained vaguely concerned at best. For all his fervor for democratic reform, for instance, Governor Patten, according to some accounts, showed little interest in helping the poor, the aged, and the needy in general. The best thing his administration did was to implement a 10 percent increase in the budget earmarked for public assistance (Zhou 1994, p.112). This was a meager gesture, considering that public-welfare budget in Hong Kong was proportionately never even half the welfare expenditures in the United Kingdom's government budget back home (Zhao 1985, p.24).

The plight of the elderly living in poverty was scandalous. According to statistics compiled in 1996, one in every four elderly residents in Hong Kong, aged 60 or above, lived below the poverty line. Those aged 65 or above accounted for 10 percent of Hong Kong's total population in 1996.[12]

One report shows that those aged 75 and above accounted for 30 percent of the elderly poor in need of help, and the majority of them had lived in Hong Kong for over ten years (Mok 1998). In the absence of a comprehensive pension system, retirees had to live off their personal savings, plus whatever meager help was available from the Comprehensive Social-Security Assistance (CSSA) scheme. The newly approved mandatory Provident Fund, finalized shortly after the handover, would not commence until the year 2000 and, even after that, it will take at least 40 years before its benefits are shown (Chang 1998). This means that most of the present generation of retirees and elderly poor will not be around to receive its benefits. The question remains: Why did the colonial government fail to do its share of the responsibility during its long tenure (1841–1997)? The reason might well be the stock answer, that it was not a problem besetting the expatriates, but only the Hong Kong natives. Hence, providing the people a social safety net was, to put it mildly, never a priority concern of the colonial government.

It is not our intention to get into the details of what may be called, for lack of a better term, an egregious "poverty of compassion" of the colonial period. But, the point here is that this part of the legacy quietly dumped an unnecessary burden of considerable weight on the SAR government after the departure of the colonial masters. Space does not permit us to go into what the SAR government needs to do in coping with the task of helping the poor, aged, and needy. Nevertheless, I think the magnitude of the problem ought to be noted. The best measure is the phenomenal increase in the SAR government's social-welfare expenditures during the very first year of its existence, responding to the plight of the needy. The total figure soared to HK$20.6 billion (or U.S.$2.66 billion) for 1997, although it might have contained spendings during the first six months under Governor Patten on the verge of his departure. But, compared with the meager HK$7 billion (roughly U.S.$903 million) five years before (1992), when Governor Patten first arrived on the job, it can be safely assumed that the nearly three-fold increase in 1997 was the result of a much needed policy change after the end of colonial rule.[13] Latest reports pointed to a 13 per cent increase for welfare and social services in the SAR budget for 1999–2000, to HK$29.1 billion (or U.S.$3.76 billion), the largest growth in public spending.[14]

Quite revealingly, the Mandatory Provident Fund, a priority project of the SAR government, is the first ever comprehensive retirement pension program, which will cover Hong Kong's entire three million workforce, including the 800,000 already covered by preexisting pension schemes.[15] It is clear from the disparity between the two figures that under colonial rule, the majority of Hong Kong's workforce, or 2.2 million, were without any pension coverage. And, it is probably not too far-fetched to assume that most of the 800,000 lucky ones covered under preexisting pension schemes were

British expatriates working for the colonial government, or government-sub-vented institutions (including public universities and hospitals), or large do-mestic and foreign corporations.[16]

Inequality was another part of the legacy from colonial times. Although there was no Chinese exclusion law per se, "insensitivity amounting some-times to brutality towards the Chinese population remained common in Hong Kong," according to a respected historian (Welsh 1993, p.165). And the existence of Europeans-only clubs had a long history. "The European ho-tels and clubs," wrote a legal authority, Peter Wesley-Smith (1994, p.91), "have routinely restricted the entry of Chinese; . . . a Chinese was not ap-pointed to the Executive Council until 1926." The lingering influence of this part of the colonial legacy, carried over into the post-reversion Hong Kong, is the controversial practice of discrimination against dark-skinned customers at night clubs or high-class restaurants. Inequality existed in many forms in the past, even in academe. In a bizarre case in 1998, two senior academics, accused of a HK$1.5 million housing allowance scam, used colonial inequal-ity in their university benefits system as a defense for their cheating. Oddly enough, the District Court Judge, Esther Toh, who tried the case, bought the argument and accepted it as a mitigating circumstance. She said the more fa-vorable terms granted to expatriates under the British administration "tempted local staff to 'maximize' their benefits" by cheating the system.[17]

In other areas, similar problems from colonial times are found, requiring remedies by the SAR government. For instance, sexual discrimination against women continued to exist even after the New Territories ordinance that deprived women of land inheritance rights was amended in 1994. Ac-cording to an authority on this subject (Wu, 1995, p.194), the practices of discrimination against women resulting from the lack of equal opportunities protection by law—i.e., practices that have been abolished in other Chinese societies—were "frozen in time by colonial ordinances" in Hong Kong.

According to Anna Wu (1995, p.190f), even after the enactment of the Bill of Rights, under the British post-1989 campaign to democratize, there were 50 existing laws that were inconsistent with the International Covenant for Civil and Political Rights, to which Hong Kong became a party as a British colony. Some of these pertained to the political censorship of films, the interception of incoming or outgoing telecommunications messages, the use of loud-hailers, and the conduct and treatment of prisoners and de-tainees. Whatever was not rectified under the outgoing British colonial gov-ernment—"in fact, the Hong Kong Bill of Rights provided a freeze period for certain of these laws (p. 191)"—devolved upon the SAR government after 1997. Hence, while the new SAR government owes it to itself to undo these injustices, it is only fair that the continuing restrictions of these and other rights, if any, in the interim should be rightfully credited to colonial neglect. Already, the United Nations was reported to have questioned the HKSAR government on its first human rights report, filed during its first year of existence. The list of 29 queries raised by a United Nations working group concerned Hong Kong's compliance with the Convention on the Elimination of All Forms of Discrimination Against Women. Specifically, the list included women's land rights in the New Territories, and the participation of women in politics and the public sector, which are unmitigated legacies from the long British rule.[18]

In addition, the officially tolerated practice of concubinage and the prevalence of mahjong parlors deserve special mention. These were also practices "frozen in time" by colonial policy because the indulgence they bred presumably would keep most ambitious men from having too much leisure in which to nurture political dreams.[19] If it had not been willful policy, the colonial government could easily have outlawed these and other similar institutions or simply taxed them out of existence. These are all part of the total legacy from the colonial era, which has devolved upon the SAR government.

When our eyes turn to these and other similar inherited problems that cry out for remedying by the SAR government, our perspective on whether the colonial legacy is an asset or a burden (an albatross around one's neck, as it were) immediately changes. The answer is, of course, it is neither, but a mixture of both.

By the same token, in a larger context, the SAR both partakes of qualities of a prisoner of colonial legacy and shows the potential of being history's bellwether. We have just seen, from the above discussion, the problems left over by the colonial legacy—albeit mostly unnoticed by a casual observer—that present a special challenge to the SAR government. It can be safely inferred that the SAR, without negating the entire colonial legacy, has to fight to stay clear of being a "prisoner" of these and other unwholesome parts of that legacy. Below, we shall have occasion to entertain the question whether, by its unique domestic "one country, two systems" model, the Hong Kong SAR holds the promise of being history's bellwether. In the run-up to that ultimate question, it is necessary to sketch briefly the outstanding, distinctly unique features of the SAR as a self-governing polity.

How Unique Is the Hong Kong SAR?

Without unduly duplicating what has been said in earlier chapters, I would single out two broad, unique features about Hong Kong in its reincarnation as a Chinese SAR. Parenthetically, claims of uniqueness can be sustained only through comparisons applying a common yardstick or in reference to a particular purpose to be served. The two broad features can be identified as

(a) a peculiar executive-legislative relationship in the SAR government, and (b) an unparalleled "unity in diversity" built into the symbiotic governance structure, known as "one country, two systems," that encapsulates Hong Kong's relationship with its Chinese sovereign. Both have been noted in different and scattered contexts before, but here I propose to discuss them schematically in terms familiar to those concerned with comparative studies of governance in human societies (that is, applying a common yardstick). In addition to its intrinsic value, the second broad unique feature holds the interest of all concerned with the practical duplicability of the "one country, two systems" model in other cases, including, but not limited to, ultimate consummation of Chinese reunification (i.e., a particular purpose to be served). The question whether the model that has proven workable in the Hong Kong SAR may hold the key to China's ultimate union with Taiwan, especially after all other models have failed, deserves to be examined in this concluding chapter. Alternatively, the model's success in Hong Kong might recommend itself to other prevailing cases of irredentism and smoldering secessionism elsewhere in the world.

I might add a third dimension to this discussion of the SAR's uniqueness as such, which is that the peculiar executive-legislative relationship, as I shall highlight below, is typical of the kind of circumstances most likely to foreshadow Hong Kong's long-term development as a polity. Hence, it bears special mention for the interest of those looking beyond the immediate postreversion transition.

Executive-Legislative Relationship in the SAR

In any modern society that claims or aspires to be a parliamentary democracy, the Chief Executive (Prime Minister) usually heads the majority party or coalition in the legislature. In a presidential system such as in the United States, the President's party may or may not control the majority seats in Congress, but at least his own party is represented in the latter and he is the titular head of that party. But, not so in the Hong Kong SAR. The Chief Executive not only does not head the majority party in the Legco, but he has no party representation at all. This is an anomaly with a number of somber implications. For one, it casts doubt on the ability of the Chief Executive to rule effectively, should the legislature be truly assertive in exercising its independent power. Secondly, it does not augur well for a balanced development toward democracy, as we understand it, which requires a built-in feature of orderly checks and balances in the distribution of power and responsibility between the two branches.

While the future is not for us to see, we can only try to understand the problem by examining the origin of this anomaly. In a nutshell, it all came about as a result of the sudden artificial shift, under the last-minute British reform, from a Legco that had always been appointed by the Governor to an elected one after 1991. During all the previous one century and a half, the Governor had had no need of party representation, as he personally headed the Legco; and his appointed "official members"—hailing mainly from the bureaucracy—would, acting as a bloc at the Governor's direction, always out-vote the "unofficial members" (likewise appointed by him) in the Legco. Besides, as the Governor was not a native but sent from London, he would not be someone heading a political party in the Legco, even if there had been parties, as there was none before 1986. The lack of any provision in the Basic Law for a political party representation of the Chief Executive in the Legco is a gross omission. It may, however, not be the result of an oversight, so much as a mere mechanical mimicry of a preexisting, albeit by now anachronistic, status quo. Besides, the unfortunate Sino-U.K. disputes over Legco elections since 1985 probably also sidetracked the issue of the future Chief Executive's party links in the legislature.

Furthermore, after the Pattenesque reform consummated in the shift to the Legco of the gravity of government, the way was paved for an assertive legislature to emerge for the first time in Hong Kong's history—but after the British departure. If no British Governor ever needed to worry about having no party affiliation in the ever subservient and party-less Legco before 1994, now the Chief Executive has no such luxury, because he has to face a totally different legislature after the May 1998 election, ten months after Hong Kong's handover. A comparison of the Chinese names for the Legco before and after the reversion is instructive. The name used in colonial times was *li-fa-ju* (literally, legislative *bureau,* suggesting it was an extended arm, at best, of the executive branch); but after 1997, it is known as *li-fa-hui* (legislative assembly or council).

Both the newly acquired prominence of the elected, and more assertive, Legco and the rise of political parties in its chamber may, in combination, threaten to make the Chief Executive's life miserable, like never before. Despite the Basic Law's intent to preserve a strong executive arm, the fact is that the Pattenesque legacy left in effect a strong legislative tail capable of wagging the executive dog. While according to the Basic Law, the Chief Executive has the nominal power to veto bills, the irony is that the Legco may not pass the bills he really wants, or the budget to suit his liking. The example of an unemployment insurance proposal favored by the executive (as noted before) that had to be scratched from the drawing board because of preemptive opposition in the Legco is a case in point. As we have seen, while the Democrats, the strongest opponents of the government, do not hold the majority seats, they nevertheless constitute the largest party in the legislature and, more important, have proven able to assemble a winning coalition.

322 • James C. Hsiung

Chief Executive Tung Chee-hwa seemed to be fully aware of his plight when he began, after the May 1998 election, to refer to himself as a politician, not just an administrator. But, in fact there is little to his claim, as he has no party representation in the legislative organ, and he himself is not elected from its ranks. This is an anachronism carried over from the recent colonial past yet, inadvertently perhaps, copied into the Basic Law. His problem is further compounded by the reality of a fragmented, but assertive, Legco after the 1998 election and, in addition, by an insulated bureaucracy not disposed to defending the Chief Executive before the legislative interrogators. A large part of the reason for the latter anomaly, besides, is that all but one of the cabinet members are holdovers from the Patten administration and, hence, do not owe their positions to the new Chief Executive.

While the Legco is unicameral, it in effect consists of two houses, one of popularly elected local-issue advocates representing geographical constituencies, and another of business-oriented individuals returned by functional groups. Only 20 of the 60 seats are open to popular vote until 2007. Many assumed that the arrangement, sanctified by the Basic Law, ensured business and wealthy elites, including the civil service, a built-in dominance over Hong Kong politics. Likewise, many assumed that the Chief Executive should have sway in the Legco, more especially if one followed the Basic Law literally. The hard reality is, while all but one of the heads of departments are civil servants, they are not known to lobby for the Chief Executive in the legislature the way the United States cabinet, for example, does for the President (DeGolyer 1998, p.47). In Britain, Cabinet members are always the solid supporters of the government because they are themselves members of the majority party in parliament, whose leader is the Prime Minister.

The lack of a ministerial system in Hong Kong, consisting of appointees of the Chief Executive and serving during his pleasure, is a legacy inherited from the colonial past. The Pattenesqe reform, in retrospect, inadvertently left behind a bizarre situation in which the SAR's Chief Executive, in effect, may very well be holding the bag almost all by himself in the Legco. The holdover high officials in the government apparently felt no particular obligation to defend his policies in the legislature. The current practice thus born of the peculiar circumstances from the past may set the tone for the future politics in the SAR, unless rectified by amending the Basic Law, which would have to go through an elaborate but cumbersome procedure.

Stemming from the same origin of circumstances during the run-up to the end of colonial rule, there is one additional identifiable peculiarity in the executive-legislative relationship, as codified in the Basic Law. While the Chief Executive has the power to dissolve the Legco (Art. 50), the latter has no statutory power to pass a "no confidence" motion against him. In parliamentary democracies, the symmetric distribution of the power to pass a "no confidence" motion by the legislature and the power of the Prime Minister to dissolve parliament is a necessary guarantee of check and balance between the two branches. If anything, it would ensure that neither will abuse its power. The Basic Law, in effect, confirms the *modus operandi* of the colonial past, in which the Governor could in effect "dissolve" the Legco without a "no confidence" vote, since he had absolute power to appoint and reappoint its members at will. And, the latter had no right to pass a "no confidence" motion against the Governor, who had appointed them in the first place.

In another sense, however, the Basic Law takes a giant step forward from colonial practice. Article 73(9) confers on the Legco the power to pass a motion of "impeachment" or a motion for investigation, which will be conducted by an independent committee to be headed by the Chief Justice of the Court of Final Appeal. The Basic Law, nevertheless, does not define the criteria for either a motion to impeach or the actual conduct of the impeachment proceedings. During the impeachment of President Bill Clinton in 1998, the criterion of "high crimes and other misdemeanors" specified in the United States constitution was found vague and even unhelpful. By comparison, the wording "serious breach of law or dereliction of duty" used in the Basic Law, as a cause justifying a motion to impeach the Chief Executive, is arguably even more vague and confusing. Since impeachment is of far more grave consequence than a "no confidence" vote, Legco members may feel intimidated to make use of this broad power that Article 73(9) confers on them, although the reverse may also be true. Conversely, as the Legco's power to impeach the Chief Executive is so loosely defined, the assurance that legislators will not succumb to the temptation to abuse their power, it seems, would largely depend on self-restraint.

Another feature incorporated into the Basic Law, but inconsistent with the general practice in parliamentary democracies, is that while unlike his British predecessors the SAR Chief Executive is no longer head of the Legco, he does not seem to have a clear-cut statutory power to initiate bills. The Basic Law confers on his government the collective right to "draft and introduce bills, motions, and subordinate legislation" (Art. 62(5)), and requires the Chief Executive to consult Exco "before . . . introducing bills to the Legislative Council" (Art 56). But, more specifically, the law provides that "[m]embers of the Legislative Council of the Hong Kong Special Administrative Region may introduce bills . . ." (Art. 74). The Chief Executive, in addition, has the power to give consent to bills "relating to government policies" (Art. 74) prior to their introduction by legislators in the Legco.

So, even when the Chief Executive's power of introducing bills can be deduced from the more generic provisions of the Basic Law, and even as it can be exercised by his government collectively (as per Art. 62(5)), the fact remains that the Chief Executive's power over legislation is a far cry from that which all previous British governors had enjoyed until 1991. Until then, the British Governor personally headed the Legco and appointed its members, who would initiate bills on his behalf. It is apparent that the vague language used in the Basic Law concerning the new Chief Executive's function in initiating bills in a Legco that he no longer heads—and whose members he no longer appoints—is an oversight precisely because this important change (reduction) in the Chief Executive's power vis-à-vis the Legco, as compared with the British Governor before, was not realistically taken into account. While future practice may change the course of political development in Hong Kong, the system it has inherited from the British post-1984 reform poses a challenge to the ingenuity of the Chief Executive. Until modified, the executive-legislative relationship described above represents an incredible departure from the typical model of what a parliamentary democracy should be and, for that matter, what the drafters of the Basic Law would like to see. These faults, however, are internal to the SAR government and not incorrigible, and, more important, in no way do they affect the SAR's relationship with its sovereign, as we shall see below.

Unity in Diversity in the "One Country, Two Systems" Model

The chapters above have, in various ways, demonstrated how the model works and, equally important, how it circumscribes the workings and functions of the SAR government. Here, at a higher level of abstraction, I would like to point out four different characteristics that assure a rare "unity in diversity" and, moreover, mark the model as unprecedented in human history, even without regard to its duplicability beyond Macau in the larger process of Chinese unification. I would first name these characteristics before discussing them separately: (a) an inherent superordinate-subordinate relationship between the two systems—one capitalist and one socialist—in the symbiotic structure encompassing Hong Kong and China proper; (b) integration and mutual non-intervention between the two systems; (c) compatibility and consistency; and (d) complementarity.

(a) Superordinate-subordinate relationship. In the most simplistic terms, this relationship is seen in the fact that the SAR itself is a creation per Article 31 of the PRC constitution. According to Article 2 of the Basic Law, the high degree of autonomy the SAR enjoys is a grant by the National People's Congress in Beijing. The same principle governs the final appointment by the Central Government of the SAR's Chief Executive after his selection through consultations held locally. The SAR government, by authorization, may "conduct external affairs" in accordance with the Basic Law (Art. 61). The Chief Executive, besides, is expected to implement the directives issued by the Central Government "in respect of relevant matters provided for in [the Basic] Law" (Art. 48(8)). While enjoying a high degree of autonomy, as Article 12 of the Basic Law clearly stipulates, the SAR is a "local administrative region" of the PRC and that it is "directly under the Central People's Government." The power of interpreting the Basic Law, the SAR's constitution, which was itself a grant by the National People's Congress (NPC) in Beijing, is vested in the NPC's Standing Committee (Art. 158). This touchy point was tested in the controversy surrounding the landmark case, *Chan Kam-gna & 80 Others v. Director of Immigration* (Final Appeal No. 13 of 1998), decided on 29 January 1999 by the SAR's Court of Final Appeal.[20]

(b) Integration and mutual non-intervention between the two systems in the symbiosis. Nothing bespeaks this principle more explicitly than the legal requirement that the Chief Executive, principal officials, Exco and Legco members, and judges of courts at all levels in the territory swear their allegiance to the HKSAR (Art. 104, Basic Law), as distinct from its Chinese sovereign, whose name is not even mentioned in Article 104. In respect of non-intervention, the sovereign shall not levy taxes in the SAR (Art. 106); and the SAR shall, on its own, formulate its own economic, monetary and financial policies and run its own economy (Chapter V, Basic Law). By the same token, the SAR shall have autonomous jurisdiction over education, science, culture, sports, religion, etc. (Chapter VI, Basic Law). In more fundamental terms, the SAR's autonomy encompasses executive, legislative and independent judicial power. The sovereign shall not interfere in the domestic affairs of the SAR, any more than shall the latter interfere in the internal affairs of the former.

(c) Compatibility and consistency. Despite the contrast between the socialist and capitalist systems embraced respectively by the sovereign and the SAR, and despite the non-intervention by either into the other, the two systems, while separate, exist in harmony. Notwithstanding Articles 5 and 6 of the PRC constitution, which provide for a socialist economic system and a socialist legal order for China proper, the Basic Law confirms the guaranteed continuance, for at least 50 years, of Hong Kong's preexisting capitalist system (Art. 5) and provides for the protection of "private ownership of property" (Art. 6). Both systems, nevertheless, coexist under Chinese sovereignty (Preamble, Basic Law).

At the empirical level, compatibility may find evidence in the administration of justice. In a concrete case involving the trial of Zhang Ziqiang, nicknamed "Big Spender," a big-time criminal sought by the SAR for sensational kidnapping and other crimes but taking refuge in the neighboring Chinese metropolis of Guangzhou (Canton), the Guangzhou Court took it upon itself to prosecute and sentence him to death after he was apprehended by Guangzhou police. The case, which involved co-jurisdiction of both Hong Kong and Mainland courts at best, and possible conflict of jurisdictions at worst, is an example of compatibility par excellence, for a number of reasons. First, although the Hong Kong court had co-jurisdiction both by the defendant's residence and the locality where the crimes were committed, the Guangzhou court had the first chance of trying him because he was apprehended in that city. It resolved the ticklish question of legality versus practicality. Secondly, many in Hong Kong, including jurists, expressed in private a preference that Zhang be tried by a Mainland court, in order to serve notice to other criminals who might be tempted to hide away in China proper after having committed crimes in the SAR. The trial of the Big Spender in Guangzhou would thus have a strong deterrent effect. And, thirdly, since the criminal law in China proper, unlike that in Hong Kong, provides for the death sentence for capital offenses, the trial and sentencing of the Big Spender in a Mainland Chinese city would send a loud and effective message to all would-be criminals who might otherwise be tempted to follow Big Spender's footsteps to Mainland hideouts.

Despite the wholesale differences between the two systems, the Basic Law takes care to ensure that they are consistent within the "one country" context. For instance, while the SAR enjoys its juridical independence, the laws enacted by the SAR Legco must be "reported" to the NPC Standing Committee "for the record," and the reporting "shall not affect the entry into force of such laws" (Art. 17, Para. 2). If, however, the NPC-SC should find any law enacted by the Legco to be "not in conformity with the provisions of [the Basic] Law regarding affairs [falling] within the responsibility of the Central Authorities or regarding the relationship between the Central Authorities and the Region, the Standing Committee may return the law in question but shall not amend it" (Art. 17, Para. 3). This provision, while it indicates the sovereign's superordinate power over the SAR, is necessary to ensure that the SAR's new laws be compatible and consistent with both the Basic Law and the wills of the sovereign. In reverse, however, other than the fixed number of national laws enacted by the NPC and listed in Appendix III to the Basic Law, the sovereign's laws shall not be applied in the territory of the Hong Kong SAR.[21]

The care with which the compatibility of the laws between the two systems is maintained, as such, adds to the unity in diversity that pervades the symbiotic structure. In addition, according to Article 7 of the Basic Law, the land and natural resources within the Hong Kong SAR shall be "State property," meaning the national assets of the Chinese nation. While the "revenues derived therefrom shall be exclusively at the disposal of the [SAR] government," the intent of Article 7 is to ensure that there shall be one *guo* (country, state, and nation rolled in one), despite the coexisting *liang zhi* (two systems).

(d) Complementarity between the two systems within the contours of the "one country" framework. A most concrete illustration of this principle is in the Basic Law's provisions on air service agreements. For example, for the provision of air services between Hong Kong and other parts of China involving airlines incorporated in the SAR and having their principal place of business in it, on the one hand, and other airlines of the PRC, on the other, the Central Government shall, in consultation with the SAR government, make the necessary arrangements (Art. 131, Basic Law). Emphasis here is on "in consultation with the Government of the HKSAR." The requirement for consultation is to ensure that no air service agreements violative of the SAR's interests will be made.

With authorization from the Central Government, the SAR government may (a) renew or amend an air service agreement or arrangement previously in force; (b) negotiate and conclude new air service agreements providing routes for Hong Kong airlines and involving rights of over-flights and technical stops, and (c) negotiate and conclude provisional arrangements with foreign states or regions with which no air service agreements exist (Art. 133).

While the coexistence of two different systems in the symbiosis would otherwise create two separate jurisdictions, the complementarity of the two jurisdictions and the care with which it is ensured, both by the Basic Law and by the sovereign's policy is impressive. More concrete examples, it is safe to assume, will unfold as the SAR's existence gains more time.

An Evaluation of the SAR's Experience Thus Far

In sum, the experience of the SAR, as summed up here and discussed in the various chapters above, we believe, can be expected to continue beyond the immediate period of post-handover transition. A number of reasons support this optimism. First, the "one country, two systems" model has proved to be working for both the interests of Hong Kong and Beijing; and this fact by itself offers a disincentive for either side, especially the latter, to want to cop out in search of an alternative *modus operandi*. Secondly, in the first two years of the SAR's existence following the British departure, Beijing has proven its ability to overcome the temptation to meddle in Hong Kong's internal affairs. This record lends credence to China's pledge, both given in the Sino-U.K. Joint Declaration of 1984 and written into the Basic Law, that the preexisting systems and way of life in Hong Kong shall not be changed for 50 years beyond the 1997 demarcation line. Thirdly, as Deng Xiaoping, the model's architect, had insisted, "one country, two systems" was ultimately designed for Taiwan in connection with China's bid to reunite with the island, a position that has since been echoed by President Jiang Zemin and other post-Deng Chinese leaders. The strategic imperative of offering the Hong Kong SAR as a show window requires the continuance of the model's success for as long as the ultimate process of winning Taiwan's return is not complete. This requirement offers a strong incentive for Beijing to continue its hands-off stance toward Hong Kong, to the extent it is the one most decisive way to assure the model's continuing success.

A few words, in this connection, about the duplicability (and its limitation) of the "one country, two systems" are in order. There is no question that what has happened in Hong Kong is going to be duplicated in Macau upon the latter's return to Chinese sovereignty in December 1999. Both former colonies share a commonality in their respective retrocession to their former motherland, instead of gaining independence, following the end of colonial rule. A crucial difference is in the contrasting lack of a functioning civil service made up of ethnic Chinese in Macau, which may cast doubt on the future Macau SAR's ability to govern itself (see Bolong Liu's chapter). However, on the other hand, Macau may have an advantage over Hong Kong in the way it exits from colonial rule. Unlike the British, the Portuguese are not known to have made last-minute changes in the name of democracy on the eve of Macau's reversion. That means, among other things, that the kind of handicaps that, as noted above, have confronted Chief Executive Tung Chee-hwa in Hong Kong are not in the cards for his Macau SAR counterpart. The lack of a Portuguese-Chinese controversy has spared Macau of the same kind of prophecies of doom heralding Hong Kong's reversion. This is a plus for Macau.

However, whether the "one country, two systems" model is similarly duplicable for Taiwan, the last piece in the jigsaw puzzle of Chinese unification,[22] is not so clear. In the first place, Taiwan is not ruled by a colonial power but by a government that owes its existence to popular will as expressed through elections. Hence, the perceptions of Taiwan's population can be reasonably expected to be very different from both Hong Kong and Macau. In the latter two cases, the key to retrocession lay in agreements reached by Beijing with their respective colonial rulers. The populace in either case was, thus, at the receiving end of decisions already made on their behalf. Besides, the ruling Governor, in either case, had no choice but to accept the deal his home government had already struck, regardless of his personal preferences or egoistic ends. In the case of Taiwan, by contrast, the aspirations of the commonfolk may have to be reckoned with.[23] Likewise, the elected political leaders may also have their vested interests and pride to defend. Unlike the last Governors in Hong Kong and Macau, the elected government in Taipei has the option of rejecting Beijing's offer of the "one country, two systems" solution to Taiwan's reunion with the Mainland. In fact, Taipei has repeatedly rejected the offer.

Even assuming all the procedural problems can be surmounted, Taiwan may nonetheless object to the "one country, two systems" solution on substantive grounds. In the discussion on "diversity in unity," above, the first feature found to seal Hong Kong's successful integration into Greater China is the principle of superordination-subordination. As such, what has worked for Hong Kong is precisely the hurdle for Taiwan, whose government leaders have no compelling reason to want to be reduced to being heads of a local government after the fashion of the Hong Kong SAR.

However, the reverse should not be ruled out, either. The continuing success of Hong Kong as a Special Administrative Region of China, if stretched out over years, not just months, may create a favorable world opinion and make its "one country, two systems" experiment a tantalizing, credible example for resolving the remaining Taiwan piece of the puzzle. If repeated reports—including the annual reports of the U.S. Secretary of State to Congress, as required by law—consistently confirm Hong Kong's success story, and more especially if that success story is widely known to result from its "one country, two systems" model, the cumulative effect might even sway the last holdouts in U.S. Congress supporting Taiwan's continued split from the Mainland. Assuming those circumstances are conducive to the process toward the final peaceful resolution of the division across the Taiwan Strait, it remains to be seen whether Taiwan will reconcile itself to accepting a relationship of subordination to Beijing as the Central Government, something Taipei now swears it cannot accept. Once this hurdle is cleared, the other three features we have identified above as decisive in generating the "unity in diversity" pervading the Hong Kong SAR's relationship with China proper will likewise assure the smooth coexistence between Taiwan and the Mainland within an ultimate Greater China framework that will have resulted from their final reunion.

The seeming tentative tones of this endorsement of the duplicability of the "one country, two systems" model for Taiwan does not imply any doubts as to the suitability of the model for the Hong Kong SAR. It is only that the totally different circumstances surrounding the Taiwan case warrant caution, which does not reflect on the merit of the model itself. By the same token, caution also requires that we not jump to the conclusion that the Hong Kong SAR's claim of being a bellwether of history is to be solely based on its successful demonstration of the workability of its symbiotic governance model and hence its suitability for Taiwan. Below, I shall come back to the question whether the "one country, two systems" model may serve as a harbinger in other cases in the world, outside the Mainland-Taiwan context.

Here, let us take note of how Hong Kongers view the question of whether Taiwan should embrace the SAR's "one country, two systems" model. In a survey conducted after the first anniversary of Hong Kong's reversion, 57.4 percent of those polled indicated their support, ranging from "strong" to "very strong," for the "one country, two systems" model to be applied to Taiwan. Some 20.4 percent, nevertheless, took the opposite view even if the model has proven to be working in Hong Kong. Only 1.9 percent, though, said Taiwan should go separatist. While 39.4 percent supported maintaining the status quo across the Taiwan Strait, the number was considerably below the 57.4 percent figure in favor of Taiwan's accepting the "one country, two systems" model as a solution to ending the cross-Strait division (Wong 1998, pp.8–9).

Before we proceed further to explore the duplicability of Hong Kong's experience elsewhere, I would pause to examine the long-term viability of the HKSAR as an exemplification of "one country, two systems" with potentials for further growth into a full-fledged democracy. This done, we shall be in a better position to ascertain the wider relevance of the SAR's "one country, two systems" as a formula for resolving conflicts due to irredentism and/or contested secessionism in other parts of the world. I shall outline, for illustration, a number of things that the SAR can do or should keep a lookout against in the short to medium term. Editorializing is not our intent; every option listed below is compelled by logic, as illuminated by the findings of this book.

Prospects for the HKSAR beyond Its Initial Success?

There is no dearth of advice on what is in store for Hong Kong or what the SAR government should do for a better future.[24] It is also true that people in many quarters, including foreign governments and the international business community, would love to know more about Hong Kong's future well beyond its initial success after the 1997 handover. I am, however, only going to outline the necessary preconditions that must be in place for the Hong Kong SAR to enjoy continued stability and to see further development toward a fuller democracy. In a real sense, they are also indispensable to the territory's economic future, as they bear on the confidence of external investors and businesspeople. I would urge that these points not be lightly discounted simply because of their brevity.

In the first place, it bears remembering, as we have shown, that Hong Kong's post-handover success thus far is, in large measure, owed to the hands-off policy of its Chinese sovereign. To sustain the success story, therefore, it is imperative that Beijing continue, and, equally important, be encouraged to continue, the same policy. But, here is the catch. Other than the Taiwan factor (i.e., the value of a Hong Kong show window for Taiwan), another decisive reason for Beijing's non-interventionism thus far is that, despite the plethora of demonstrations against China, often on the human rights issue, Hong Kong after reversion has not turned out to be a bastion of anti-Communist activities as was feared by some in Beijing, such as to warrant intervention by the Central Government. In a large sense, the credit should go to everyone in the territory, including even the demonstrators, who apparently acted within the bounds of the legally acceptable. Part of the credit should also go to the SAR government, which showed great prudence and finesse in dealing with challenge. For instance, the police resorted to the playing of Beethoven in an attempt to soothe the nerves as well as to drown out the noise of the demonstrators, instead of forcibly dispersing or arresting them, during President Jiang Zemin's visit to Hong Kong on 1 July 1998, the first anniversary of its reversion.

Critics have made hay of Article 23 of the Basic Law, which requires that the HKSAR "enact laws on its own to prohibit any act of treason, secession, sedition, subversion against the Central People's Government." These anti-subversion laws would also be aimed at "theft of state secrets," foreign political organizations conducting "political activities" in Hong Kong, and local political organizations maintaining illicit liaison with foreign political organizations, presumably seeking to subvert the Chinese government (Art. 23). Despite the jitters this Basic Law provision had given to many bona fide democracy advocates before 1997, the fact remains that at the time of this writing, two years after the handover, the SAR government is in no hurry to introduce such bills in the Legco. Nor is Beijing known to be in such a hurry. The reason is that despite the worst fears of the NPC, which enacted the Basic Law in 1990—that Hong Kong might fall prey to purveyors of terrorism and subversion against its Chinese sovereign—nothing has taken place that would even come close to it in real life. It is only logical that, to ensure the continuance of Beijing's hands-off policy, people in the SAR owe it to themselves not to allow this situation to be reversed. And, the world at large should respect this ticklish but crucial point.

Secondly, Hong Kong after reversion can function efficaciously only if it learns how to adjust itself to being no longer an "administrative state," which it once was under colonial rule. It was called an "administrative state," because the bureaucracy in coalition with the business community always dominated public policy-making (Harris 1976). While the Chief Executive has not inherited the nearly absolute power of the British Governor, and while the Legco enjoys more clout than at any time during colonial rule, the fact remains that high officials, most of them holdovers from the Patten era, are inseparable from the civil service they hailed from. In the SAR government, these high officials make policies and run the show as their predecessors used to before. But, they are too often above accountability like never before—not even for the fiasco of the Chek Lap Kwok Airport of July 1998. Besides, unlike cabinet members elsewhere, these high officials in the SAR government are not known to be keen on defending the Chief Executive's policies before the legislature, as already noted. During the colonial era, the Governor could make his high officials accountable to him, and he had no need for vicarious defense by his cabinet before a docile Legco, whose members, like the officials, owed their jobs to his appointment.

All this has changed, thanks to the post-treaty British reform. The SAR government has yet to figure out how to deal with the anomalies spawned by the anachronistic parts of the British legacy, as also noted above. One dilemma faced by the Chief Executive is that to redress the anachronisms, he could face charges of tampering with an inherited system supposedly not alterable for 50 years. But, without redressing them, as we have pointed out, the Chief Executive may find himself often holding the bag all by himself, in dealing with an assertive Legco that finds no precedent during the colonial era.

It is plain that the existing executive-legislative relationship, as we have seen, has the potential for trouble unless the imbalance in effect is redressed. Likewise, the tradition of inducting cabinet members (high officials) from the civil service has to be reexamined, if the government is going to function efficiently, which in turn requires, *a priori*, an accountability system. As the investigations of the new Chek Lap Kwok Airport fiasco have shown, failures by high officials also damage the credibility and long-standing reputation of the civil service. But, these much-needed adjustments will not be easy unless the Chief Executive can overcome the stigma of changing the existing system before the 50-year moratorium is up, despite the fact that the system inherited in 1997 was not what had existed in 1984, when the U.K.-Chinese agreement was signed on Hong Kong's return. Thus, a precondition for the continued smooth functioning of the SAR government is, *a priori*, a recognition by the public and the world media that making the kind of necessary adjustments spoken of here, to the extent they do not violate the Basic Law and the UK-Chinese Joint Declaration of 1984, does not constitute structural alteration of the existing institutions. This is all the more so, if the necessary adjustments apply to post-1984 changes made by the last two British Governors which, aside from unilaterally tampering with the "current" system as of 1984, have turned out to be anachronistic and inconsistent with the Basic Law and violative of the Joint Declaration.

Thirdly, despite the general "unity in diversity" prevailing in the SAR's relationship with China proper, there are admittedly crevices or loopholes that lend themselves to exploitation by the unscrupulous. One example is the scandal about a ship belonging to Pan Nation Shipping. The ship was boarded by PRC customs officers while it claimed to be in Hong Kong waters. But, the Hong Kong police, arriving afterwards in response to calls for help, estimated that the ship was actually in Mainland waters, hence properly subject to PRC jurisdiction, not Hong Kong's.[25] The incident simply illustrates a potential source of friction, either when the proper jurisdiction is not readily ascertainable or if a claimant should claim to be in one jurisdiction while in fact he/she is in another. We noted the Big Spender case before. It is imperative, in view of just these two cases alone, that rules and procedures be worked out between the SAR and Chinese authorities so that in the event of a similar fuzzy case or even possible conflict of jurisdiction, a speedy answer and solution can be readily found. This condition is necessary for the assurance of a continuing peaceable and cooperative relationship between the SAR and its sovereign over the long term.[26]

Fourthly, the SAR must not, out of expediency or good will, relinquish its own legal-judicial responsibility to Beijing, under normal circumstances. However, equally important, in making a decision regarding the interpretation of the Basic Law, care must be taken so that Hong Kong courts do not contravene, or appear to contravene, the authority of the National People's Congress (NPC), which has the ultimate power to interpret the Basic Law, if in doubt, and even to modify that Law. For example, a controversy surrounded the question of whether the SAR's judges should, as the SAR government had initially suggested, refer issues like the right of abode to the National People's Congress Standing Committee (NPC-SC) for a definitive answer.[27] The Basic Law confers on the SAR an "independent judicial power, including that of final adjudication," which means that the SAR's courts "have jurisdiction over all cases in the Region" (Art. 19). The determination of such legal questions as the right of abode for immigrants of questionable legal status, a technical and relatively narrow issue almost exclusively concerning Hong Kong, should properly fall within the jurisdiction of Hong Kong courts. Referral to the NPC-SC, as such, would be a self-abnegation, an unnecessary surrender of judicial power by the SAR to the NPC. It would make a bad precedent. Under the common law doctrine of estoppel, such an act of surrender would deprive the SAR of its proper right to exercise judicial interpretation of similar legal questions in the future, if the surrender leads to *reliance* by the NPC that the SAR does, by its referral, intend to signify relinquishment of its judicial power of interpretation. Rather than letting that happen, possibly leading only to more future disputes, it is imperative that the SAR government is on guard not to yield the rights and power the Basic Law grants to it, including judicial independence. Again, this is a necessary precondition for sustaining the smooth give-and-take between the two polities joined together by the "one country, two systems" formula.

The SAR Court of Final Appeal made an epoch-making decision, on 29 January 1999,[28] when it granted the right of abode to thousands of young immigrants—including children born out of wedlock on the Mainland to

Hong Kong fathers—whose rights had been in limbo following restrictions clamped down by the NPC on migration from the Mainland within days of the handover. Much controversy has surrounded the decision. The background of the dispute was about the application of Article 24 of the Law regarding the "right of abode," or who has the right to immigrate and reside in the SAR. The now defunct Preparatory Committee, at the time of the handover had urged a strict construction of Article 24 in a report to the NPC, which would deny the right of abode to children born out of wedlock on the Mainland to Hong Kong fathers. In terms of legal niceties, the ultimate question boils down to whether the NPC's approval of the said report constitutes an act of the national legislature, whose wishes no SAR court can contravene.[29]

Substantively, the Court's decision opened the door to a consequence that the SAR may not be able to live with, for it paved the way for the influx of an estimated 600,000 people waiting to crash the gate into Hong Kong,[30] which was later revised to over 1.67 million. Immediately, the decision created jitters for the SAR government, and many in private quarters worried about the strains on housing and other social services. Procedurally, however, the Court's move may set a precedent that Hong Kong courts have autonomous jurisdiction in matters of concern to the territory. In effect, if allowed to stand, the Court's decision laid down the crucial principle that Hong Kong courts can even diverge from acts of the NPC if they breach the Basic Law in its present form, as interpreted by no other than these local courts. It would mean that interpretation of the Basic Law is to be carried out ultimately by the SAR's Court of Final Appeal, not the NPC, if and only if the main issue in a case is within Hong Kong's autonomy—unless, of course, the Court should opt to canvass the NPC's views on its own accord.

The SAR government's claim that migrants in the territory were subject to Mainland emigration laws was found by the Court of Final Appeal to be "untenable." The Court, it seemed, gave a most liberal construction to Article 158 of the Basic Law, which allows Hong Kong courts the maximum freedom in interpreting the Law at the expense of the NPC.

Beijing's reaction was mixed. Officially, a spokesman for the Foreign Ministry swiftly commented that, since the decision concerned an issue—that of immigration—that according to the Basic Law fell properly within the jurisdiction of the Hong Kong SAR, the Central Government would therefore not interfere.[31] On the other hand, some legal experts in Mainland China reacted differently. Their criticisms, carried by the official Xinhua agency on 7 February 1999, were that the Hong Kong Court's decision contravened the power of the National People's Congress. In Hong Kong, on the other hand, most legal experts came to the defense of the Court's move. Local experts, including some former Basic Law drafters, hailed the Court's ruling as a decisive move in maintaining the integrity of the "two systems," in both essence and spirit.[32]

One day later, however, official Beijing reversed itself. Zhao Qizheng, head of the State Council's information office, said the Central Government endorsed the criticisms by Mainland experts. At an appropriate time, he said, the government will express its opinion, adding: "The [Hong Kong] Court's decision is a mistake and against the Basic Law."[33] While it was odd that someone from the State Council, rather than the NPC-SC, should have come out with such a statement, two things are clear. First, this was the first time since the handover in July 1997 that Beijing came out openly criticizing the SAR (or its judiciary, in this instance). Secondly, and more important, the controversy confirms our point that, while the Hong Kong judiciary should not surrender its independent judicial power, it should also exercise due diligence not to contravene, or give the appearance of contravening, the NPC's wishes. For the NPC Standing Committee has the ultimate power over the Law's interpretation, and the NPC is also vested with the power to amend the Law if necessary (Art. 158). This dictum (i.e., judicial independence without overstepping the proper bounds), in the final analysis, is imperative for the continuance of the smooth functioning of "one country, two systems" for the benefit of the SAR.

Hong Kong's Court of Final Appeal may make independent decisions according to law, and in cases exclusively concerning the SAR's autonomy, the Court may diverge from the wishes of the NPC. But, as we have learned from the *Chan Kam-nga* case, which in the NPC Standing Committee's view simultaneously involved the Mainland's emigration law, the divergent part of the ruling by the SAR's Court of Final Appeal could be treated as *obiter dictum* as a way out. In layperson's language, an *obiter dictum,* in the common law tradition, offers counsels of value for possible judicial cognizance, even though, like dissenting opinions, it is not controlling or directly binding for the case on hand. This was the gist of a resolution agreement finally reached after Secretary of Justice Elsie Leung's trip to Beijing.[34] I would suggest that this *obiter dictum* solution means that the NPC-SC's ultimate authority as stipulated in Art. 158 is upheld, while the ruling by the Hong Kong Court of Final Appeal, in the *Chan Kam-nga* case, was allowed to stand. The Court's subsequent clarification, in response to a peculiar application by the SAR government, to the effect that its jurisdiction to interpret the Basic Law in adjudicating cases was "derived by authorization from the Standing Committee [of the NPC] under Articles 158(2) and 158(3)" of the Basic Law. The Court's original judgment in the disputed case, it continued, "did not question the authority of the Standing Committee [of the NPC] to make an interpretation" of the Basic Law, as pursuant to Article 158, which "would have to be followed by the courts" of the Hong Kong SAR.[35] Despite some raised eyebrows both locally and abroad, including in the U.S. Congress,[36] this seems to me the only sensible way of resolving a deceptively simple question concerning the SAR's judicial independence, which, at its very root, is whether a Hong Kong court has the statutory power to challenge the wills and wishes of its sovereign, as expressed through its parliament.

In the end, the SAR's judicial independence was not compromised by what in retrospect was a storm in a teacup, which could have been avoided. The lesson, nevertheless, should not be lost if the SAR is to continue to live in peace with its sovereign. In Chapter 4, Daniel Fung, whose term as Hong Kong's Solicitor General straddled both the last years of British rule and the first year and a half of the HKSAR, has more to offer on the paradoxes on the legal and judicial dimension in the SAR's experience thus far.

Fifthly, as a sequel to the above point, courts of Hong Kong should be mindful of the crucial fact that in the "one country, two systems" design, two strictures are equally important: (a) that there are two jurisdictions in the symbiotic model, and (b) that both jurisdictions are within the "one country" that is built on the predicate of a "unitary system." Simply put, without the "one country," there cannot be the "two systems." I do not say this lightly, for it seems the SAR's courts may not be always able to overcome the temptation of pronouncing their independence at the expense of Hong Kong's sovereign. One more recent case was the controversial ruling by the Court of Appeal, dated 23 March 1999, that burning or defacing the HKSAR's *and* China's national flags was no criminal offense.[37] On the surface, the appellate court was hardly on assailable ground when its ruling, written by Mr. Justice Michael Stuart-Moore (under Art. 92 of the Basic Law, Hong Kong SAR courts may have non-Chinese judges recruited from "other common-law jurisdictions" as well), invoked the freedom of expression guaranteed by the Basic Law and by the International Covenant on Civil and Political Rights (ICCPR).[38] For the same reason, the ruling attacked the constitutionality of the laws that made the burning of regional and national flags a criminal offense. I do not question the particular part of the Court's ruling that pertained to the burning of the SAR's regional flag. But, I am concerned about the sweep of the ruling as it touched on a prerogative of the sovereign (the legality of burning the Chinese national flag in the SAR). I wish to emphasize that the Court apparently succumbed to the temptation of asserting its judicial independence at the sovereign's expense. The question concerning the legality of burning China's national flag, it is plain to any reasonable person, should properly belong to the sovereign's jurisdiction. Analogizing from the American experience may, I hope, serve to illustrate this point, *mutatis mutandis* (allowing for necessary differences). In the United States judicial system, a state court would invariably shun jurisdiction, and defer to the federal courts, in a case that clearly involves a "federal question."[39] To translate it into our language as regards the Hong Kong-China relationship ("one country, two systems"), a local Hong Kong court should refrain from pronouncing itself on an issue that clearly falls within the bailiwick of its sovereign, more especially if the latter has laws governing the issue on hand. To put it another way, the question regarding the legality of Hong Kong residents (within the SAR's territory) defiling China's national flag is not an issue about which an SAR court should arrogate jurisdiction to itself. China's unitary system, it bears noting, makes this judicial self-restraint by local SAR courts all the more crucial than in a federal system such as that of the United States. Moreover, the question of whether or not China's own laws making flag-burning a crime are valid is not for a local Hong Kong Court to decide. They may be unconstitutional, abominable, even offensive. But, the point is that a local Hong Kong Court is the wrong forum to pass a judgment thereon, since they are "national laws," especially if they are among those incorporated into the Basic Law, as per its Annex III. At issue is a sensitivity absolutely required for the sustained workability of the "one China, two systems" governance structure, which although unprecedented in human history, as explained, does not require a prior firm grasp of Chinese political culture to fully comprehend. Instead, the question can be dealt with by using language familiar to a non-Chinese speaking judge, although it may require a fundamental knowledge of a little bit of modern Western political thought. One may begin by reference to the Hegelian formulation of "state and civil society." While Hegel made much of the state-civil society distinction, it is worth recalling that before Hegel, "state and civil society" was considered together as on the opposite end of natural law, as Neocleous (1996, p.1) notes. I would urge that, in thinking about the "one country, two systems" model, state and civil society not be viewed either in the Hegelian (opposing) sense or in the pre-Hegelian (concomitant) sense. Instead, they should be viewed as vertically complementary in this symbiotic governance structure, in the sense that one is superordinate to the other. The "state" in this instance is the sovereign, under which the "two systems" are subsumed, and which is endowed with a necessary *constitutive power*[40] that both justifies the "one country" and sustains the "two systems." The paradigm, while different, can be better understood by analogizing from what prevailed in Western political thought before Hegel, when in practice, it meant that to be a member of civil society was to be a member of the state.[41] This dual membership smacks of the rationale of the "one country, two systems" model as it applies to Hong Kong's relationship with its sovereign. I would simply add that, as a derivative of the fact that the SAR is an inalienable part of its Chinese Sovereign, it is incumbent upon the SAR's courts not to try to take on the sovereign and pass on the legality of its acts and stakes, if merely to demonstrate their judicial independence, however big the temptation to do so may be. To do otherwise merely makes a bad case of ultra vires.

The bottom line here is that, while Hong Kong courts should be encouraged to maintain their judicial independence and integrity vis-à-vis their sovereign, they should likewise exercise judicial self-restraint from transgressing into territories that properly fall under the sovereign's jurisdiction. To use Deng Xiaoping's pithy language, while "river water" (China) does not cross "well water" (Hong Kong), it is equally incumbent upon "well water" not to challenge "river water" without provocation. Now that Beijing has shown great restraint in its hands-off policy toward the Hong Kong SAR, well into the second year of the latter's existence, it is incumbent upon the SAR's courts to be on constant guard, lest they succumb to the temptation of encroaching upon the sovereign's territory, thereby risking the possibility of wrecking the "one country, two systems" dictum, as encroachments, however unwitting, are bound to invite counter-encroachments. To reiterate, whether "one country, two systems" will prove workable in the long run does not just depend on Beijing alone, as we have seen one more good reason why it is so. As Danny Paau poignantly points out in his chapter, SAR judges should refrain from trying to test the outside limit of Beijing's tolerance.

Sixth, now, well into the second year after reversion, it is imperative that Hong Kong's democratic advocates come to the realization that the political fault-lines are totally different. As we have noted, before 1997 these same advocates were allied with the British rulers in a common fight against a perceived common threat to the territory's political and legal tradition and democratic aspirations. They now should be able to work together with the SAR government and find peace with the formerly assumed bully, who has turned out to be a harmless patron (if not exactly a pussy cat) thus far. For the local political parties themselves, the loss of a specifically targeted bully, a rallying point for intramural unity before, has resulted in the erosion of their internal cohesion. This can be evidenced in the challenge posed by the faction led by legislator Lau Chin-shek (Liu Qianshi) to the leadership of Martin Lee within the Democratic Party, following the May 1998 election.[42] Similar feuds have plagued some of the others of the eight political parties in the Legco. The best hope for Hong Kong's democratization is, as a first step, to see a fully elected Legco materialize after 2007, as the Basic Law envisages. In order to have a well-greased modern legislature functioning normally, a precondition is that there be a sound system of political parties, with minimum cohesion within each of them. On the other hand, the future Chief Executive, who may have to contend with greater odds in the Legco due to more sophisticated party politics, is most likely to need his own party affiliation and support in the legislature.

Seventh, with the expectations for a fully elected Legco from geographical constituencies, doing away with the functional constituencies, there is also widespread expectation among the Hong Kong population that the Chief Executive will also be popularly elected.[43] While the Basic Law provides for an eventual popularly elected Legco, it does not similarly provide for a popularly elected Chief Executive. Thus, for that to happen, the Basic Law would have to be revised accordingly. Again, as was shown before, it probably will not be enough to have an elected Chief Executive who has no party representation in the Legco. In view of the unitary (as distinct from federal) system of China, in which tradition the "one country, two systems" concept is steeped, however, the introduction of a popularly elected Chief Executive might present a problem for the unitary principle. But, it is not irreconcilable. For example, a two-step arrangement, whereby the Chief Executive is first elected by the Hong Kong electorate at large and then affirmed by appointment from the sovereign, may offer a solution. In this case, the sovereign's appointment would be like the British monarch's *assent* to the result of parliament's decision, following an election, to make the majority party's leader the next Prime Minister.

Last, but not the least, in the interest of a full flowering of democracy in the Hong Kong SAR, a precondition is that the territory should have a proper civic culture supporting it. In order to have that happen, there must be, to paraphrase Alexis de Tocqueville, an "urge for democracy" on the part of the citizenry in general. This is something not to be taken for granted, considering the over 150 years of colonial rule, during which the commonfolk were encouraged to value, even worship, materialistic idols, not to mention indulge in corrupting practices such as concubinage, mahjong playing, etc. Suen Ming-yeung, Secretary for Constitutional Affairs of the SAR government, astounded a visiting U.S. Senate team when he remarked, in terms too crude though honest for politics, that the Hong Kong people did not really prize democracy as much as economic liberty, or the license and opportunity to make money.[44] But, by coincidence or not, in the same issue of the *South China Morning Post* that carried the news about Suen's apparent slip of tongue on 11 January 1999, a letter to the editor by a certain Frank Kwong explicitly stated that for him and his fellow Hong Kong brethren, economic liberty always came before democracy. If Kwong's view is representative of the Hong Kong populace in general, then the "revolution of rising expectations" regarding Hong Kong's democracy, whose creation we have ascribed to the departing British rulers, was probably largely limited to the elite, certainly the principals in the emergent political parties. Nowhere is this lingering lack of a civic culture supporting participatory democracy better demonstrated than the increasing difficulty experienced by the political parties, all of them having shallow roots, in trying to recruit young, new members. The under–30 age group cohorts are just not interested in politics, and all political parties face the harrying problem of trying to fill their depleting ranks.[45] While this trend confirms our speculation that the politicians of Martin Lee's generation were encouraged and egged on by the departing British after 1984, in the latter's post-treaty disputes with Beijing, it does not augur well for the democratization and self-governance for Hong Kong. A democracy without a civic culture supported at the grass-roots is like a plant, no matter how beautiful, grown on shallow soil. How to foster a popular "urge for democracy," at least a greater interest, especially among the young, in politics, is an urgent task for the SAR, requiring the systematic reeducation of the population, to reverse the effects of long colonial neglect. In addition, more or less can be said of the would-be politicians, who need a crash course on "civic competence," or the ability to articulate and aggregate political interests, to campaign and lobby for votes professionally in elections, to translate ideals and aggregated political interests into policy and legislations, and in sum to live up to the electorate's mandate, once elected into office. It is a precondition necessary for the future of Hong Kong's democratization beyond its present stability and successful post-colonial transition. Mere noise making about instant democratization or negative politicking against certain targeted officials of the SAR government (such as Elsie Leung, Secretary of Justice) or even bad-mouthing about the SAR or China abroad (including before the U.N.), as some Legco members are wont to do, is no substitute for adequate civic competence. It can only be a worrisome sign of the prevalence of a political culture deficient in "civic virtue."[46] Even worse, such practices by the incumbent politicians in the Legco, while demonstrating a bleak lack of civic competence, could unduly shake the world's confidence in the SAR's future. The fact that civic culture, civic competence, and civic virtue do not come automatically with any governance structure installed anywhere, such as the "one country, two systems" model in Hong Kong, is self-apparent and needs no further elaboration. I shall only add that all of it has to be nurtured purposively.

The International Relevance of the Hong Kong Model?

In this chapter we have raised the question whether the Hong Kong SAR is a prisoner of legacy or history's bellwether. We have followed a concededly convoluted discourse to reach a definitive answer, duly reflecting the complexity and, at times, fuzziness of the matter. It is easier to answer the first part of the question. In search of a truly efficient and democratic future for itself, the SAR, as we have demonstrated, has to free itself from some unwholesome parts of its inherited legacy from the past, the obscurity of which only makes the needed effort more difficult. Its success or failure in doing so would determine whether post-colonial Hong Kong can free itself from being a mere "prisoner of legacy." On the other hand, however, the SAR's laurels as "history's bellwether" are an elusive issue. As we have seen, there is no automaticity in the transferability to Taiwan of the "one country, two systems" solution, despite its successful experiment in Hong Kong. Such variables as Taiwan's volition and the international climate are beyond Hong Kong's control. Under the circumstances, its claim to being "history's bellwether" may alternatively have to depend on proof that its "one country, two systems" structure can serve as an example for the resolution of similar cases elsewhere involving divisive irredentism or contested secessionism. A convincing answer as such would, concededly, depend on whether this unprecedented symbiotic model can survive the test of time in Hong Kong itself. This very question we have addressed in the section above, where we touched on certain preconditions for the model's sustainability and the prospects of a full flowering of democracy. I would argue that the Hong Kong SAR model, minus the "albatross" effects from the unwholesome parts of its colonial bequest, is the unadulterated "one country, two systems" structure, which can be made an example for possible emulation elsewhere. We are now ready to explore if there are possible "candidates" in the rest of the world, outside the Chinese unification context, for the possible transplant of Hong Kong's "one country, two systems" governance structure.

Candidates for a "one country, two systems" solution. While the world's decolonization process has largely played out, there are still many cases in which a component part (or parts) of a territorial state, either due to of historical reasons or because of a fundamental dissension due to ethnic, cultural, linguistic, or other origins, is engaged in activities of protest in search of a solution, including a separatist existence. A study by Halperin and Scheffer (1992) compiles a list of 51 countries, in four continents (Africa, Asia, Europe, the Americas) plus the Middle East, that have an irredentist problem of various sorts and with varying degrees of severity, ranging from one to three or four dissenting entities in a territorial state. Other than the China Mainland-Taiwan case, some of the more celebrated and better known cases are: Canada-Quebec; Spain-Catalonia; and Mauritania-Western Sahara. Likewise, similar disputes exist between India and Assam and Punjab; between Britain and Northern Ireland; and between France on the one hand and Brittany, Corsica, and New Caledonia, on the other. The United States has a remotely similar problem with Puerto Rico, not to mention the native Americans (American Indians), as do a number of countries in the Middle East, including Iran, Iraq, and Turkey, with pockets of Kurds within their national boundaries. Similar tensions exist in the successor states to the Soviet Union and Yugoslavia. One should not forget there is the separatist Chechnya question in Russia. The Kosovo situation, which arrested world attention in spring 1999, may be *sui generis* because of the ethnic-cleansing problem involved. But the complexity can nevertheless be reduced to one of a tortured secession by a former province (Kosovo) from what remained of Yugoslavia, in which Serbia reacted with unusual terrorism in an attempt to interdict and abort the secession bid. Also in Asia, Indonesia has the East Timor problem. Then, there are others identified as "trans-states," such as Jammu-Kashmir, sandwiched between India and Pakistan, and France's Basque regions. Thus, the list is much longer than one realized.

Out of the numerous cases of irredentism and secessionism identified by Halperin and Scheffer, I have singled out only those that, in my view, are most amenable to a transplant of the "one country, two systems" solution. And, I have added a few that gained importance more recently, such as East Timor and Kosovo. In each, there are two sides of disproportionate size in territory, population, and power; and the larger of the two is known to be intolerant of, and in extreme cases even deadly opposed to, the insubordination or secessionism of the smaller party. Likewise, the latter in each case is invariably fearful of being engorged by the larger party but is unable to translate its separatist inclination, if it has one, into reality. At least in one case (Quebec), attempts to find a way out by referendum have not yielded a definitive answer thus far.

The only, though crucial, difference between all these (and other) cases and Taiwan is that none of them is an existing entrenched polity with an international personality recognized by any significant number of foreign sovereign powers. Although Beijing refuses to recognize Taiwan as a country-*qua*-state, the fact is that the island is still diplomatically recognized by 27 sovereign states and, furthermore, it participates in many international institutions under different names, usually "Chinese Taipei," to avoid Beijing's objection. Besides, most important of all, its continuing existence is supported by the United States.

Considering the material conditions of these cases—such as the disparity in size between the disputant components, and the inability of the dissenting party to successfully pursue a separatist course despite its fear of being engulfed by the larger component—I think the "one country, two systems" model might just be an equitable way out of the morass they are in, if only as a last resort. In view of the outstanding features we have identified above for the model's successful experiment in Hong Kong, i.e., those that have made for its unusual "unity in diversity," the solution might work in some of these cases as well. The fact that none of the minor disputants in the cases mentioned above has an existing political identity and status such as Taiwan's should actually be a plus. It should make it easier to break down the minor party's resistance (as its leaders, unlike in Taiwan, have less vested interest and pride to defend), in the event the larger disputant in each case offers "one country, two systems" as a solution. It might just happen, for instance, that Canada and Spain will be able to persuade Quebec and Catalonia, respectively, to accept the symbiotic model, taking a leaf from Hong Kong's book. In that event, the latter's contribution, stemming from its successful demonstration of the workability of the unprecedented "one country, two systems" model, integrating two disproportionate and otherwise incompatible socio-political systems into a symbiotic governance structure, is destined to earn for post-colonial Hong Kong the distinction of being "history's bellwether."

It bears reiterating that we are talking about the duplicability of the *authentic* "one country, two systems" model, minus the "albatross" burdens of the unwholesome parts of Hong Kong's anachronistic colonial bequests including the fortuitous side-effects of Patten's reform and long-standing colonial neglect (of course keeping the other parts of that inherited tradition intact), as explained before. Hence, in the final analysis, whether post-reversion Hong Kong can be a bellweather of history depends largely on whether it can free itself from being a mere "prisoner" of an unmitigated legacy from its past. Here lies the ultimate paradox.

Notes

1. "A Case of Legal Misunderstanding," op-ed page, *SCMP,* 9 February 1999, p.15.
2. *Wen Wei Po,* 11 December 1998, p.16.
3. "Basic Right Intact, Says EC Report," SCMP, 10 January 1999, p.1. See also "The Commission Publishes Its First Annual Report on Hong Kong," in *EU News,* Winder 1998–99, p.5.
4. I am indebted to Dr. Ka-ying Timothy Wang of the Institute for making available to me a set of statistics from the monthly opinion polls conducted by the Institute.
5. I am grateful to the United States Information Service (U.S.I.S.), U.S. Consulate-General/Hong Kong, for providing me with a copy of the report, known as: "Quarterly Report by Task Force on The Hong Kong Transition, U.S. House of Representatives," submitted to the Speaker of the House, U.S. Congress., dated 4 August 1998.
6. Cf. "A Firm Grip on Elections," *SCMP,* 14 December 1998, p.17.
7. Critics from the Human Rights Monitor attacking the SAR government's 500-paragraph report to the United Nations, on the post-handover human rights record, echoed a common criticism that business interests were overemphasized in the SAR's functional constituency elections. See "Anger at Official Rights Report," *SCMP,* 13 January 1999, p.1.
8. As reported on TV Pearl in Hong Kong, on 17 December 1998.
9. "Social security" in Hong Kong is a modest public-alms handout program for applicants for emergency assistance and is not contribution-financed.

10. "Poor Left Behind as Gap Grows," *SCMP,* 6 December 1997, p.1

11. "Hong Kong's Poor Upped by 80 percent in 10 Years," *SCMP,* 6 December 1997, p.1.

12. I am indebted to the Asia Pacific Institute of Ageing Studies for providing me much of the information that goes into this discussion.

13. *Hong Kong, 1997: SAR's First Year,* published by the Information Services Department, HKSAR (1998), p.170.

14. "Social Assistance Bill to Rise," *SCMP,* 27 February 1999, p.4.

15. *SCMP,* 3 October 1997, p.4.

16. I used "assume" here for the sake of intellectual honesty, because very little has been written on this subject, and practically no comparisons on social-welfare policies before and after the end of colonial rule in Hong Kong, of a sort relevant to this discussion, are known to me, despite an initial search.

17. "Colonial Bias 'led to flats fiddle'," *SCMP,* 22 July 1998, p.1.

18. "UN Seeks Answers on Rights of Women," *SCMP,* 22 January 1999, p.2.

19. Here I am relying on a keynote lecture I gave at an international symposium at Franklin College (Franklin, Indiana), 7 November 1997, entitled: "Post-Colonial Hong Kong: The Aftermath of Reversion." The ubiquitous mahjong parlors, the official tolerance of concubinage, and even the craze about horse racing—all carried over from colonial times remind one of a comparable device of imperial control by the Czarist rulers in traditional Russia. The latter's exhortation to libation created a folk culture whose effects are still shown today (witness Boris Yeltsin's drinking habit). The original rationale, in encouraging the Russian people to uninhibitedly consume vodka (manufactured by state-owned breweries) under the pretext of patriotism, was allegedly to keep the people forever intoxicated. For intoxicated people would make poor revolutionaries, until of course a sober Lenin came along. Those interested in British colonial legacy in Hong Kong should consult Mitchell 1998, and Lau 1997.

20. While I have more to say on this in a different section below, the controversy arose when the Court, in an *obiter dictum,* alleged that Hong Kong courts could interfere with acts of the National People's Congress (NPC) if they breached the Basic Law. It ended with a clarification by the same Court, issued on 26 February 1999, that its 29 January judgment "did not question the authority of the Standing Committee [of the NPC] to make an interpretation under Article 158 [of the Basic Law] which would have to be followed by the courts of the region," referring to the Hong Kong SAR. See "Justices Clarify Ruling: We Were Not Challenging NPC," *SCMP,* 27 February 1999, p.1.

21. Appendix III has since been amended twice, deleting one from the original six but adding six other national laws, with the result that the list now consists of 11 such national laws that are applicable to the HKSAR. See *Wen Wei Po,* 11 February 1999, p.11.

22. Some people have raised the question of the huge tracts of Chinese land lost to Tsarist Russia during Manchu era, the last dynasty that ended in 1912, and

asked whether these should be the really last piece of the jigsaw puzzle of Chinese reunification. My view is that since the return of these territories is so iffy, we should not consider them in the same breath with Hong Kong, Macau, and Taiwan. Besides, these lost lands are not known to have a coherent governing structure as a separate polity.

23. This is not suggesting that the future of Taiwan is to be determined by a plebiscite, in the sense as used by the advocates of the island's separatist independence. China's Vice Premier, Qian Qichen, as recently as 28 January 1999, reiterated Beijing's opposition to a plebiscitary solution precisely for that reason. See *Wen Wei Po,* 29 January 1999, p.1.

24. In a special "Series on Hong Kong's Future," carried by *SCMP* starting on 25 January 1999, covering diverse subjects such as rule of law, creating more jobs, press freedom, and "conflicting signposts on political path," there were as many views as pundits being interviewed. The beginning article set the tone for all pontifications with the title, "Challenge of Defining Our Tomorrow."

25. *SCMP,* 1 August 1999, p.3.

26. A team led by the SAR's Secretary for Security, Regina Ip, was in Beijing in late March 1999, on a mission to work out an agreement on details regarding the transfer of suspects between Hong Kong and the Mainland. See "Security Chief in Talk on Transfer of Suspects," *SCMP,* 23 March 1999, p.3.

27. Cf. *SCMP,* 11 January 1999, p.16.

28. *Chan Kam Nga and 80 Others v. Director of Immigration* (Final Appeal No. 13 of 1998). There were a number of other similar cases on appeal involving children born out of wedlock in Mainland China to Hong Kong fathers, in some cases before the father even acquired Hong Kong legal resident status, such as *Ng Ka Ling and Ng Tan Tan v. the Director of Immigration* (FACV No. 14), and Tsui Kuen Nang v. the Director of Immigration (FACV No. 15), but decided on the same date. And, the Court of Final Appeal arrived at an identical ruling in each. Hence, I use the Chan Kam Nga case as representative, in fact an example of what in the United States would have been a class suit.

29. See "Secretary for Justice Off to Beijing on Court Ruling," *Xin Bao* (*Hong Kong Economic Journal*), 12 February 1999, p.5.

30. "Future Shock for SAR," *SCMP,* 7 February 1999, p.2.

31. "Foreign Ministry Comments on the Final Appeal Court's Decision," *Wen Wei Po,* 3 February 1999, p.1.

32. *SCMP,* 31 January 1999, p.1; and *SCMP,* 8 February 1999, p.1. See also editorial in *Xin Bao* (*Hong Kong Economic Journal*), 8 February 1999, p.1.

33. "Beijing Says Abode Ruling Was Wrong and Should Be Changed," *SCMP,* 9 February 1999, p.1.

34. "Heads Cool on Bode, Say Officials," *SCMP,* 15 February 1999, p.1.

35. "Justices Clarify Ruling: We Were Not Challenging NPC," *SCMP,* 27 February 1999, p.1. See also Beijing's reaction, in "Beijing Hint of End to Abode Row," *SCMP,* 28 February 1999, p.1.

36. Senator Jesse Helms, Chairman of the U.S. Senate Foreign Relations Committee, on 29 March 1999 urged the State Department to draw particular attention

to "threats to the SAR's judicial independence" in the wake of the Chan Kam-Nga and other cognate cases. *SCMP,* 30 March 1999, p.6.

37. "Defiling Flags No Crime: Judges," *SCMP,* 24 March 1999, p.1. The case involving two Hong Kong young men, Ng Kung-siu, 25, and Lee Kin-yun, 19, who were convicted by a lower court in May 1998 for deliberately defiling in public the Chinese national flags and the SAR flags. They were bound over for a year "to keep the peace." After the Court of Appeal ruling, Ng said he was surprised but pleased.

38. The ICCPR, according to Article 39 of the Basic Law, continues in force in Hong Kong beyond the handover.

39. In the United States, "federal question" jurisdiction is granted to federal courts (and denied to state courts) over civil actions arising under the constitution, federal laws, or treaties of the U.S. For an illustration of how the "federal question" principle works, see, e.g., U.S. Supreme Court: *Kerr-McGee chemical Corp. vs. Illinois,* 459 U.S. 1049 (1982). I am grateful to Dr. REN Yue, of Lingnan University, for bringing this case to my attention.

40. I am borrowing this concept from Mark Neocleous 1996, p. viii and ch. 1.

41. Cf. J. Jeane, "Despotism and Democracy," in J. Jeane (ed.) 1988.

42. *SCMP,* 11 January 1999, p.4.

43. In two separate public-opinion surveys, conducted by the Hong Kong Asia Pacific Research Institute at the Chinese University of Hong Kong, 71.6 percent of the respondents said they thought the Chief Executive ought to be popularly elected in October 1997, and 72.2 percent supported the same view in August 1998. In addition, 70.5 percent and 72.3 percent agreed that the Legco should be fully elected in the two surveys, respectively. These findings were made available to me by courtesy of Dr. Wong Ga-ying, of the Institute.

44. As reported in the *SCMP,* 11 January 1999, p.1.

45. "Search for Successors Proves an Uphill Job," *SCMP,* 5 April 1999, p.13.

46. For a treatise on the development of civic virtue and its linkage to citizenship in the American democracy, see Glendon and Blankenhorn (eds.) 1995, esp. ch. 5: "Re-institutionalizing virtue in Civil Society," by William M. Sullivan. While this book considers the family, union, and the church as the seedbeds of virtue, I think Hong Kong needs a greater role played by the school system in inculcating such civic virtues as empathy, public spirit, participation in collective action, public service above self, and "habits and beliefs that hold human appetites in check" (p. 4).

References

Boucher, Richard A. 1998. "U.S. Policy and the Asian Economic Crisis." English text of an op-ed piece by the United States Consul-General in Hong Kong, carried in a local Chinese-language newspaper; courtesy of the United States Information Service (U.S.I.S.), Hong Kong.

Chang, Chao-wei. 1998. "A Government that Did Not Heed to the Welfare of the Old." *Poverty Watch* (Hong Kong) (June). (In Chinese)

Chan, Ming K. 1997. "The Legacy of the British Administration of Hong Kong: A View from Hong Kong." *China Quarterly* no. 151 (Septmber): 567–82.

DeGolyer, Michael E. 1998. "Having Their Say in Hong Kong." *China Business Review* (September-October): 44–7.

Glendon, Mary Ann, and David Blankenhorn. 1995. *Sources of Competence, Character, and Citizenship in America.* Lanham. MD: Madison Books.

Halperin, Morton H., and David Scheffer, with Patricia Small. 1992. *Self-Determination in the New World Order.* Washington, D. C.: Carnegie Endowment for International Peace.

Harris, Peter. 1976. *Hong Kong: A Study in Bureaucratic Politics.* Hong Kong: Heinemann Asia.

Hsiung, James C. 1997. *Anarchy and Order: The Interplay of Politics and Law in International Relations.* Boulder, CO: Lynne Rienner.

Jeane, J., ed. 1988. *Civil Society and the State.* London: Verso.

Kuhn, Thomas S. 1970. *The Structure of Scientific Discovery.* Chicago: University of Chicago Press.

Lau, Chi Kuen. 1997. *Hong Kong's Colonial Legacy.* Hong Kong: Chinese University of Hong Kong Press.

Liu, Shuyong. 1997. "Hong Kong: A Survey of Its Political and Economic Development over the Past 150 Years." *China Quarterly* no. 151 (September): 151–592.

Lui, Samuel Hon Kwong. 1997. *Income Inequality and Economic Development.* Hong Kong: City University Press.

Lung, Alan Ka-lun. 1998. "From Blaming the Brits to Accepting Responsibility." *Hong Kong Democracy Foundation Newsletter,* issue 7 (August).

Mitchell, Robert Edward. 1998. *Velvet Colonialism's Legacy to Hong Kong.* Hong Kong: Hong Kong Institute of Asia-Pacific Studies, Chinese University of Hong Kong.

Mok, T. K. 1998. "Don't Dodge the Question of the Old Poor." *Poverty Watch* (Hong Kong) (June). (In Chinese)

Neocleous, Mark. 1996. *Administering Civil Society: Towards a Theory of State Power.* London: Macmillan Press.

Welsh, Frank. 1993, *A Borrowed Place: History of Hong Kong.* New York: Kodansha International.

Wesley-Smith, Peter. 1994. *Constitutional and Administrative Law in Hong Kong.* Hong Kong: Longman Asia Ltd.

Wong, Ka-ying (Timothy). 1998. *A Retrospective and Prospective View on "One Country, Two Systems" after One Year.* Hong Kong: The Hong Kong Research Center on Cross-Strait Relations. (In Chinese)

Wu, Anna. 1995. "Hong Kong Should Have Equal Opportunities Legislation and a Human Rights Commission." *Human Rights and Chinese Values.* Edited by Michael Davis. London: Oxford University Press.

Zhao, Weisheng, ed. 1985. "The Fable of Economic Prosperity and the Social Change of Hong Kong." *Xianggang shehui gongzuo de tiaozhan* (The Challenge of Social Work in Hong Kong). Hong Kong: Jixianshe Publishers.

Zhou, Yongxin. 1992. *Xianggang shehui fuli zhengce pingxi* (An Analytic Commentary on Hong Kong's Social Welfare Policy). Hong Kong: Tiandi Books.

Zhou, Yongxin. 1994. *Shehui baozhang yu fuli zhengyi* (The Disputes over Social Protection and Welfare). Hong Kong: Tiandi books.

About the Editor and Contributors

JAMES C. HSIUNG, Ph.D. Columbia University, NYC, is Professor of Politics and International Law at New York University, where he teaches international relations theory and international law. During 1997–99, he served in a visiting capacity as Chair Professor and Head, Department of Politics-Sociology, Lingnan University, Hong Kong. He observed at first hand the transition of Hong Kong into a Chinese Special Administration Region (SAR), from the handover ceremony of 1 July 1997, which he attended by invitation, through the end of the summer of 1998, after the SAR's second anniversary. He is author and editor of 16 well-received books, all published in the United States, not to count his numerous articles and book chapters. His most recent books are: *Asia Pacific in the New World Politics (1993)*, and *Anarchy & Order: The Interplay of Politics and Law in International Relations* (1997). His ongoing research project, "Hong Kong in the International Community: Capacity to Act and Effect Change," shows his continuing interest in Hong Kong as a post-colonial polity and nonsovereign international actor.

FRANK CHING is a journalist who has written and commented on developments in China and Hong Kong for more than two decades. He worked for the *New York Times* in the 1960s and 1970s before joining the *Wall Street Journal*. He opened the Journal's bureau in Beijing in 1979, shortly after the normalization of Sino-U.S. relations, and became one of the first four American newspaper correspondents to be based in China since 1949. He remained in Beijing for four years. Mr. Ching is now a Senior Editor of the *Far Eastern Economic Review*. Among other things, he writes a weekly column called "Eye on Asia."

DANIEL FUNG, QC, SC, JP, served successively as the last Solicitor General of Hong Kong under British rule from 1994 to 1997 and as the first Solicitor General of the Hong Kong Special Administrative Region (HKSAR), from 1997 to 1998, prior to taking up consecutive appointments as Visiting Scholar at Harvard Law School in September 1998 and Senior Visiting Fellow at Yale Law School in February 1999.

HO LOKSANG, Ph.D. Toronto, is Head of the Economics Department and Director of the Center for Public Policy Studies at Lingnan University, Hong Kong. He has authored about two dozen journal articles and over 30 chapters in various scholarly books. His research interests span a broad range of public policy issues, including housing, transportation, health policy, social security, labor market policy, and education.

He is also deeply interested in macroeconomic issues and policy, particularly monetary and exchange rate regimes. He has served on various public advisory bodies, including the Central Policy Unit of the SAR government, Hong Kong Committee of Pacific Economic Cooperation Council, and Health Services Research Committee. He is the Managing Editor of *Pacific Economic Review* and a Vice-President of the Hong Kong Economic Association.

Y. Y. KUEH, Ph.D. in Economics, Marburg University, Germany, is Chair Professor of Economics at Lingnan University, Hong Kong. Before joining Lingnan in September 1992, he was Professor and Foundation Director of the Center for Chinese Political Economy at Macquarie University, Sydney. He also taught at the Chinese University of Hong Kong for many years. He is author and co-author of several books, including *Industrial Reform and Macroeconomic Instability in China* (1999), *The Political Economy of Sino-American Relations: A Greater China Perspective* (1997), *China and the Asian Pacific Economy* (1997), *The Chinese Economy Under Deng Xiaoping* (1996), *Agricultural Instability in China, 1931–1991 (1995), and Economic Trends in Chinese Agriculture* (1993). He has also published numerous articles on various aspects of Chinese economic policy and economic changes in China in a wide range of leading international journals.

LAU SIU-KAI, Ph.D., Minnesota, is Professor and Chairman, Department of Sociology, and Associate Director, Hong Kong Institute of Asia-Pacific Studies, both at the Chinese University of Hong Kong. Professor Lau is a long-time student of Hong Kong and comparative politics. In addition to his academic work, he is active as a political advisor and commentator on politics. During 1996–97, he served as a member of the Preparatory Committee of the Hong Kong Special Administrative Region. His publications include: *Society and Politics in Hong Kong* (1982); *The Ethos of the Hong Kong Chinese* (1988; coauthored with KUAN Hsin-chi); *Hong Kong Politics in the Transitional Period* (in Chinese, 1993); and many articles in scholarly and popular journals.

BEATRICE LEUNG, Ph.D., London School of Economics and Political Science, is an Associate Professor in the Department of Politics-Sociology, Lingnan University, Hong Kong. Her research interests are in China's domestic politics and international relations. She has published widely in the area of state-society relations, church-state relations, and center-periphery relations involving China, Hong Kong, and Macau. Her most recent work, "Sino-Vatican Negotiation: Old Problems in New Context," was published in the *China Quarterly* (no.153, 1998). She was co-editor (with Joseph Cheng) of an earlier work, *Hong Kong SAR: In Pursuit of Democratic and International Order* (1997).

BOLONG LIU, Ph.D. in International Relations, is Associate Professor and Dean of the Faculty of Social Sciences and Humanities, University of Macau. His research interests are in Macau's political and cultural development. He has published in these areas and on Macau's external relations. His other publications are on the public administration and foreign relations of China. Besides following Macau's transitional problems in its retrocession to China, due to take place on 20 December 1999, he is writing a book on public policy-making in China.

ANTHONY NEOH, Senior Counsel, Hong Kong Bar and member also of the English and California Bar, is a member of the Hong Kong SAR Basic Law Committee. Now a Visiting Professor at Peking University, he was the Chairman of the Hong Kong Securities and Futures Commission and Chairman of the Technical Committee of the International Organization of Securities Commissions.

DANNY SHIU LAM PAAU, Ph.D. and M.A., Georgia, B.A. Hons. Chinese University of Hong Kong, is Senior Lecturer and formerly Head of the History Department; Director, Sino-Western Relations Research Program, David C. Lam Institute for East-West Studies, Hong Kong Baptist University. His publications include: *Beyond the Taiwan Independence Movement: American Advocates and Policies* (1992); and *Visions of Civilization: The Search for National Models in Modern China* (1999), plus dozens of articles in refereed journals. He is also on the editorial board of: *Comparative Civilizations Studies Review* (U.S.); and *Sino-Humanitas* (U.K.). He teaches, among other courses, Current Issues in Hong Kong and China, which is also one of his research interests.

TING WAI, Ph.D., University of Paris (Campus 10), is an Associate Professor in the Department of Government and International Studies, Hong Kong Baptist University. His research interests include Chinese foreign policy, China-Hong Kong relations, and external relations of Hong Kong. Currently he is working on a research project on Sino-American relations and the issue of nuclear proliferation. His latest publications include a monograph on *The External Relations and International Status of Hong Kong* (1997), and a book chapter on "China, the United States, and the Future of Hong Kong" (1997).

Acronyms

ADPL	Association for the Development of People's Livelihood
AGC	Attorney General Chamber
APEC	Asia Pacific Economic Corporation
ASEAN	Association of Southeast Asian Nations
ATV	Asia Television
BBC	British Broadcasting Corporation
BIS	Bank of International Settlements
CCP	Chinese Communist Party
CE	Chief Executive
CFA	Court of Final Appeal
CFTC	Commodity Futures Trading Commission
CITIC	China International Trust and Investment Corporation
COCOM	Coordinating Committee for Multilateral Export Controls
CPPCC	Chinese People's Political Consultative Conference
CSSA	Comprehensive Social Security Assistance
DP	Democratic Party
EIB	Environmental, Institutional-Organizational, Behavioral
EU	European Union
Exco	Executive Council
FAO	Food and Agriculture Organization
FDI	Foreign Direct Investment
GDP	Gross Domestic Product
GITIC	Guangdong International Trust and Investment Corporation
HIBOR	Hong Kong Interbank Offered Rate
HKMA	Hong Kong Monetary Authority
HKSAR	Hong Kong Special Administrative Region
HOS	Home Ownership Scheme
ICAC	Independent Commission Against Corruption
ICC	International Chamber of Commerce
ICCPR	International Covenant on Civil and Political Rights
ICESCR	International Covenant on Economic, Social, and Cultural Rights
IMD	International Institute for Management Development
IMF	International Monetary Fund
IPO	Initial Public Offering
Legco	Legislative Council

MFN	Most Favored Nation (treatment)
MOP	Macau pataca (unit of currency)
MTCR	Missile Technology Control Regime
NAFTA	North American Free Trade Association
NCNA	New China News Agency
NGO	Non-governmental Organization
NPC	National People's Congress
NPC-SC	National People's Congress Standing Committee
NRI	Nomura Research Institute
NSG	Nuclear Supplies Group
NTD	New Taiwan Dollar
OECD	Organization for Economic Cooperation and Development
PLA	People's Liberation Army
PRC	People's Republic of China
QEF	Quality Education Fund
RMB	renminbi (unit of Chinese currency)
RTHK	Radio and Television Hong Kong
SAR	Special Administrative Region
SCMP	South China Morning Post
SFC	Securities and Futures Commission
SOE	State-owned Enterprises
TDC	Trade Development Council (Hong Kong)
TKP	Ta Kung Pao (Ta Kung Daily)
TMD	Theater Missile Defense
TPS	Tenant Purchase Scheme
TVB	Television Broadcasts Limited
UAR	United Arab Republic
UN	United Nations
UNESCO	United Nations Economic, Scientific, and Cultural Organization
USIS	United States Information Service
WTO	World Trade Organization

Index